ISBN 978-1-332-19534-3
PIBN 10296659

1 MONTH OF FREE READING

at

www.ForgottenBooks.com

By purchasing this book you are eligible for one month membership to ForgottenBooks.com, giving you unlimited access to our entire collection of over 700,000 titles via our web site and mobile apps.

To claim your free month visit: www.forgottenbooks.com/free296659

English
Français
Deutsche
Italiano
Español
Português

www.forgottenbooks.com

Mythology Photography **Fiction**
Fishing Christianity **Art** Cooking
Essays Buddhism Freemasonry
Medicine **Biology** Music **Ancient
Egypt** Evolution Carpentry Physics
Dance Geology **Mathematics** Fitness
Shakespeare **Folklore** Yoga Marketing
Confidence Immortality Biographies
Poetry **Psychology** Witchcraft
Electronics Chemistry History **Law**
Accounting **Philosophy** Anthropology
Alchemy Drama Quantum Mechanics
Atheism Sexual Health **Ancient History**
Entrepreneurship Languages Sport
Paleontology Needlework Islam
Metaphysics Investment Archaeology
Parenting Statistics Criminology
Motivational

SHIPYARD PRACTICE

AS APPLIED TO

WARSHIP CONSTRUCTION

BY

NEIL J. McDERMAID

MEMBER OF ROYAL CORPS OF NAVAL CONSTRUCTORS

LATE INSTRUCTOR ON PRACTICAL SHIPBUILDING AT THE ROYAL NAVAL COLLEGE,
DEVONPORT

WITH DIAGRAMS

LONGMANS, GREEN, AND CO.

39 PATERNOSTER ROW, LONDON

NEW YORK, BOMBAY, AND CALCUTTA

1911

PREFACE

THIS work is intended to provide Students and others with a knowledge of the actual operations performed in the Shipyard during the construction of a warship.

Great care has been taken to include as many as possible of the important operations performed both during the construction and fitting-out of the ship, and embraces a course of lectures given to Cadets of Naval Construction when Instructor at the Royal Naval College, Devonport.

N. J. McDERMAID.

BARROW-IN-FURNESS,
1911.

223975

SHIPYARD PRACTICE

THE BUILDING SLIP.

IN the Royal Dockyards these are constructed of masonry, the slope of the floor being about $\frac{41}{64}$ inch to a foot. At intervals of 5 feet to 6 feet down the slip balks of timber, termed "land ties," are placed. These are of English oak 12 inches by 6 inches, their upper surface being flush with the floor of the slip, and are for the purpose of securing ways, keel blocks, shores, etc.

The keel blocks are laid midway between the ways, and are spaced the same distance apart as the land ties, and are vertical, and not at right angles to the floor of the slip. A set of these blocks is shown on p. 8. At the back of each set of blocks two tripping or spur shores are placed to guard against risk of the blocks capsizing while the ship is building. Only those pieces of timber below the wedges are fastened to one another, since those above the wedges have to be removed for riveting flat keel and painting bottom prior to launching.

The first work to be done when commencing a new ship is to arrange keel blocks and launching ways, and it is assumed that the groundways are being prepared for the first time, as it is not usual to alter these for each subsequent ship.

Considering first the groundways, these will be of least depth at their lower end, and to guard against the risk of them being crushed should the ship travel so far before being waterborne that the fore poppet will pass over the outer end, it is not desirable to have a less depth than about 12 inches at this point. Hence, set up a distance of 12 inches above the slip at the after end of groundways, and from this point draw a line at a slope of $\frac{49}{64}$ inch to a foot, which will be the tangent line from which the camber of the ways will be set off (Fig. A, p. 4). The amount to be set off can be readily calculated when the amount of the

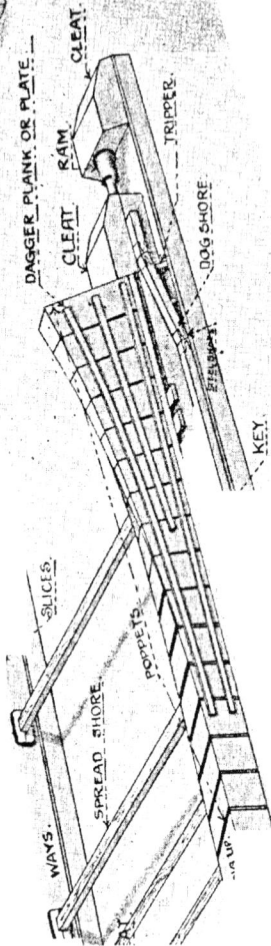

STANDARDS or PUDLOCKS.
STAGING.
DERRICK.
Lᵉ PIECE RAM.
RAM.
SLICES
WAYS.
SPREAD SHORE.
POPPETS
KEY.
BAND SHORES.
DOG SHORE.
TRIPPER.
CLEAT.
RAM.
CLEAT.
DAGGER PLANK OR PLATE.

camber has been decided upon, and these are given on a sketch from which the groundways can be built up to their correct height.

The groundways are laid directly on the slip, and are built up of timber laid tier on tier crosswise, the upper surface being completed by the "sliding plank." The surface of the sliding plank is not a plane one, but "cambered," that is, a longitudinal section of it is an arc of a circle of very great radius, the curve being convex upwards. The idea of "cambering" the ways, as it is called, is to guard against the possibility of them becoming hollow as the ship passes over them when launching. Its effect is to lessen the initial velocity and increase the final velocity as compared with ways having no camber, and further, it permits of the groundways being shorter, as will be explained later.

The camber is measured by the amount the arc of the circle rises above its chord, and is usually about 9 inches in 400 feet. The curve is constructed by drawing a tangent line at $\frac{49}{64}$ inch to a foot, the touching point being at about the point where the after perpendicular of the ship will come on the slip. The spread of the ways is usually one-third the beam of the ship.

The points to consider in laying the keel blocks are the following :—

(1) To allow sufficient height between the keel of the ship and the slip for men to work, taking care at the same time to avoid excessive height of blocks, or there will be risk of tripping. The groundways when once built up are not generally altered unless it be found necessary to extend them farther up the slip for any particular ship.

(2) To allow sufficient space between the bottom of the ship and the top of the groundways to get the bilgeways, etc., in place.

(3) To arrange that the forefoot of the ship shall not touch the slip at any point in its passage down the slip, taking account of the fact that after the stern lifts the forefoot will approach it a certain amount over and above that due to the difference between the slope of the slip and groundways.

(4) That the ship shall clear the sides of the slip.

(5) That the after end shall clear the ground, *i.e.*, the depth of water must be greater than the maximum draft aft, which will be when the stern begins to lift.

As the ways are at a greater slope than the slip, it will be clear that the forefoot approaches the latter at a rate of $\frac{49}{64} - \frac{41}{64}$ inches

HEIGHT OF FOREMOST BLOCK.

for every foot the ship travels. The travel is obtained from the launching curves and the amount the forefoot approaches the slip calculated, and this added to the "clearance" will give the "minimum" height of foremost block. This takes no account of the amount the ship approaches the slip due to the stern lifting and the ship hinging about the fore poppet, and it is therefore necessary to consider the question in greater detail.

Considering (2), first take a profile of the ship and lay off a buttock line at a distance out from the middle line equal to the half spread of the groundways, that is, to their inner edge (this drawing should be on tracing paper), and place this on top of a drawing of the slip on the same scale. Place the top drawing with the L.W.L. at $\frac{5}{8}$ inch to a foot slope, and so that the buttock just touches the surface of the groundways. Now move the tracing up a distance equal to the depth of bilgeways, plus stopping-up, plus slices, plus about $\frac{1}{2}$ inch for the grease and an amount to take account of the crush of the blocks while building (about $\frac{3}{4}$ inch), and mark off the keel line, which will clearly give the least height of blocks for getting cradle in place (Fig. C, p. 4).

The next process is to arrive at the height of blocks required to obtain any given clearance. This can be done as follows :—

The estimated weight and trim at launching is calculated, and the water line corresponding to this drawn on the profile of the ship. Take this tracing, and place the water line to coincide with the tide line, so as to give the minimum clearance between forefoot and slip. Next assume a position for the fore poppet, then it will be clear that the point on the keel where the fore poppet comes will move along a curve parallel to the top of groundways when launching. Hence measure the distance of this point from the top of ways, and set it off at the position the fore poppet will occupy with the ship at rest on the slip. The point so obtained will be a point in the actual keel line, which can then be drawn in (see Fig. A, p. 4).

As regards the position of fore poppet, if this be too far aft, there will be a considerable portion of the fore part unsupported, whilst on the other hand if a position well forward is selected the poppets will be of excessive height, due to the ship fining rapidly. Thus a mean position for the fore poppet should be selected.

Thus the height of keel blocks has been obtained from point of view of clearance of forefoot and getting cradle in place, and whichever gives the higher line must be the one selected for line of keel blocks.

It will be clear that the greater the amount of stopping up, the farther the ship will have to travel before being waterborne, and this alone will reduce the clearance ; but as the proportionate gain in clearance through raising the ship is much greater, on the whole the clearance has been increased.

Should the trim of the ship come out less than estimated, this again will reduce clearance, because the ship will have to travel farther before being waterborne.

When considering the travel of ship before being waterborne, it is clear that increased travel brings the C.G. nearer the end of the ways, and risk of tripping is increased, and it is necessary to consider this point. In some ships ballast has been placed in the fore part to bring C.G. further forward to prevent tripping, but this again alters trim and reduces clearance.

Thus it will be seen that in fixing up height of keel blocks a number of opposing considerations have to be dealt with.

As regards (4), sections of slip and ship at her fullest part, showing bilge and docking keels, are made, and it can then be seen if there is the necessary clearance.

If the estimated water line is above the tide line when fore poppet reaches the end of the ways, then the ship will drop off the ways, and it must be ascertained that there is no possibility of the forefoot striking the ground.

Appliances. Page 2 shows the masts, derricks, etc., used round slip for dealing with material to be hoisted on board, and also the uprights for the staging. These uprights are of the single type amidships and of the double type at the ends, as in figure. In the double type no diagonal strut will be required for the bearers on which staging is laid, unless these bearers are of very great length, when they should be fitted. In the single type diagonal supports will clearly be required. Test loads of about $\frac{1}{2}$ ton are applied to the ends of the bearers before they are allowed to be used.

Laying Blocks. The information supplied consists of a drawing showing the height of blocks at intervals down the slip and a declivity base, a board about 20 feet long, one edge of which is inclined at the same angle as the line of keel.

The method is to erect a set of blocks forward, amidships, and aft to the heights given on the drawings. Midway between these a base is erected, and its lower edge sighted in with the top of existing blocks, and a set built up to this height (see p. 8). The sighting-in is necessary, as it takes account of any unevenness in the floor of the slip, which working to the height given on drawings would not.

From these sets of blocks a series of four or five blocks is erected and trimmed to the declivity base as shown on p. 8, and the whole finally tested by stretching a line along the upper surface, and sighting through from end to end. The top block is called the "cap" block, and is of wood which can be easily split out and not crack, such as teak; it is of less width than the other pieces to permit of the flat keels and garboards being riveted without disturbing the blocks. The middle line of the ship is cut in on the blocks.

At the cut-ups at bow and stern, flight moulds are made to give the form to which the blocks must be trimmed at these parts (see p. 9).

While the work on the slip is in progress the flat and vertical keels are prepared and put together temporarily near the slip.

Preparation of Keel Plates.

The information supplied for this purpose consists of (1) *keel battens* giving position of butts of inner and outer flat keel, and vertical keel plates; (2) *keel section moulds* for plates which require dishing. A mould is supplied for each butt of the outer keel plate, the middle line of ship and the plate edges being marked on it. This mould is made to the inner surface of outer keel, and can also be used for the inner flat keel plate by taking off the thickness of inner plate from its edge; (3) *scheme of riveting,* giving details of all riveting; (4) *sketch of middle line work,* giving positions of butts of all plates and angles of flat and vertical keels for the midship or flat portion of the ship's bottom (see also p. 12), and similar information for the longitudinals.

Flat keel plate (outer).—The length of bottom plate which can be worked is determined by the length of bending rolls, and must be a multiple at the same time of the frame spacing, *e.g.* with a

A- ang e blocks.
C- cap   "
W- wedge   "
G- ground   "
D- declivity base.
T- tripping or spur shores.
S- sights.

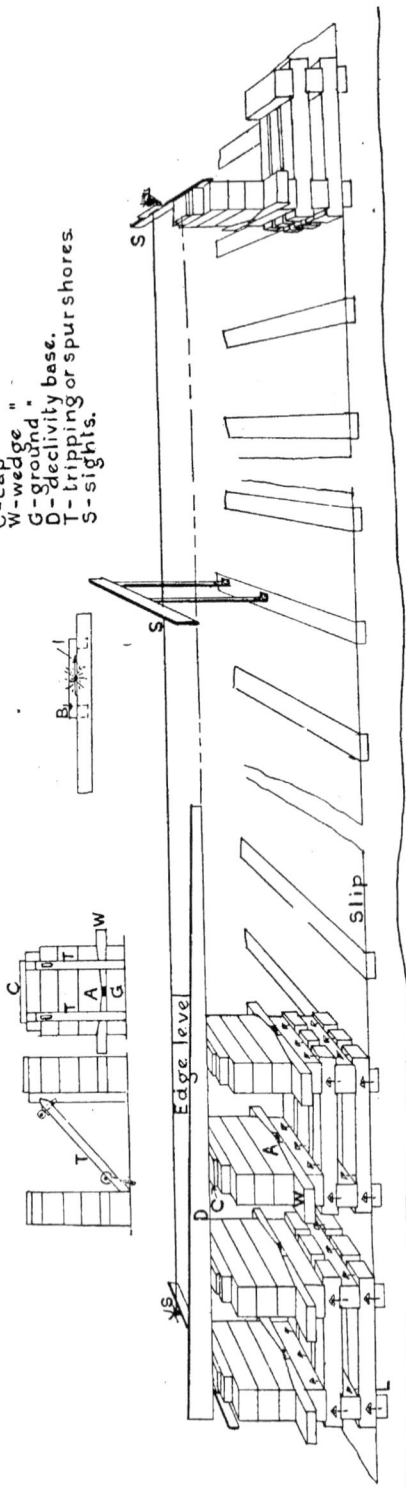

LAYING KEEL BLOCKS.

spacing of 4 feet and 25-feet rolls the limit is 24-feet plates. For plates in the flat portion of the bottom no dishing will be necessary ; longer lengths of plates can be worked if desired, and we proceed thus :—Mark off middle line, vertical keel plate, and angles of transverse frame, then mark off rivet holes along edges and butts, in the former case disposing those for the frames first, as the rivets have to be kept clear of the frame angles, and their spacing will depend on the type of frame, *i.e.* whether watertight or not ; after which, mark off the holes for garboard butt straps, and those for butts of inner flat keel. The edge rivets are arranged with a watertight or oiltight spacing, as per scheme of riveting, and a few tack rivets worked in plate. The marking of the plate is completed by writing on it the size of rivet and " C.K. other side." The plate is then taken to the machine shop, where the edges and butts are machined, holes punched or drilled according

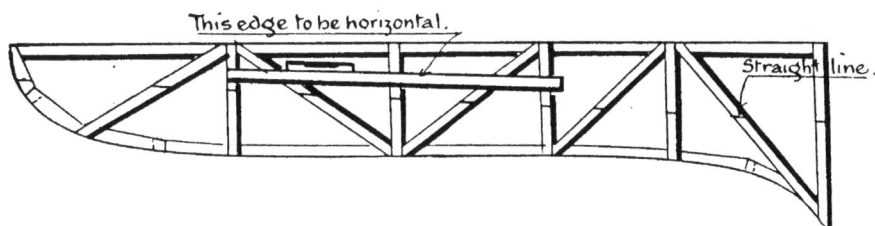

This edge to be horizontal.

Straight line.

FLIGHT MOULD FOR AFTER KEEL BLOCKS.

as to whether it is of mild or high tensile steel, plate turned over and countersunk, and finally put through the straightening rolls to relieve the state of stress and ensure its being a plane surface. In the case of a plate which has to be dished, a line is given on it showing where this is to be done, and no holes must be placed near this, or they will become elongated when the plate is dished. The dishing is the last operation carried out on the plates The method of checking a dished plate is shown on p. 10, and consists in placing the section moulds at the butts and outwinding the edges of the battens, which will show if the plate is twisted.

Inner flat keel.—It will be seen that the inner and outer flat keels form edge strips and butt straps to one another, and in addition butt straps are fitted to both outer and inner flat keel plates extending to the edge of the particular keel to which it serves as a butt strap. No countersunk holes are required in this plate, and the plate must be marked off from the outer keels after the latter have been dished, the section moulds being altered as before described for this purpose. The thickness of inner and outer flat keels is decreased towards the ends of the ship. This can be

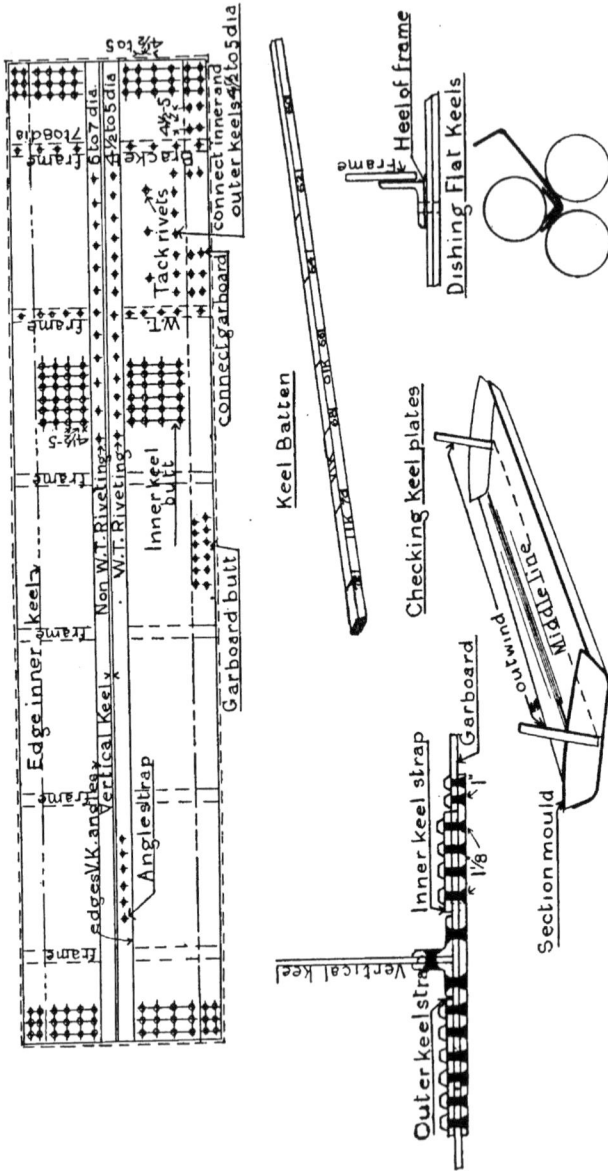

LINING OFF AND CHECKING FLAT KEELS

arranged in the following ways:—(1) Upper surface of inner keel flush with liners between the two keels at the butts of inner keel; (2) faying surface flush with liners at the inner keel butt straps; (3) underside of flat keel flush with liners at butts of both keel plates and vertical keel angles. In (1) the heel of frame line is level, but the underside of flat keel will be stepped, which is a disadvantage as regards docking, since the blocks at the ends would require to be specially trimmed. (2) The underside of flat keel is stepped, but less than in previous case, and by this method a better result is obtained, as the edge of inner flat keel can be efficiently caulked against outer keel, there being no liners. (3) This is an advantage as regards docking, but necessitates fitting liners and lowers the heel of frame. The best result is obtained by arranging the plates as in (2), the exposed butts of outer keel plates being chamfered off to obtain a flush caulk, and also reduce resistance.

The flat keel plates which connect on to the stem and stern castings are generally required to be curved in a fore and aft direction, as well as being dished. The information supplied for bending these consists of cradle moulds which give the exact form of the inner surface of the plate. The final fitting of these keel plates is made after the stem and stern castings are erected, and as it may be necessary to remove them for furnacing, no riveting can be done until the fitting is satisfactory.

Cradle moulds are also supplied for the *contour* or *fashion* plates, which are continuations of the stem and stern castings.

A cradle mould for a contour plate is given below :—

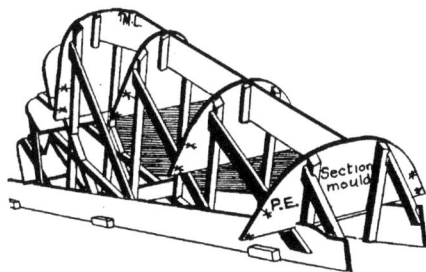

CRADLE MOULD.

TRANSVERSE FRAMING.

In large ships, as the longitudinal strength is of more importance than the transverse, the longitudinal framing is continuous and the transverse intercostal.

The type of frame to be worked in any part of the ship

TYPES OF FRAMING.

Vertical Keel outside Double bottom

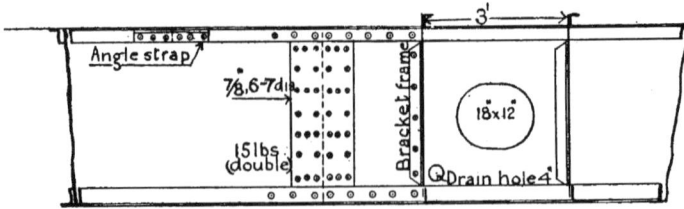

Angle strap

7/8, 6-7 dia

15 lbs (double)

Bracket frame

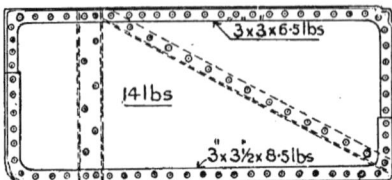

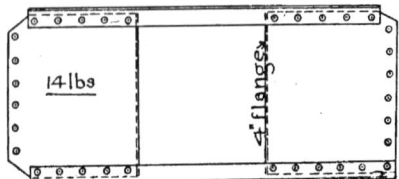

18"x12"

3'

Drain hole 4"

W.T. Longitudinal

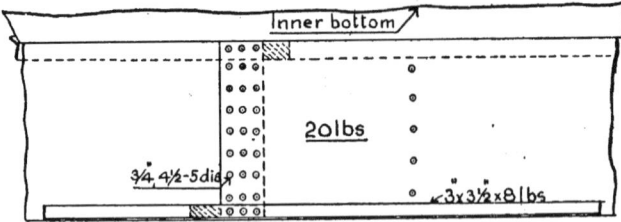

Inner bottom

20 lbs

3/4, 4½-5 dia

3"x3½x8 lbs

Non W.T. Longitudinal

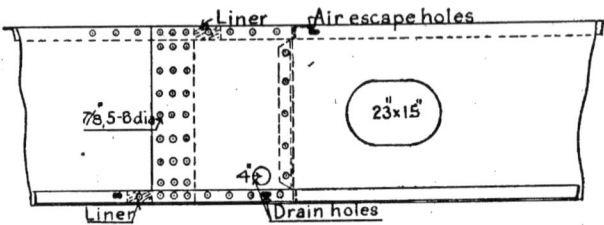

Liner

Air escape holes

7/8, 5-8 dia

23"x15"

Liner

4"

Drain holes

Vertical Keel within Double bottom

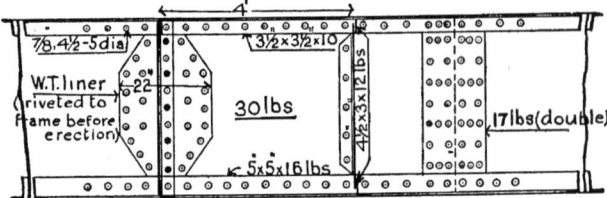

7/8, 4½-5 dia

3½x3½x10 lbs

W.T. liner (riveted to frame before erection)

22"

30 lbs

4½x3x12 lbs

17 lbs (double)

5x5x16 lbs

●-through liner and angle bar only

Lightened Plate frames

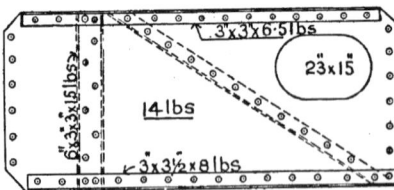

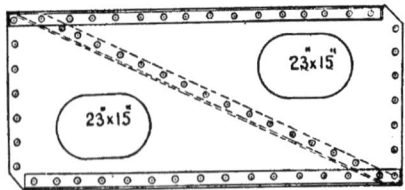

3x3x6.5 lbs

23"x15"

6x3x15 lbs

14 lbs

3x3½x8 lbs

23"x15"

23"x15"

Watertight frame

3x3x6.5 lbs

14 lbs

3x3½x8.5 lbs

Bracket frame

14 lbs

4" flange

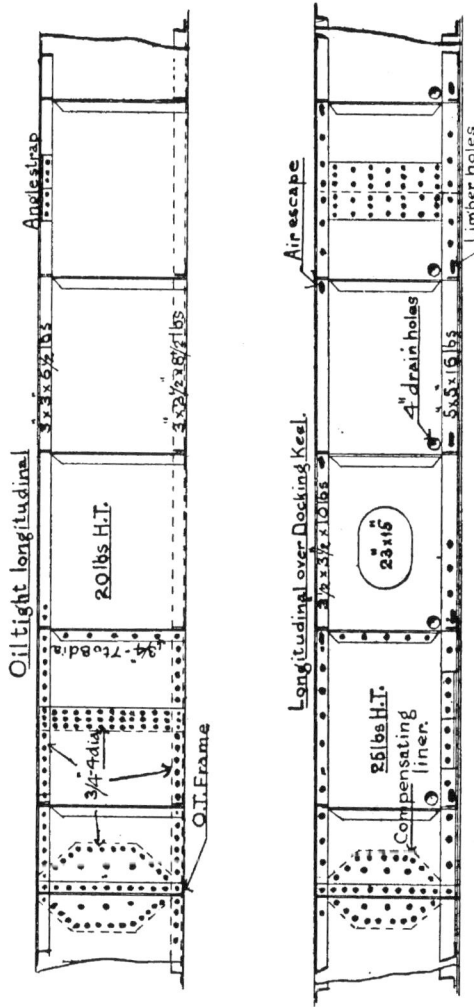

will depend on what function it has to fulfil ; for example, under a heavy concentrated load, as a barbette, the framing must be rigid. Again, as the double bottom space is divided up into watertight and oiltight compartments, solid frames will have to be provided, and between any two of these the framing must be arranged to provide access to the frame spaces.

The types of transverse framing fitted in double bottoms are bracket, lightened plate, watertight, and oiltight (see pp. 12 and 14).

The first is the ordinary framing ; the second is fitted under heavy weights, as barbettes, and holes for access must be made in these ; the third and fourth for bounding W.T. and O.T. spaces.

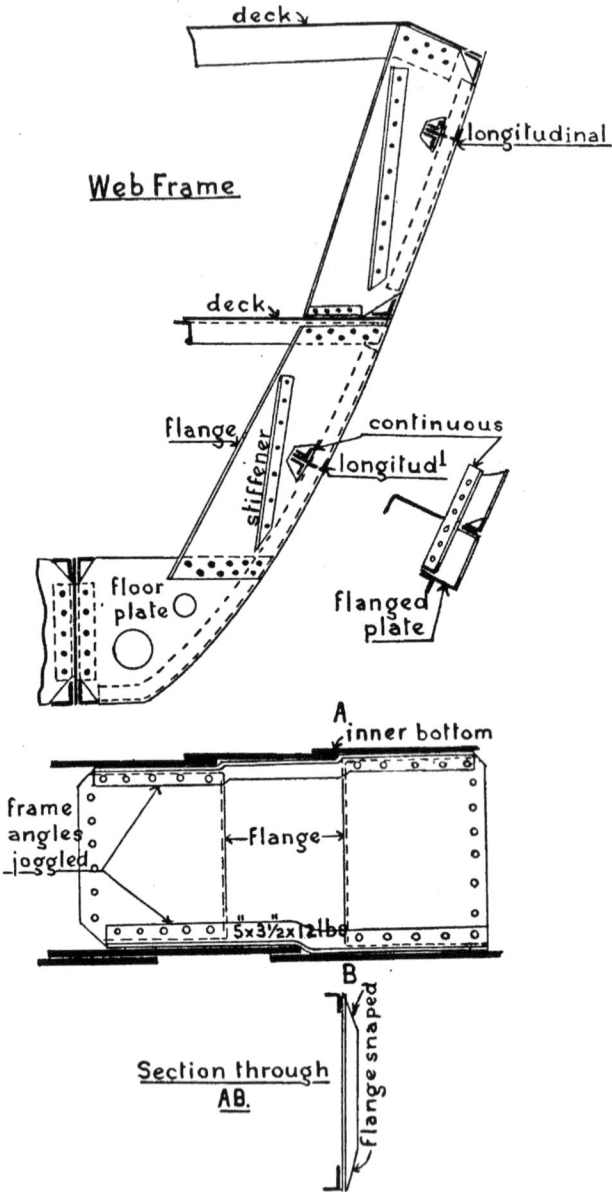

deck

Web Frame

longitudinal

deck

flange

stiffener

continuous

longitud¹

floor plate

flanged plate

A inner bottom

frame angles joggled

flange

5x3½x12lbs

B

Section through AB.

Flange snaped

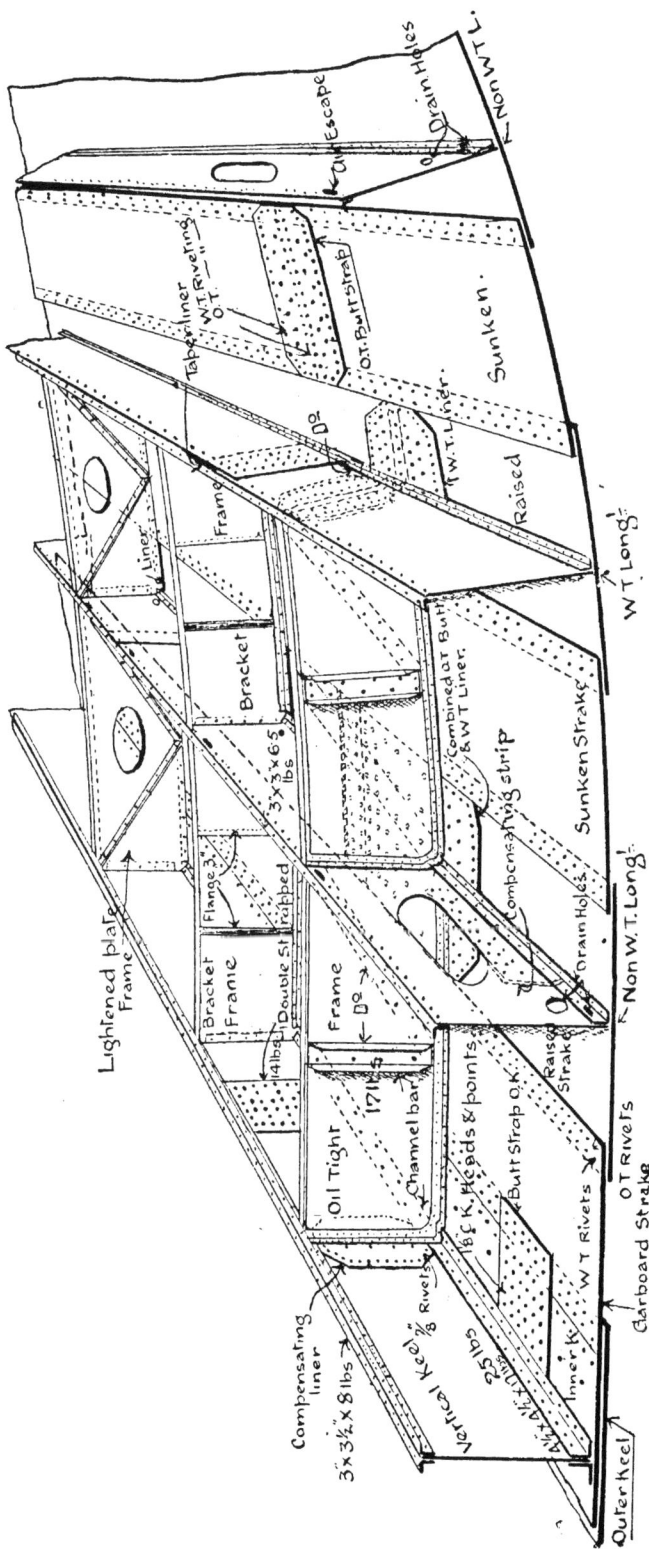

FRAMING WITHIN DOUBLE BOTTOM.

At the ends of the ship the type of transverse frame alters to that of a Z bar, with a floor-plate at its lower end.

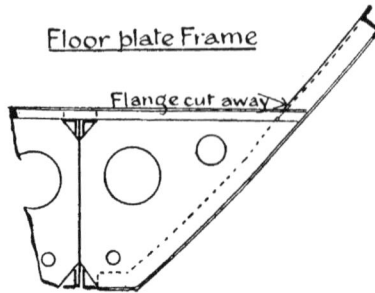

Floor plate Frame

Flange cut away

At the extremities of the ship specially strong transverse frames are sometimes required, and these are known as web frames, a type of which is shown (see p. 14).

The general arrangement of transverse framing can be seen on the midship and other sections, and on p. 15, which shows a portion of the frames in double bottom.

LONGITUDINAL FRAMING.

Vertical Keel.—This is continuous from end to end of the ship, the portion made watertight or oiltight depending on the arrangements for storage of water and oil fuel, but it is generally watertight throughout double bottoms. The plates are worked in as long lengths as can be conveniently handled, and are connected together by double straps, the butts coming between the transverse frames.

The longitudinals consist of watertight and non-watertight, the former with the W.T. and O.T. frames dividing the double bottom into W.T. and O.T. spaces for storage of oil fuel, and water. These longitudinals are continuous throughout the double bottom, and some are continued to the ends of the ship, either in the form of intercostal flanged plates worked between the transverse frames riveted to bottom plating, and with continuous angles on the free edge, or as the ordinary plate longitudinal, this depending on how far forward and aft the bracket-frame system is carried ; but where the transverse framing changes to Z bars, the former method has to be adopted. The method of continuing the angles of longitudinals through the web frames is given on p. 14.

All the longitudinals cannot be carried right forward and aft owing to the diminishing girth of the ship, and some have to be stopped short. Further, the longitudinals worked as intercostal plates with continuous angles are arranged so that there shall be no sudden loss of strength at this junction, as follows:—

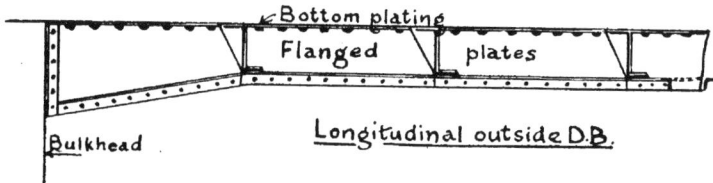

The butts are usually lapped, but in some cases where a longitudinal serves as a docking keel for side-docking blocks the butts are strapped as for vertical keel.

In some cases of non-watertight longitudinals compensating liners have been worked below the holes for access, as shown on p. 15.

The information supplied for lining off a longitudinal is a mould giving exact shape of plate with position of centre of lap, frame stations, and of the lightening holes, etc., in the case of a non-watertight longitudinal. The sketch of middle-line work also gives position of butts of angles and plates for those longitudinals in the middle portion of ship. For method of obtaining mould, see under Laying-off (p. 228).

In the non-watertight longitudinals, drainage and limber holes are cut, and also holes at the top of the longitudinal for escape of air when flooding, otherwise air-pockets would be formed in the corner and prevent compartment being completely filled. The edges and butts are sheared, all rivet holes punched, those for access and lightening purposes stamped out, and the plate then put through the straightening rolls, as the work done on it will distort it. In the case of the longitudinals being of high tensile steel all holes would be drilled and the butts and edges planed. The butts of angles to longitudinals are strapped.

In the double-bottom spaces under the engine-rooms extra framing is worked in connection with the engine seatings to locally stiffen up this part of the structure.

Special framing has to be worked round shaft swell and

recess to take the stern castings in wake of shaft brackets, etc. (see sketches on pp. 18 and 19).

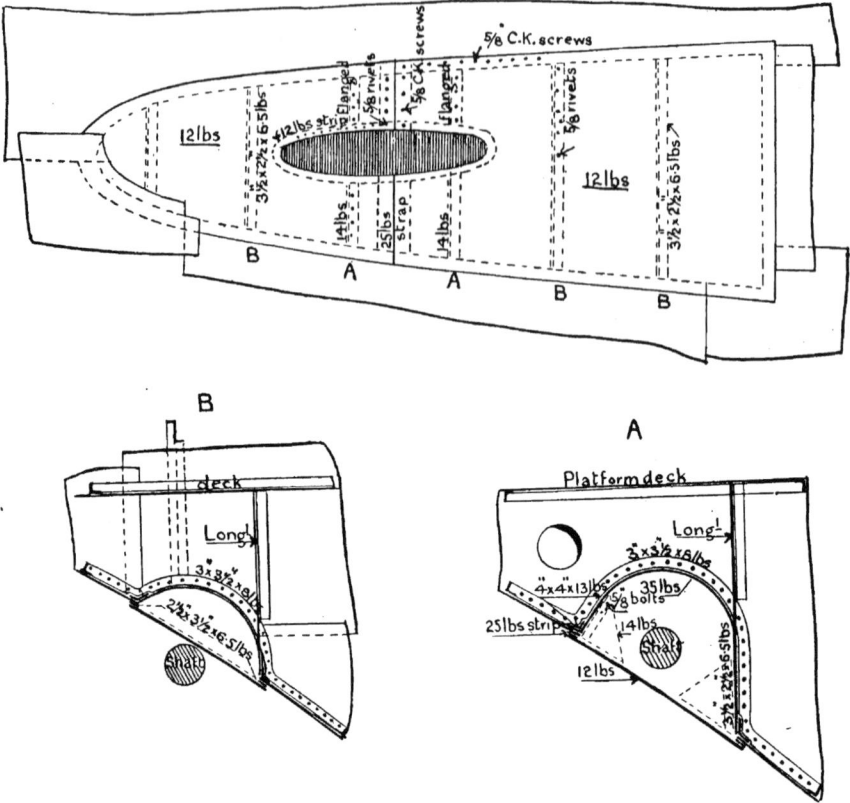

WASHPLATES AND FRAMING OF SHAFT RECESS.

SCRIEVE BOARD.

On p. 20 a fore body is shown on the scrieve board, and the lines drawn on it are: inner and outer bottom frame lines; inner and outer (sight) edges of longitudinals, beam at middle and beam at side lines; protective deck in full; flats; plating behind armour; bilge and docking keels; plate edges; lines of bracket frames.

The plate edges of inner and outer bottom are painted in distinctive colours to avoid confusion. They are shown in the figure as broad black lines, and only those for outer bottom are shown to avoid confusion.

Before describing the lining-off of frames in detail, it is

necessary to consider the methods adopted, usually two, viz. the *mould* and the *scrieve board* systems. In the former a mould is made to each frame from the body on the mould loft floor, while in the latter a copy of the body on the mould loft floor is transferred to a specially prepared floor near the bending slabs, the frame lines being drawn in on both sides of the middle line, the forward and after bodies being kept separate

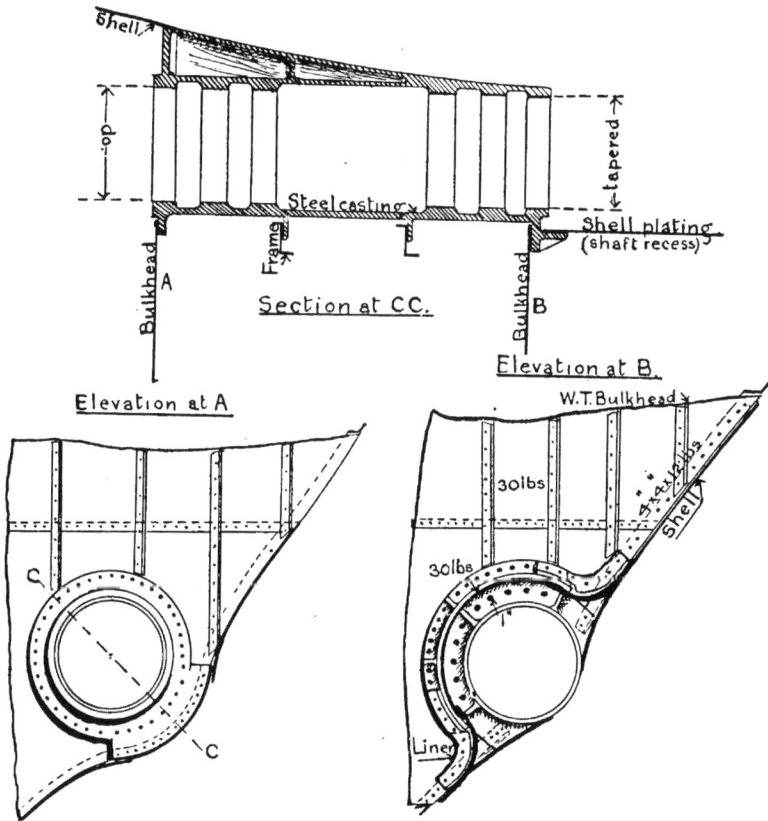

FRAMING IN WAY OF PROPELLER SHAFT.

With the *scrieve board*, work can be carried on very rapidly, the board being near the building slip, and information can readily be obtained from it, without the necessity of going to the mould loft. It would appear that to obtain the best results a combination of the methods is advisable, moulds being made to all watertight frames at the mould loft as soon as this body has been faired, so that work can be proceeded with at the moulding sheds while the scrieve board body is being prepared, the remainder of the frames being lined off on the scrieve board.

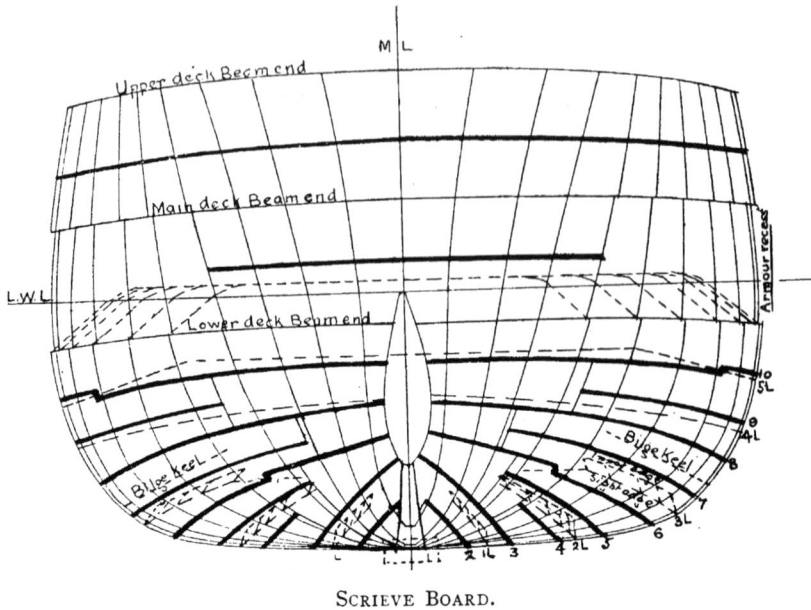

SCRIEVE BOARD.

LINING-OFF FRAMES IN DOUBLE BOTTOM (see p. 21).

The frame selected is a lightened plate frame, the procedure on the scrieve board being as follows : Pick up the inner and outer bottom frame lines and join the points where the inner and outer edges of longitudinal cut the frame lines. These lines will be the side of longitudinal nearest middle line, so it is necessary to draw in lines giving thickness of longitudinal. As the frame lines will be covered by the plates when laid down, it is necessary to have a means of picking up the lines, which is done by producing the lines and setting out spiling spots S for the frame lines.

The frame angles are first punched, omitting holes in wake of the plate edges, and then bent on the slab, the shape being given by sett irons made to the scrieve board lines (see p. 23).

As the process of bending may alter the shape of the sett irons, it is necessary to provide a check, and this is done by drawing a line on the slab giving the shape of the sett iron before the bending is commenced. The chalk line H is shown on p. 23.

The lines of longitudinals are struck across the plate, and

then lines parallel to them about $\frac{3}{8}$ inch inside. A line is also drawn at a given distance, usually 6 inches from it, which is marked on the plate, for use when erecting the frame. The frame angles

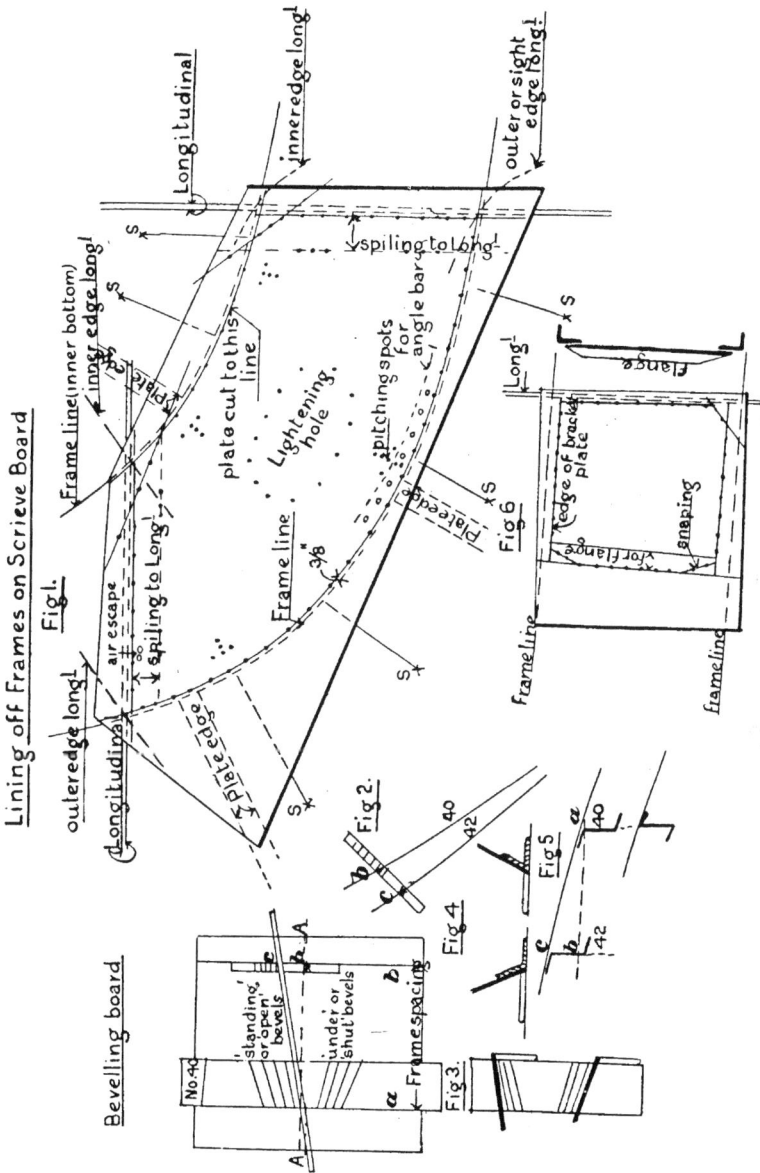

are now placed on the plates, being pitched correctly by means of the spiling spots. Centre punch marks are placed on angle and plate to ensure their coming to same position again after holes have been punched in plates, and the holes in angles then transferred to plates and positions of plate edges marked. The angle is then moved in about $\frac{3}{8}$ inch and marked by and the centre line punched in, also the holes for angles connecting frame to longitudinal. Any lightening holes or stiffeners are also marked off, and then the plate is sent to shop to be punched, etc. The frame is completed by riveting frame angles to plate, taking care when doing so that the centre punch marks in angles and plate agree.

If it is found that some of the holes are blind, they must be rimed, and the angle bars must not be shifted to give good holes, or it is clear the frame angles will not agree with frame lines, and the frame will not be the correct shape.

The necessity for the 6 inches spiling will be clear when it is considered that the plates are $\frac{3}{8}$ inch short of longitudinals on each side, and if the angles connecting them to longitudinals were marked off with plates in contact with longitudinals, the latter would be forced out of place.

Before the frames are erected the angles which connect them to the longitudinals are first riveted to the longitudinals, and the heels of the bars caulked. This applies to non-W.T. frames.

BEVELLINGS.

The bevellings are obtained by measuring the normal distance between consecutive frames as *bc*, Fig. 2, p. 21 ; set off on the bevelling board the "*frame spacing*" equal to *ab*, Fig. 5, and from any line AA drawn across the bevelling board set up *bc* as shown, then it is clear that the angle between AA and the batten on the bevelling board is equal to *cab*, Fig. 5, the required bevelling for frame No. 40. The distances as *bc* are conveniently taken off at the plate edges and longitudinals on a stick, and the bevellings for each frame are set off on a board as that marked No. 40.

Fig. 4 shows how the bevelling is applied on the slab, and consequently the smith will take the bevelling as in Fig. 3.

Figure on p. 23 shows how angle and Z bars are bevelled by hand, the frames being first bent to their correct form.

The sett irons F of square section, shown under Z bar, are pieces about 6 inches long, and only serve to keep the zed to its proper height.

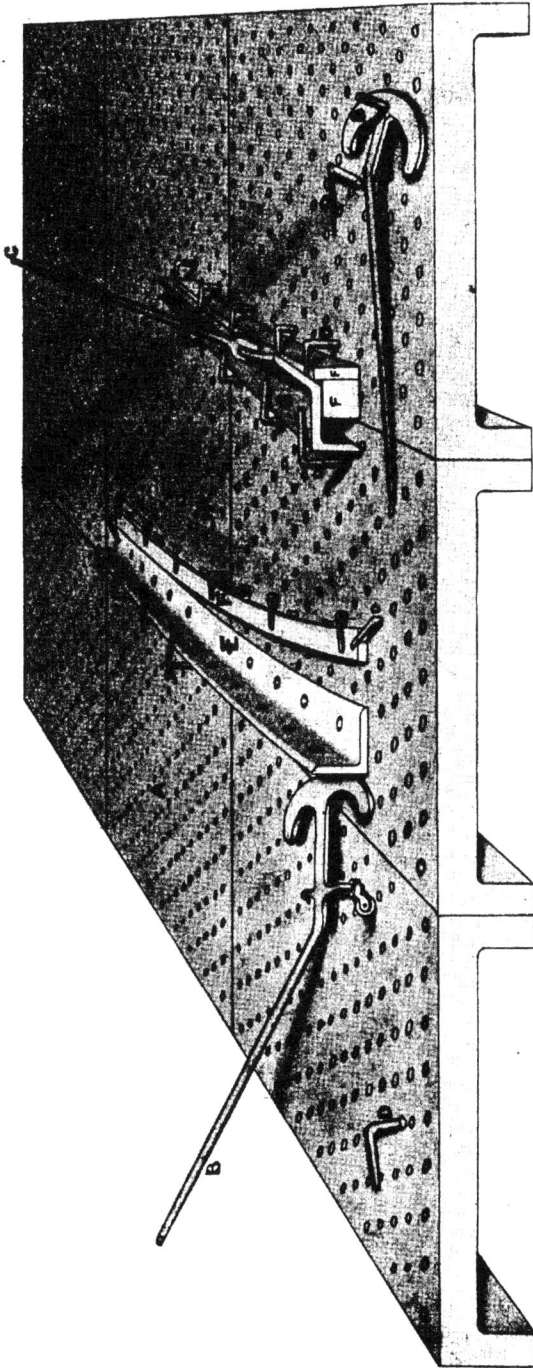

BENDING AND BEVELLING FRAME BAR

A—Slab.	C—Beveller.	E—Angle bar.	G Zed bar.
B—Squeezer.	D –Dogs.	F—Sett irons.	H Check on sett iron.

When angles and zeds are bevelled in the machine the bevelling must be done on the straight bar and the bending afterwards.

Angle and Z frames which have large bevelling are slightly hollow at the heel of the bar, owing to the greater thickness at this part, and for proper contact, to ensure watertightness, the heel should be chipped flush. The flanges should be slightly hollow, so that the riveting will thus close the heel and toe of the bar.

If the flange has a round on it, the bar will only be close at the rivet and the heel, and will stand off, making it difficult to obtain efficient caulking without resorting to liners.

This more particularly applies when the rivet is large compared with the thickness of the flange of the bar.

INNER BOTTOM.

The method of working this is shown on p. 25, the plates usually being lap-jointed, the edges being arranged so that water shall not lodge at the laps. The moulding of the plates is similar to that described for outer bottom plating, all the plate edges having been cut in on the frames when lining them off on the scrieve board.

The inner bottom is faired on the mould loft floor by means of buttock lines for the flat portions, and by diagonal lines for the other portions.

The riveting depends on whether oil or water is to be carried, and if the former, the two rows of rivets will be differently spaced, that nearest the middle line being watertight, and the other oiltight.

Access is provided to all double compartments through the inner bottom by fitting raised manholes as shown on p. 110.

Pockets have to be formed in the inner bottom to catch the drainage from wings, machinery spaces, etc. (see also under Pumping, etc.).

In some cases inner bottom plating is joggled and liners dispensed with.

LINING-OFF MAIN TRANSVERSE BULKHEAD.

As the principal decks are continuous, the bulkheads are worked intercostally between them.

The information supplied consists of a working sketch giving disposition of plating, stiffeners, riveting, and marks on each plate.

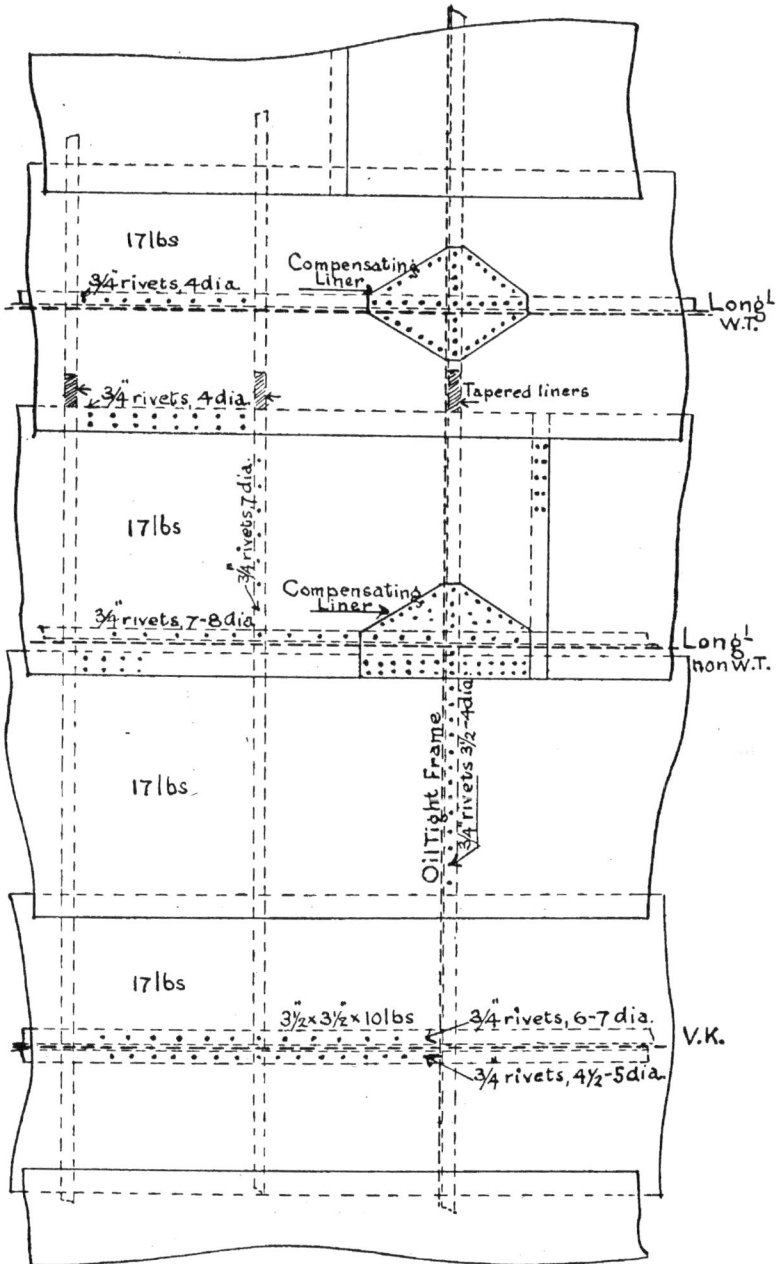

INNER BOTTOM.

Referring to p. 26, pick up from the scrieve board the frame line, and by the means of level lines, etc., transfer the curve to the place where bulkhead is to be laid out.　Next set out spiling marks to enable frame line to be picked up where plates cover it, and lay down the plates in accordance with the working sketch ; the plates being worked in horizontal strakes, edges and butts of plating generally being lapped for transverse bulkheads.　In laying down the plates, arrange so that it shall be necessary to shear one edge only of the plates, as the plates, as received from

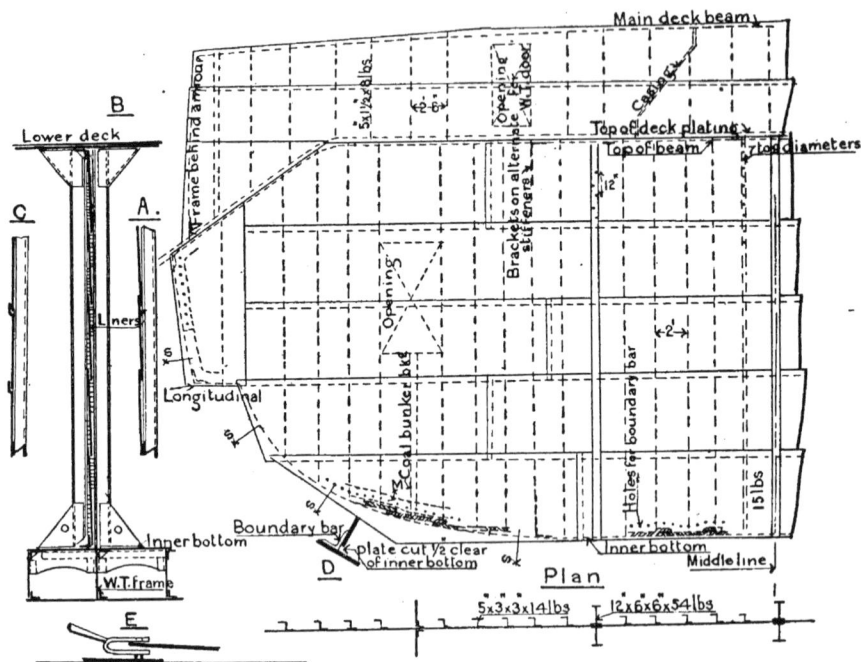

LINING-OFF MAIN TRANSVERSE BULKHEADS.

steel makers, may not always be rectangular.　While this is in progress the boundary angles are punched—omitting holes in wake of plate edges—the bending and bevelling being carried out on the slab, as described for frames.

The boundary angle is made up of several lengths, and for those portions which have sharp turns moulds are supplied to the smiths, and no holes are punched in the bends, as they would close up or elongate when forming the bend.　The boundary angle is placed on the plates, being correctly pitched by measuring from the spiling marks, and the holes in it transferred to the plating, and centre-punch marks are placed on plate and angle to ensure their being brought to correct relative positions before marking off holes in plates from boundary angle. The boundary bar is now shifted in about ½ inch from frame line, the angle marked by and the plating cut to this line to allow the heel of the boundary bar being accessible for caulking against the bottom plating.

The holes in edges and butts are marked off and transferred to the lower plate by means of the tools shown. Next arrange rivet holes for stiffeners from template, the spacing for these being non-watertight, *i.e.* 7 to 8 diameters. The Z stiffeners are so placed that the upper edges of laps of plating can be caulked for their whole length, as in Fig. A. If the edges of plates are joggled, then the stiffeners are placed on the flush side, as in Fig. C, and liners will have to be worked where the boundary bar crosses the plate edges.

The stiffeners are generally on the after side in after body and *vice versâ.*

Any openings in the bulkheads are lined off before plates are disturbed, the plates punched, sheared, rolled, and then the bulk head is ready for erecting.

The stiffeners are erected as soon as possible, as they keep the plating together and the bulkhead to shape. The whole is temporarily bolted together and plumbed, that is, the plane of the bulkhead must be square to the load water line. This is done as described for frames, by means of the declivity base, or a plumb-bob and base. Tapered liners will have to be fitted under every stiffener where it crosses the lap, but the liners need only be about 6 inches long and not the whole width of a plate. The stiffeners are generally angle and Z bars, with H bars in the place of the former at intervals. To avoid the necessity of fitting long liners along the inner bottom, this portion of the boundary angle may be joggled over the laps, as shown.

The main transverse bulkhead between engine-rooms will be seen to be very strongly stiffened both horizontally and vertically, the former being worked intercostally and also attached to the latter in some cases. To ensure getting the full amount out of the stiffeners, they must be efficiently connected at head and heel to the deck and inner bottom, which is done by brackets flanged on the free edge to further resist buckling (see p. 108).

At the ends of the ship the transverse bulkheads have the plating worked vertically to avoid having a number of short lengths to rivet off, which would be the case if the plates were worked horizontally. These bulkheads are completed and hoisted into the ship whole.

In addition to the ordinary stiffening of the bulkheads in machinery spaces to withstand water pressure, local stiffening is necessary to prevent the plating buckling where steam pipes are attached to them. Such stiffeners have often to be stopped short of the top of the bulkhead, and the bulkhead stiffened by fitting doubling plates and brackets.

The behaviour of the bulkhead is noted on the steam trials and additional stiffening fitted if necessary. This stiffening is in addition to the results of the water test.

The load water line is marked on all bulkheads, as it is required for a reference line for taking measurements during the construction of the ship.

Taking Account of Bottom Plating.

This term is applied to the process of obtaining the shape of a bottom plate. Before actually making the moulds the plate edges are faired after all the frames are in place. This is done by means of a fairing batten placed with its edge coinciding with the plate edges, and allowed to bear fairly against the frames without any edge set, when it can be seen if the plate edges lie on a fair curve or not, and any small adjustments necessary can be made. This also gives a check on the form of the ship.

Having faired the plate edges, a surface mould is made, the top and bottom battens of which coincide with the plate edges. The battens are held in place by clips as shown, and then cross battens are placed at every frame and at the butts.

For plates without twist this surface mould will be all that is necessary; but if the plate has twist, additional moulds are made at the frames, generally two, the edges of which outwind. These cannot be used when checking the plate, as it will lie with its concave face upwards, as in the figure, so that it is necessary to make reverse section moulds to the former, the edges of which likewise outwind.

The surface mould is laid on the plate and clipped to it.   The
holes on the cross battens, which indicate those in the frames, are

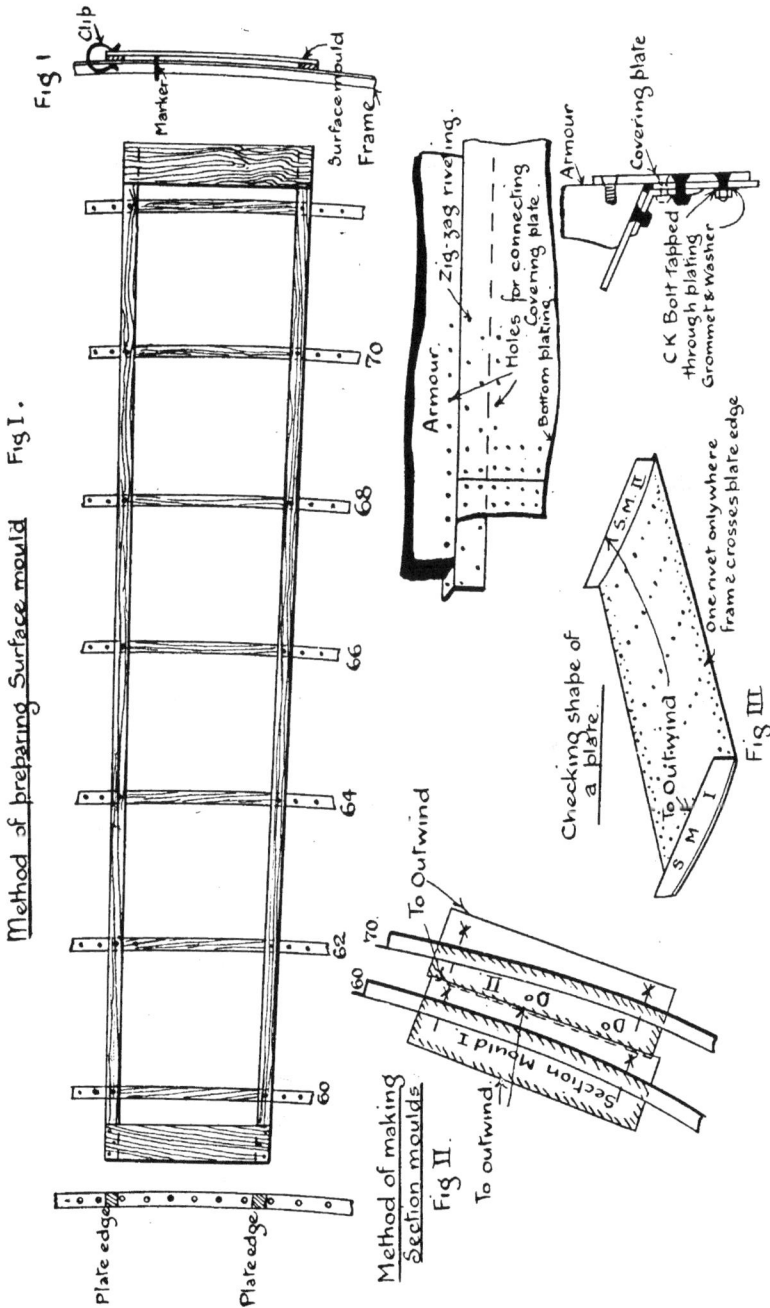

Method of preparing Surface mould   Fig I.

Fig I

Clip

Marker

Surface mould

Frame

70

68

66

64

62

60

Plate edge

Plate edge

Zig zag riveting.

Armour.

Holes for connecting
Covering plate

Bottom plating

Covering plate

Armour

C K Bolt tapped
through plating

Grommet & Washer

Checking shape of
a plate

To Outwind

S.M. II

S M I

One rivet only where
frame crosses plate edge

Fig III

Method of making
Section moulds
Fig II.

To Outwind

To outwind

To outwind

Section Mould

I

II

70.

60

first marked on the plate, after which the outline is marked by
from surface mould.   The surface mould is now removed, and the

rivets for edges and butts disposed, the size and spacing being obtained from the scheme of riveting, remembering that only one rivet is placed at the edge where the frame crosses it, as shown on p. 32. It will be remembered that when lining-off the frames, the holes in wake of the plate edge were omitted, these being drilled off after the plate is in place. In the case of a sunken strake, the rivet holes along the edges will clearly not have to be countersunk, while those at the frames and also at the butts must be, and the plates must therefore be marked to show this difference. The plate is sheared to the lines shown on it, after which it is punched from the faying surface, and the butts planed.

Sight edges only of plates are planed, the remainder being rough sheared and chipped if necessary to remove sharp edges. The plate is now put through the bending rolls to give it the correct twist. When the plating is of H.T. steel the holes for the edge connections in the sunken strakes are not drilled until the raised strake is worked, when the latter is used as a template for drilling sunken ₛtrake·

Below are shown a ring punch and a double centre punch, the former being for the purpose of providing a check on the correctness of shearing or planing the edges of a plate, and the latter as a check on the correctness of drilled holes.

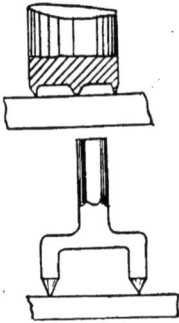

Raised Strake.—There are a few slight departures in this case from the previous ; for example, as the holes are already in the sunken strake, they as well as the holes in the frames must be copied on the edge battens of the surface mould, and further, the cross battens at the frames must be blocked off from them by pieces the thickness of the plate.

It will be obvious that all the holes in this plate will have to be countersunk.

The wide liners in wake of W.T. frames at all the raised strakes are sometimes riveted to the frames before erecting, alternate rivets only passing through frame angle, liner, and bottom plate. Hence the number of holes in bottom plate for attachment of a watertight frame will be less than at the other frames, these of course being copied from the frame itself.

The plates are temporarily secured by bolts, and it is of the greatest importance that they should be properly "closed," that is, made to fay close against the frames, so as not to bring undue strains on the rivets on cooling.

The holes in the shell plating in wake of bilge keels, and the portions of docking keels formed of forgings, should be omitted until the positions of these fittings have been lined off at the ship. The fastenings for these can be then disposed with a view to avoiding an unnecessary number of holes, especially at the butts of shell plating where the bilge keel angles cross them, and so avoid weakening the shell.

The strakes of bottom plating immediately below and above the side armour have an additional row of holes (tapped) for securing the covering plates. These plates must be riveted off and caulked complete before covering plates are fitted.

The holes for securing edges of covering plate to the armour are all formed in the armour before it is delivered, and as it may happen that the holes in armour and those in the bottom plating for securing the covering plates are not vertically over one another, the butts of the covering plate will have to be cut diagonally.

The details of the connections are as on p. 29.

The intention in having bolts instead of rivets for the lower edge of the covering plate is to do away with the possibility of the caulking of the edges of deck and bottom plating being affected by the hammering if the latter were used, and also if the armour is struck the bolts would probably sheer at the plating, the holes remaining filled.

The bottom plating is generally double riveted, except the sheer and garboard strakes, which are treble riveted. Where high tensile steel is used, the butts are treble riveted, and the holes kept $\frac{1}{8}$ inch further from edges.

Liners are worked on the raised strakes at every frame, those at the bracket frames and lightened plate frames being the width of the angle only, and lightened between the rivet holes. If the frame angles are joggled over the inner and outer bottom plating, no liners will be necessary.

At the watertight and oiltight frames special wide liners are worked to compensate for the loss of strength due to the close spacing of the rivets, and where such liners come near a butt of bottom plating the liner and strap are combined as shown below. It should be noticed that the liners are divided by the

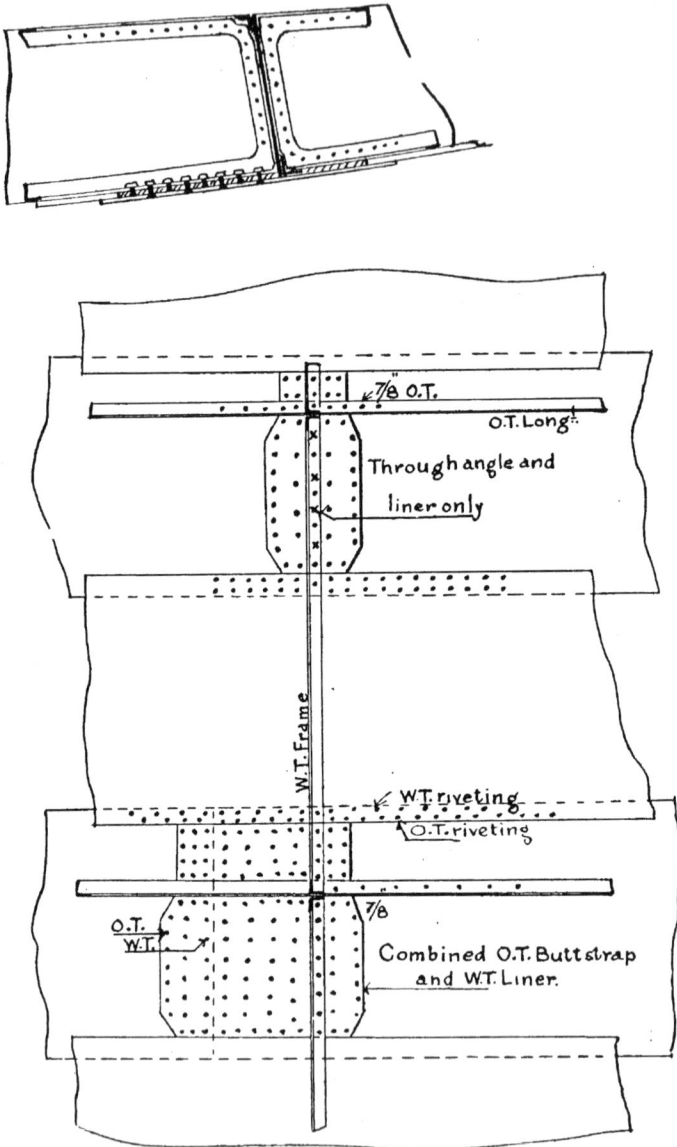

longitudinal into two portions, otherwise the lower angles of the longitudinals would have to be joggled over them.

The bottom plating is connected to the stem and stern castings as shown on the sketches of those castings on pp. 33–36. Through

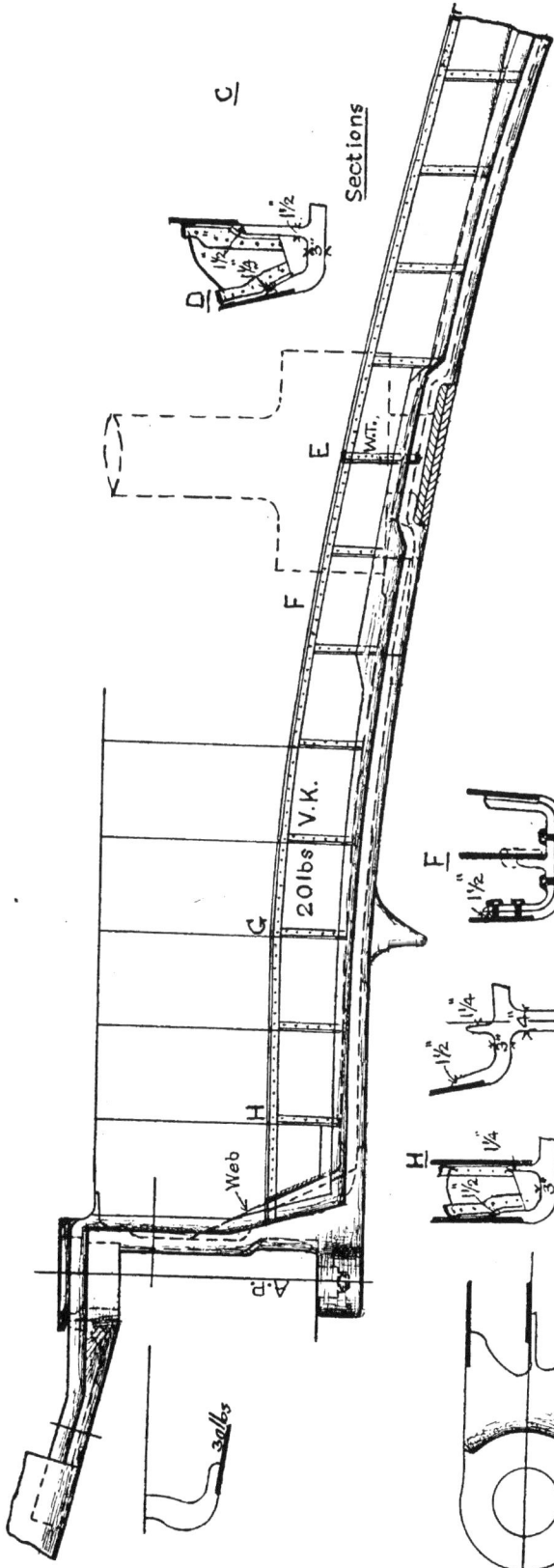

Sections

C

D

1½"

1½"

3"

E

W.T.

F

20 lbs V.K.

G

H

Web

A.P.

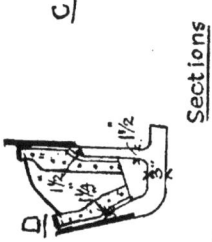

B A

W.T.

C

D

20 lbs
25 lbs

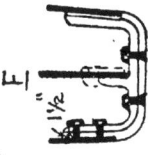

F

1½"

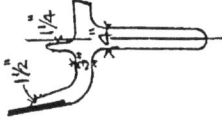

1½"

1¼"

¾"

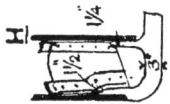

H

1½"

1¼"

3"

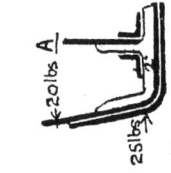

A

20 lbs

25 lbs

Sections

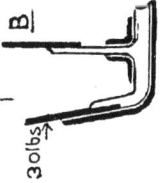

B

30 lbs

30 lbs

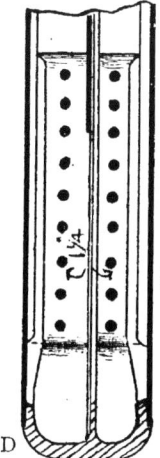

Plan of Scarph.

1¼"

D

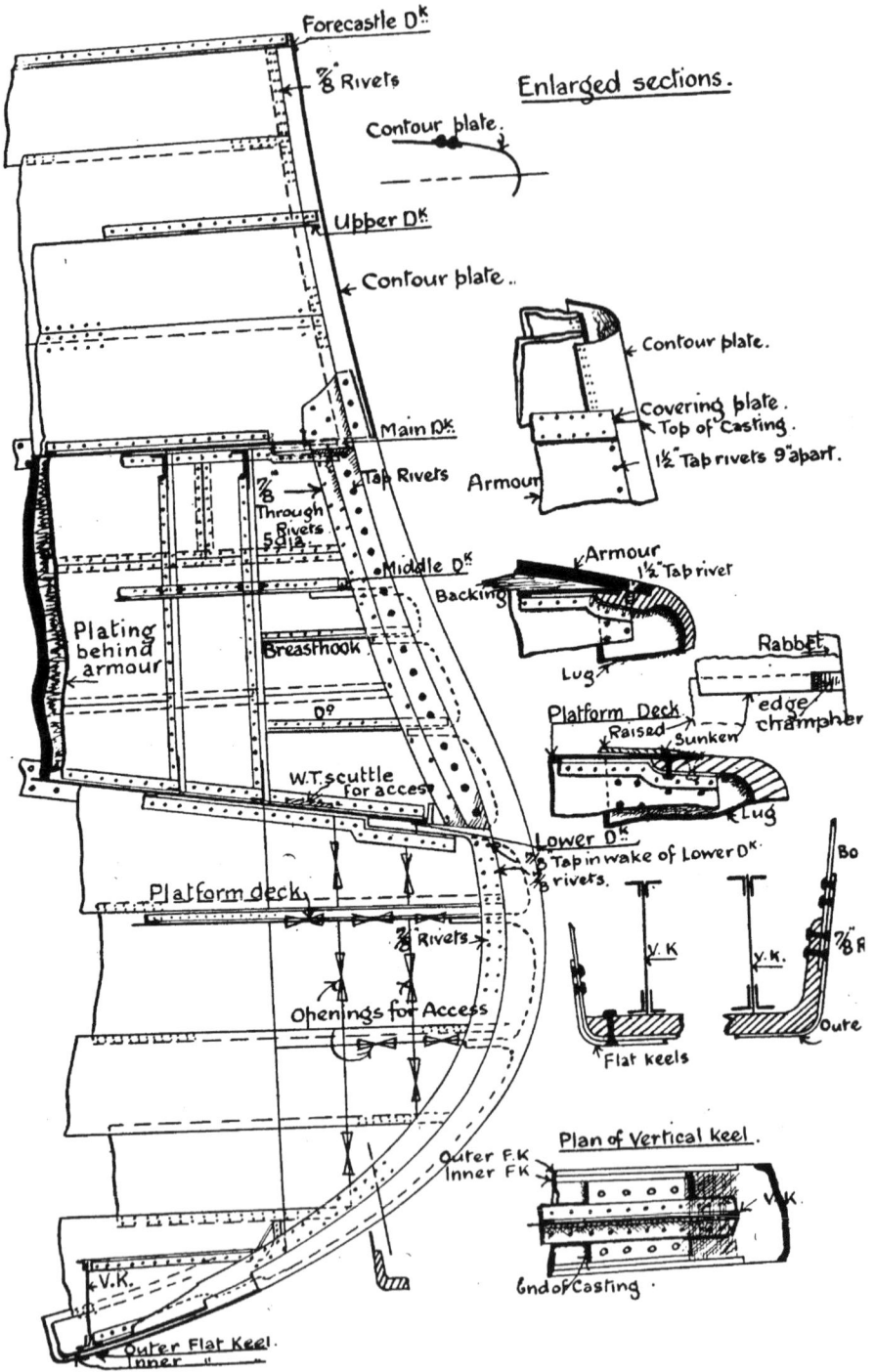

Forecastle D^K.

⅞" Rivets

Contour plate.

Upper D^K.

Contour plate..

Main D^K.

Tap Rivets

⅞" Through Rivets 5 dia

Middle D^K.

Backing

Plating behind armour

Breasthook

D^9

W.T. scuttle for access

Lower D^K.
⅞" Tap in wake of Lower D^K.
⅞" rivets.

Platform deck

⅞" Rivets

Openings for Access

V.K.

Outer Flat Keel.
Inner "

Enlarged sections.

Contour plate.

Contour plate.

Covering plate.
Top of Casting.
1½" Tap rivets 9" apart.

Armour

Armour
1½" Tap rivet

Lug

Rabbet

edge chamfer

Platform Deck
Raised   Sunken

Lug

Bo

⅞" R

Outer

V.K.   V.K.

Flat keels

Plan of Vertical keel.

Outer F.K
Inner FK

V.K.

End of Casting.

STEM FOR BATTLESHIP.

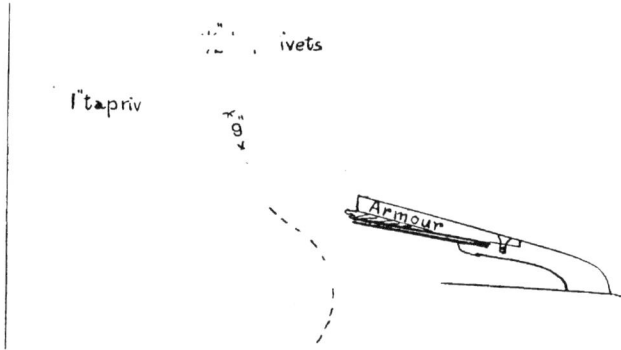

ivets

l"tapriv

Armour

Thro
ri

Web

Section at 1.

Wet

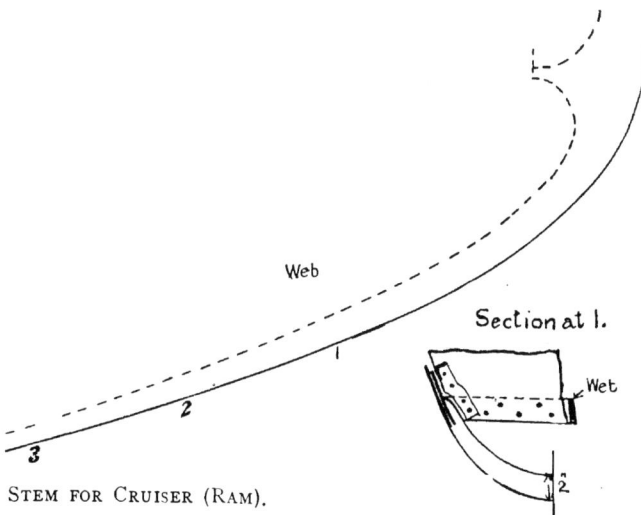

STEM FOR CRUISER (RAM).

rivets are always worked where possible to connect the plating to the casting, but in the way of deck lugs and webs on the casting tap rivets must be used.

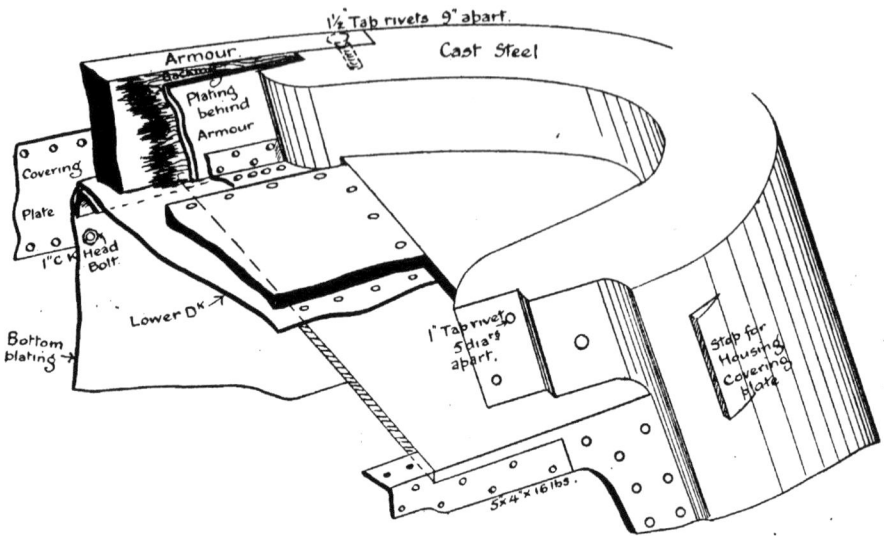

The method of arranging the rabbets on a casting for two thicknesses of plating are shown. Two cases arise (1) in which the inner surface is flush and the doubling on the outside, *e.g.* the

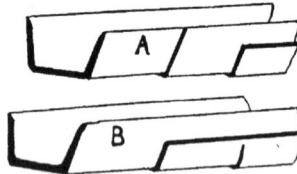

two flat keels on to the stem (Fig. A), and (2) in which the outer is flush and the doubling on the inner side, *e.g.* the two thicknesses of contour plating on the stern casting (Fig. B).

Two methods of working stealers are given on pp. 37 and 39. The first requires the fitting of tapered liners, as the plating will be clinker fashion beyond the point where stealer stops. In the second, the end of stealer is arranged to come at a butt of adjacent plating which gives parallel liners, but edge strips have to be worked.

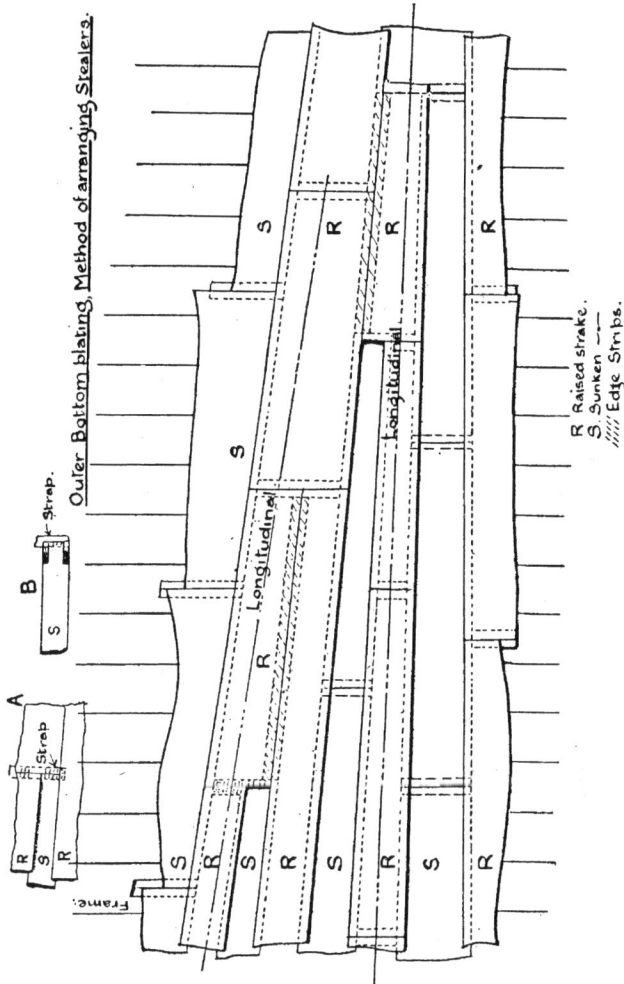

Outer Bottom plating. Method of arranging Stealers.

R Raised strake.
S. Sunken ——
//// Edge Strips.

A slight modification to the second method is given in (A), where the upper edge of sunken strake is chased away to give a flush surface to the plating, thus permitting of the butt strap being continued across the stealer as well as the adjacent plate, the plates being worked clinker beyond the butt, no edge strips are required.

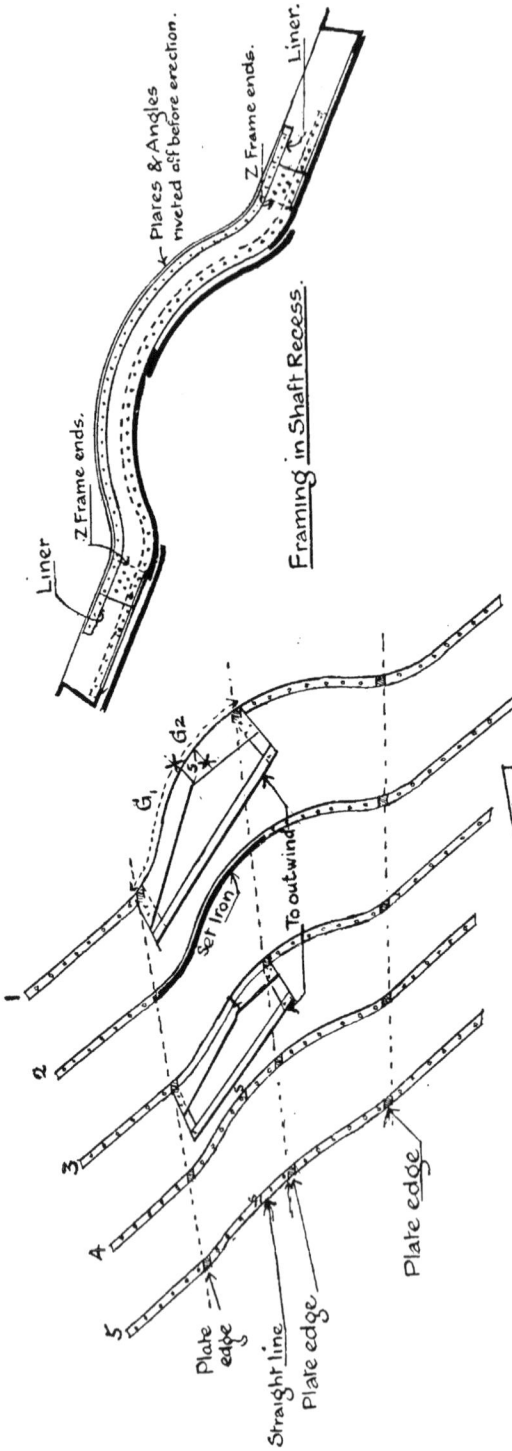

Framing in Shaft Recess.

Taking account of Plating in Shaft Recess.

PLATING SHAFT RECESS.

This is taken account of as on p. 38. The plate edges are marked on the frames, and moulds made to give the curvature

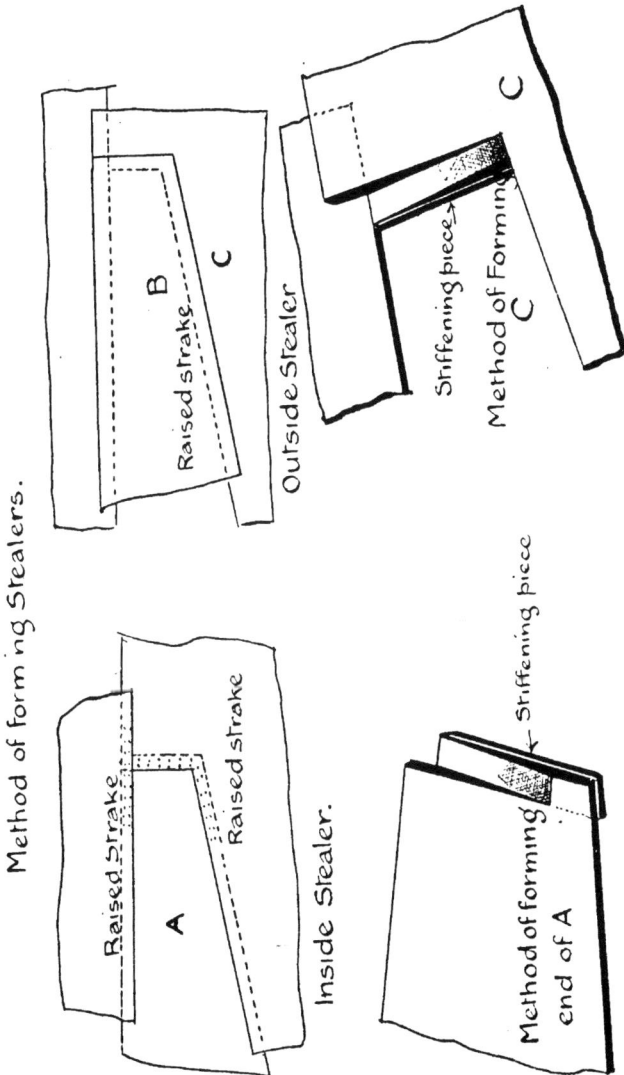

of the plates, the edges of these moulds being made to outwind. A straight line is marked on the frames, and this line together with the plate edges are marked on the moulds.

The moulds are girthed, and the distances set off on the plate as shown from the straight line, making due allowance for the dishing. Sett irons are made for checking the plate while being furnaced, the final check being made with the moulds referred to previously.

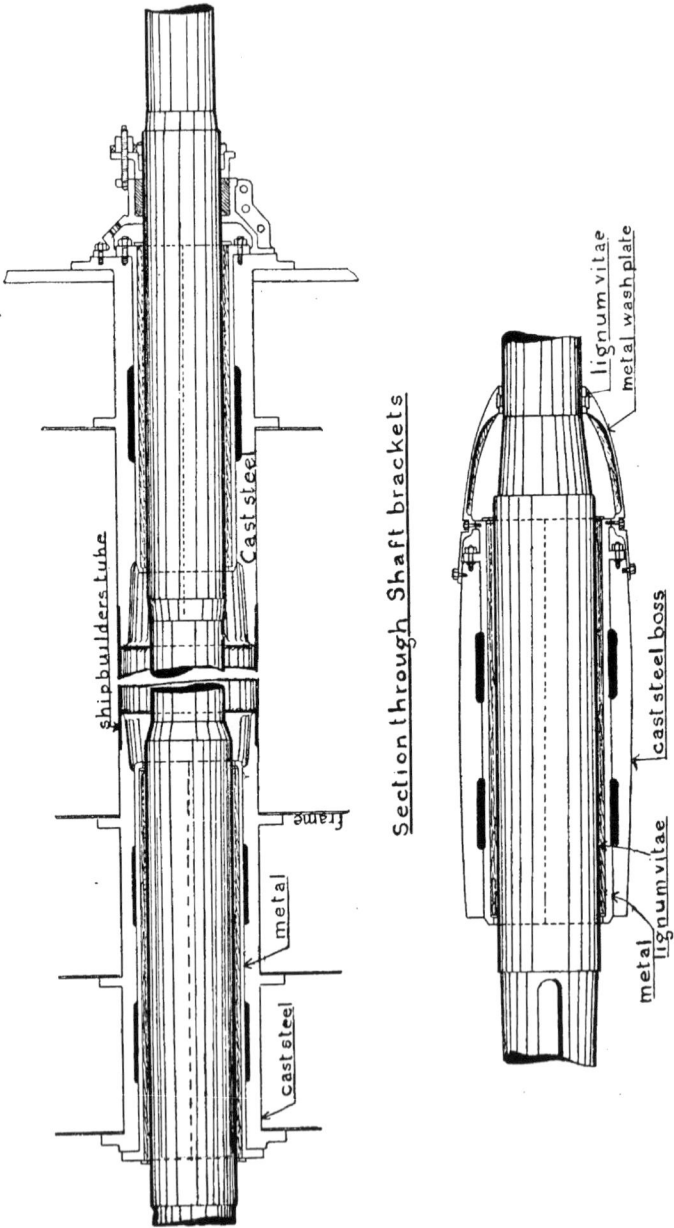

The shipbuilder's tube through which the propeller shafting passes secures watertightness at this part of the ship, and also forms a bearing for the shafting. It is formed of steel castings, one at each end, which form the bearing for the shaft, and plates bent to form and riveted to the castings, being connected to the boss frames, etc., through which it passes, as shown on p. 19.

The tube is made watertight, and at the forward end a gland is fitted around the shafting to prevent water entering the ship at that part. For cleaning and painting purposes manholes are provided, to the covers of which zinc protectors are attached, the latter having perforated plates fitted over them to prevent the shafting being scored should the zinc protectors become detached owing to the corrosion of the securing bolts.

BEAMS.

The majority of the beams are of angle bulb section, and have a round-up on all decks above the protective deck. This

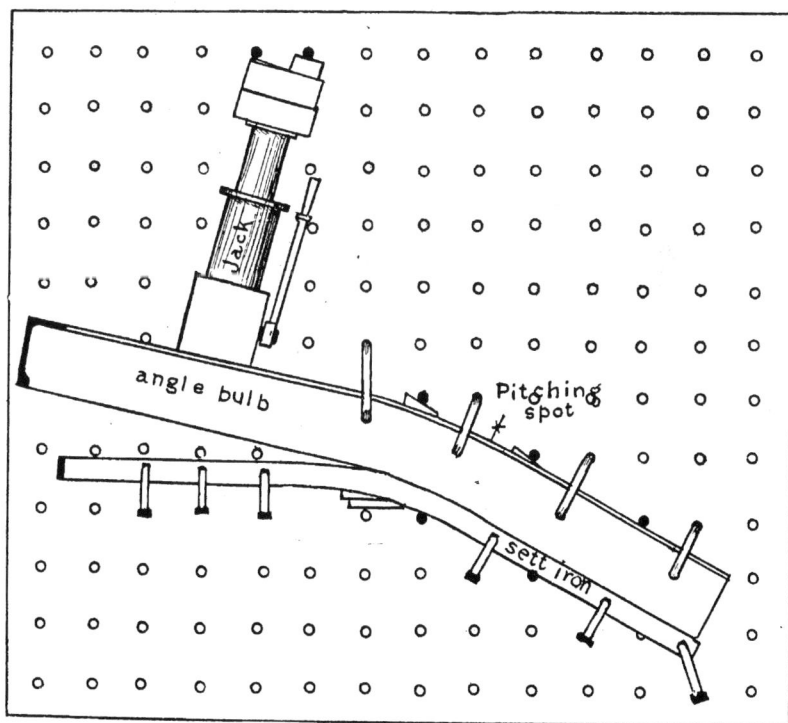

BENDING LOWER DECK BEAMS.

curve of the beam is an arc of a circle, the round-up being the amount the arc rises above its chord, and varies with the beam of

Checking & Maintaining form of Ship while building & permanent shoring of Beams.

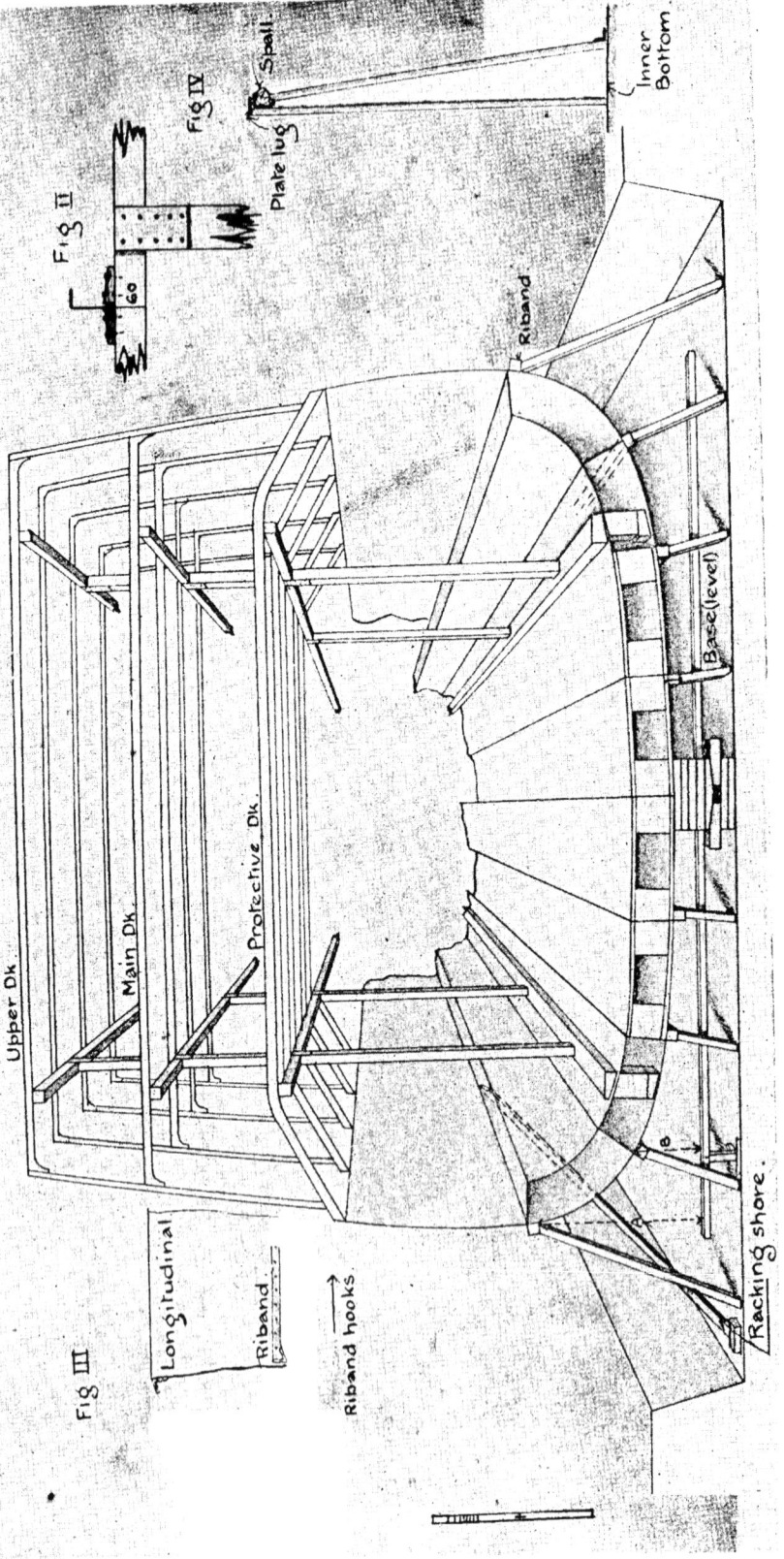

Fig III

Fig II

Fig IV

Plate lug   Spall.

Inner Bottom

Riband

Base (level)

Upper Dk.

Main Dk.

Protective Dk.

Longitudinal

Riband.

Riband hooks.

Racking shore.

the ship, the usual amount being from 9 to 12 inches. The method of bending lower deck beams is shown on p. 41. The holes in the flange of the beam for connecting it to the deck plating are first punched, omitting those in wake of plate edges. The beam is bent cold to the beam mould, and two holes are made in each bracket or knee, for temporarily connecting it to the frames, the remaining holes in the bracket being drilled off

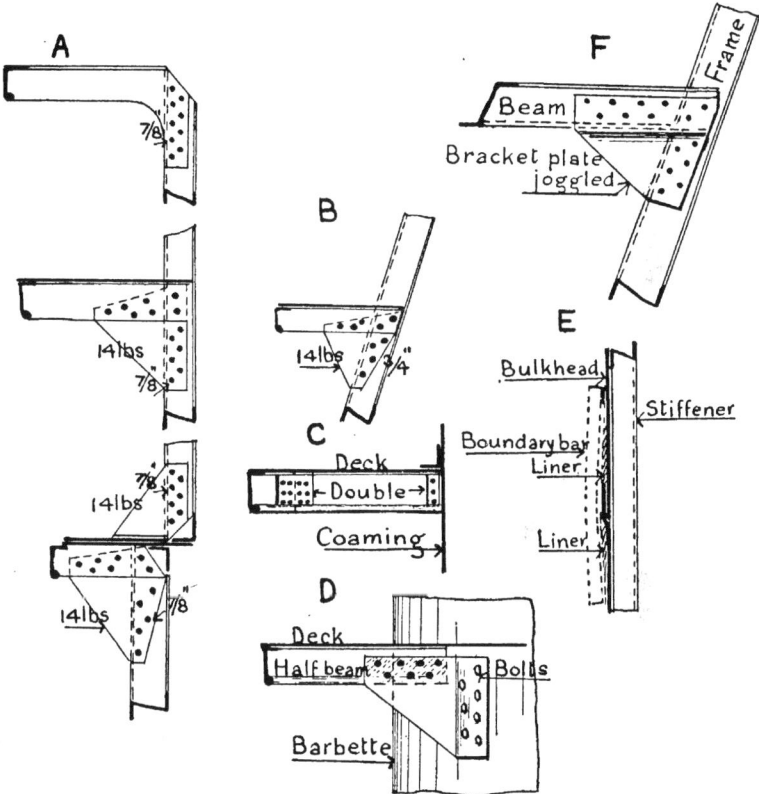

in place after the beams have been faired and set to their correct heights by means of battens supplied from mould loft.

As the protective deck at side forms a shelf for the armour, especial care is necessary in fairing this, and to permit of this being done the heels of the frames immediately below this deck are left loose, and the angles connecting their lower ends to the longitudinals are not riveted off until the armour shelf is fair.

Long beams have spalls lashed to them to prevent them altering shape under their own weight when being hoisted in.

The beams after being bolted to the frames are temporarily shored as shown on p. 42, and after a number have been erected the permanent shoring is got in and the temporary ones removed. The permanent shoring is shown in Fig. p. 42, and consists of quartering running fore and aft under the beams supported on struts reaching to the inner bottom for the protective deck beams, and to the deck below for beams above this deck, all the struts,

Making an angle bulb W.T.

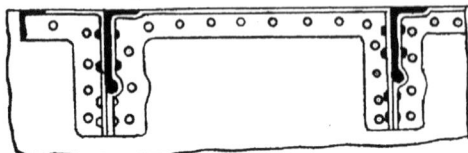

FIG. A.

etc., being in the same vertical plane to give rigidity. To prevent damage to the inner bottom, short struts are placed between the inner and outer bottom as shown, wedges being placed under each strut for making any required adjustments. The spacing of the frames is kept by means of chocks on the quartering, as shown, in which the bulbs fit. Where deck plating has to be left loose for shipping machinery the beams must also be cut in wake of the openings, the remaining portion being securely connected to exist-

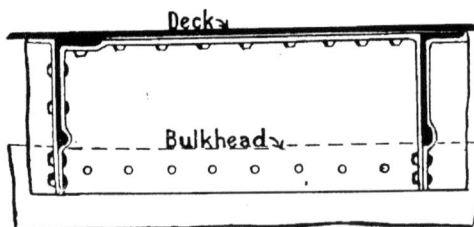

Deck

Bulkhead

FIG. B.

ing ones after the machinery is in place. Beams in wake of barbettes are connected as in Fig. D, those at funnel hatches as in Fig. C, p. 43.

It is necessary to fair decks on which wood is laid before commencing to work the planks, to ensure getting a good surface and to avoid the necessity of working planks of different thicknesses to make up for any unfairness in the deck plating.

Page 43 shows methods of connecting beams and half beams

to frames, etc., and p. 44 shows how they are made watertight where they pass through watertight bulkheads or casings.

An alternative method consists of stopping the bulkhead plate at underside of beams, and working a plate collar as shown by Fig. B.

BILGE KEELS.

The method of fitting bilge keels for a steel ship is shown below, and that for a sheathed ship on p. 46. After the position of the bilge keel has been determined at the ship, the lower angle is got in place and riveted off together with the

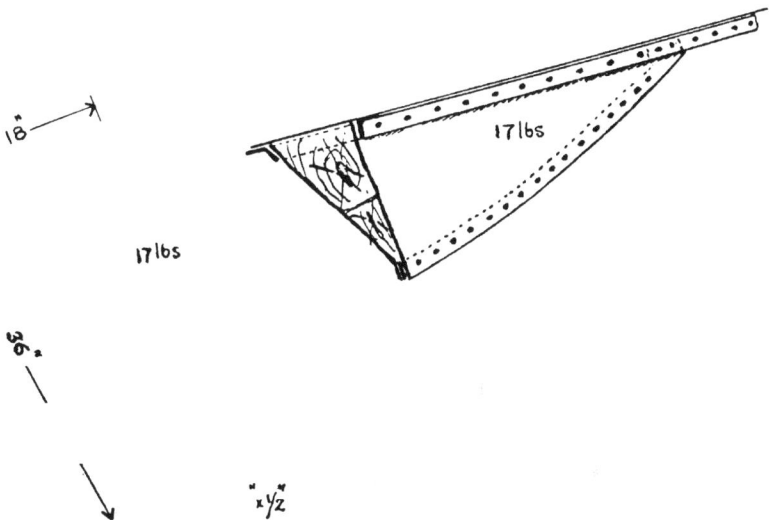

BILGE KEEL.

lower plates and their butt straps. The fir filling is then prepared and placed in position, and the upper plates and angles fitted. The upper plates and angles are then removed in sections as convenient, riveted off, and then again placed in position on the top of fir filling, when the angle is riveted to the ship's side, and the butt straps connecting the several sections of the keel are secured by means of tap rivets.

The bilge keel for a sheathed ship is formed of a plate connected by brackets spaced about 12 feet apart, worked alternately on opposite sides of the plate, and the whole covered with teak, the latter being secured to the plate by $\frac{3}{4}$-inch clenched bolts as shown. The copper sheathing is continued round the bilge keel, the copper at the lower end being 60 ozs. per square foot, and secured to it by means of screws.

Docking keels are fitted to ships of great beam with a full underwater portion for docking purposes, and at the flat portion of the ship consists of a doubling plate worked on the shell similar to the outer flat keel. At the extremities, where the form of the ship begins to rise, the keel is constructed as shown on p. 47, the forging being worked for a few feet in length, as the teak cannot be carried down to a fine edge.

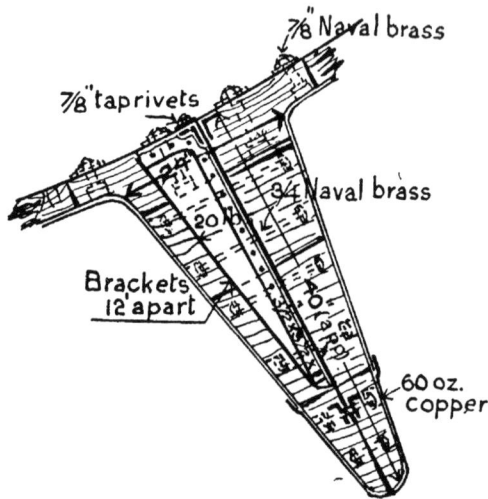

BILGE KEEL FOR A SHEATHED SHIP.

CHECKING FORM OF SHIP WHILE BUILDING.

The structure of a ship in the early stages consists of a large number of parts temporarily bolted together prior to riveting, and means must be taken to ensure that the form is being adhered to. The work of checking commencing with the—

Keel. The blocks have to be constantly removed from under the keel for riveting to be done, and when they are replaced it is necessary to ensure that the keel is straight. For this purpose the heights of underside of flat keel above floor of slip are measured before any blocks are disturbed, and after replacement the blocks must be brought to the same height. The wedge blocks enable this to be done.

If the floor of the slip is not of masonry, pegs are driven into the ground, and the measurements taken from the top of these.

Frames. The frames are at right angles to the middle line of ship, and also to the load water line. The method adopted is to strike lines on the floor of the slip square to the middle line at

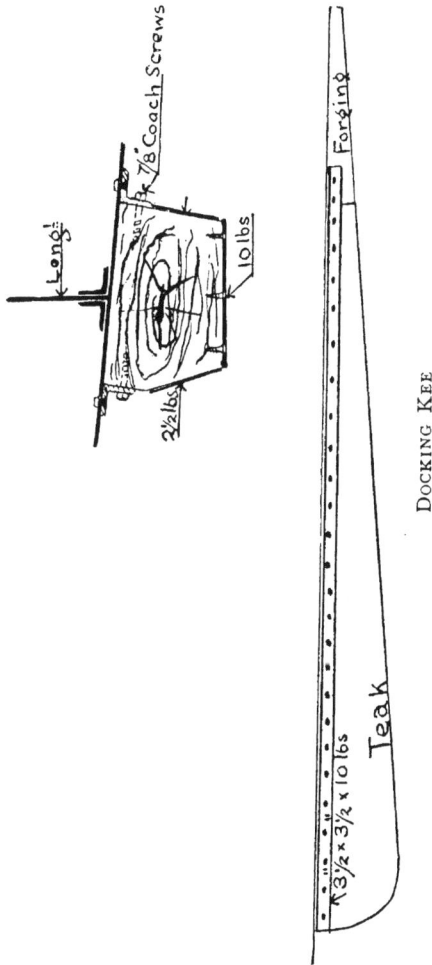

convenient intervals and cut these in, as they will be constantly referred to. The slope of the load water line being known, say $\frac{5}{8}$ inch to a foot, a plumb bob is held against the heel of frame line (see p. 48), and the distance it should swing out from the heel of frame marked on the floor of slip can readily be calculated when the height of the point of suspension is known. It should be noticed that the heel of frame line on the floor of the slip is not that plumbed down from the heel of frame at the keel. It is the line obtained by using a declivity base with its edge coinciding with the heel of frame at the keel, the other edge of the base being vertical.

To ensure the different sections of a transverse frame being in the same plane, the plumb bob is suspended at different positions round the frame. Instead of having the reference lines on the floor of the slip, a levelled base which is square to the middle line of the ship can be used, as shown.

The correct spacing of the frames is maintained by means of "ribands," which are pieces of quartering about 6 inches square, placed just below the lower edges of longitudinals, and attached

Plumbing Frames

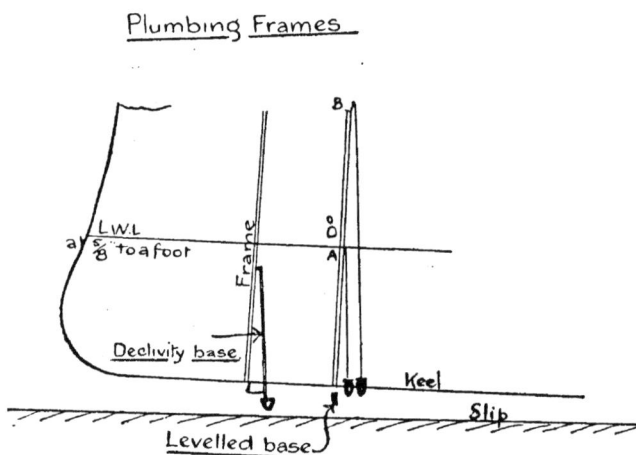

to the frame angles by riband hooks, the position of the heel of every frame and the number of the frame being marked on them. The ribands are shored as in Fig. p. 42, both transversely and longitudinally. Ribands high up in the ship would require very long fore and aft shores, and instead of these it is convenient to have wire hawsers with cat-screws attached at their upper ends to the ribands, the lower ends being carried forward and secured to the slip. The above process is called "plumbing," and the correct form is got by "horning" the frames, erecting a base square to middle line, levelling it, and making the heights at A, B, C agree with those obtained from the scrieve board. Particular care is required in maintaining the form of the bottom at docking and bilge keels. The half breadths of the ship are also checked, more especially at the fullest parts of the ship, to make sure the beam is not being exceeded.

DECK PLATING.

This is commenced after the beams have been faired. The arrangement of deck plating varies for the various decks, being worked in two thicknesses on some decks and one on others. The protective deck is generally in two thicknesses, and there are certain conditions regarding the method of arranging the plating.

In all cases the butts of the plating come between the beams, but with two thicknesses the butts of upper and lower thicknesses in the same strake must not come in the same beam space. Butt straps are sometimes fitted to both thicknesses of the deck, those for the upper thickness being on top of the deck, and for the lower thickness on the under side, as in Fig. p. 50, but no butt straps are fitted on top of a deck on which corticine has to be laid. The edges and butts are arranged as shown, and the riveting of these should be carefully noticed. The first plates laid are the stringer plates, the moulds for which are prepared as described on p. 236.

Protective Deck. A model is made of this deck on which the plating is lined off. It shows all openings in the deck, and the position of all bulkheads immediately above and below the deck, also openings required for shipping machinery. The first thickness of plating is worked directly on the beams, and as the holes are already in the beams they must be correctly copied on to the plates, so that when they go in they will correct the spacing of beams. The upper thickness is worked as on p. 50, and account must be taken of the holes for lower thickness by means of moulds and copied on to the upper thickness. The edges and butts of both thicknesses are planed.

Certain of the plates of both thicknesses are closers, and moulds are made for these. As each plate of the upper thickness is laid, its edges are caulked against the lower thickness, stopwaters being fitted only where absolutely necessary, *e.g.* in wake of bulkheads, deck angles, hatchways, etc., and when the upper thickness is complete the edges and butts are all caulked.

The method of connecting protective decks at different levels to avoid sudden discontinuity is shown on p. 51.

Lower Deck Plating.

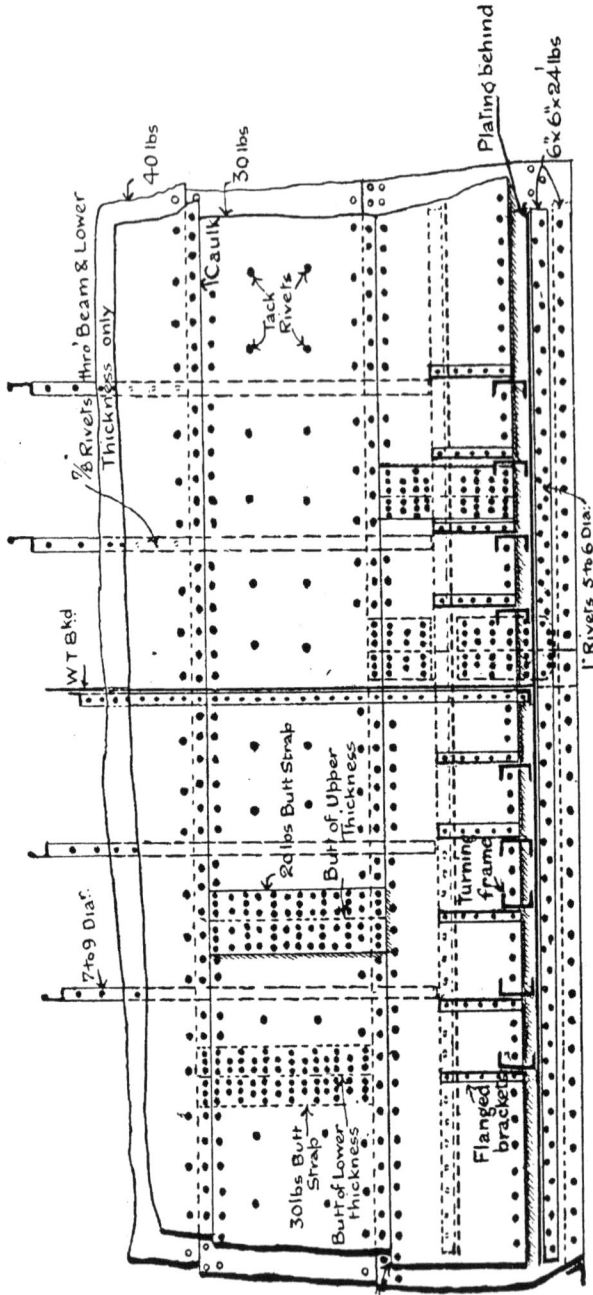

Labels within the figure:
- 40 lbs
- 30 lbs
- Plating behind
- 6"×6"×24 lbs
- ⅞" Rivets thro' Beam & Lower Thickness only
- 1" Caulk
- Tack Rivets
- Plating behind Armour 27 3 lbs (Single thickness)
- 12×3½×3½×30 lbs.
- Butt Strap
- Rivets 5 to 6 Dia.
- W.T. Bkd
- 20 lbs Butt Strap
- Butt of Upper Thickness
- Turning Frame
- 7 to 9 Dia.
- 30 lbs Butt Strap
- Butt of Lower Thickness
- Flanged brackets
- Edges lap caulked against lower thickness also flushed caulked after upper thickness is laid.

Some of the plates have to be left loose for shipping machinery, the size of necessary openings being given by the machinery contractors, and sufficient plating must be left loose to give these. These plates are lined off at the same time as the other plates, so that after machinery is shipped they have only to be put in place and riveted off. All loose plating, both in decks and bulk-

Connection of lower and middle deck

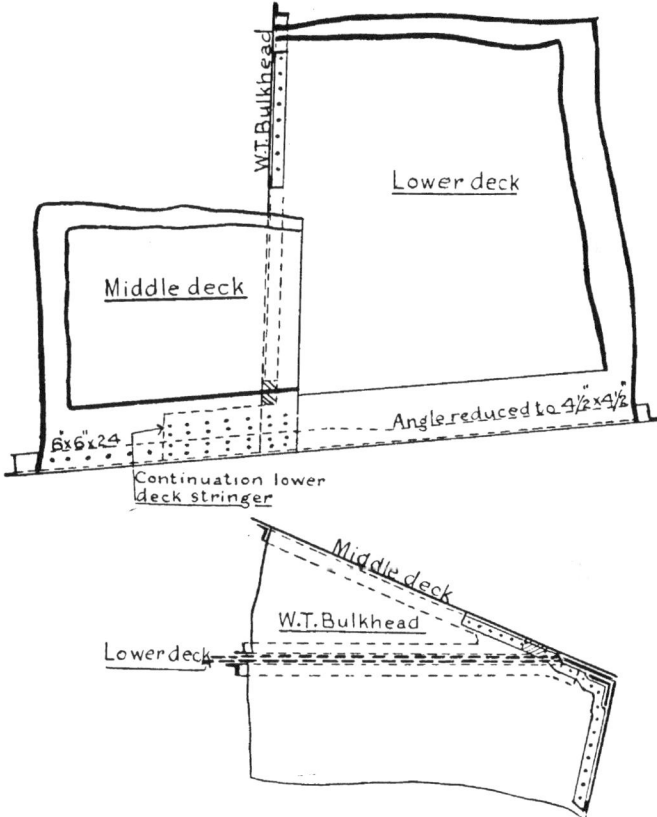

heads, should be clearly marked to prevent the plates being riveted off. The arrangement of loose plating for shipping machinery may determine the shift of butts of deck plating in the vicinity, as the openings required may include portions of plating which cannot conveniently be left loose without interfering with the progress of other work, so it is necessary to look into this question carefully.

Main Deck. This deck is covered with corticine, hence the butt

Method of working
Deck Stringers.
External View.

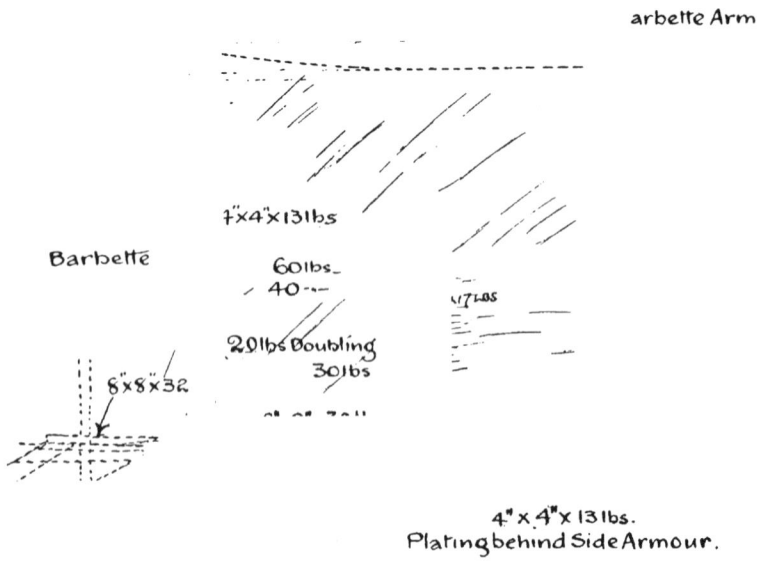

Barbette Armour.

40 lbs H.T.

30 lbs

8"x 8"x 32 lbs

120 lbs Upper
Deck Stringer

Skin plating continu
-ous.

Skin plating

Barbette Armour.

4"x 4"x 13 lbs.

6"x 6"x 26 lbs.

1" Rivets

40 lbs H.T.

8"x 8"x 32 lbs

8"x 8"x 32 lbs

Main Deck

6"x 6"x 26 lb

30 lbs H.T.
120 lbs stringer N.S

8"x 8"x 32 lbs

Side Armour.

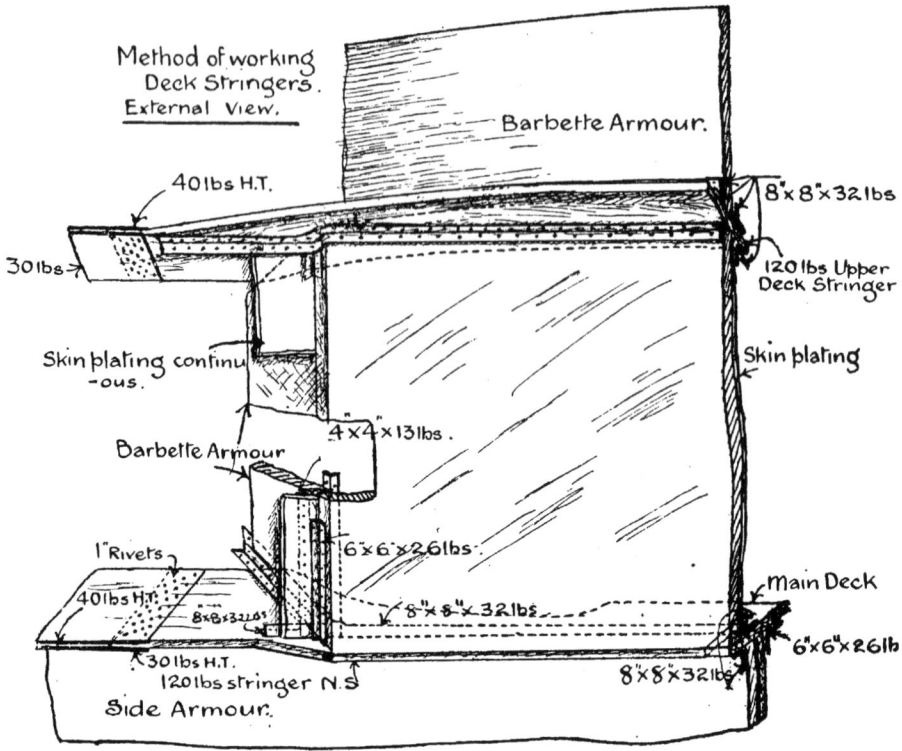

Internal View.

arbette Arm

7"x 4"x 13 lbs

Barbette

60 lbs
40

17 lbs

20 lbs Doubling
30 lbs

8"x 8"x 32

4"x 4"x 13 lbs.
Plating behind Side Armour.

straps and edge strips must be placed on the underside of the deck, the edge strips being intercostal between the beams. The stringers for this deck are laid off as described on p. 236. Portions of the deck in two thicknesses are worked as for protective deck, and no edge strips will be necessary.

To compensate for the weakening of decks by barbettes, special stringers are worked in wake of barbettes and connected to the ordinary stringers as on p. 52.

Upper Deck. The portion of this deck outside the forecastle is planked, and where the plating is of a single thickness the edge strips can be worked continuously on top of the deck. When the deck is doubled the butt straps can be placed either on top or underneath the deck, or both, if special strength is required.

Under the forecastle corticine is laid, and the plating must be flush, edge strips and butt straps being placed on the underside. As in the case of main deck, special stringers are fitted in wake of barbettes.

Forecastle Deck. This plating is worked lapped with thicker plates for stringers and is covered with teak planking. In the wake of capstan gear and hawse pipes doubling is worked.

The stringer plates at all decks follow the run of the side of the ship, the remaining strakes being worked with edges parallel to the middle line, and this necessitates the plates at the ends of ship being properly ended against the stringers, as shown on p. 54.

The decks are connected to the stem and stern castings at the ends of the ship, as shown on the sketches of these castings.

The topmost thin deck, *e.g.* boat or shelter deck, is not continuous, but has a break in it to avoid the risk of its being fractured by the strains which would come on it in a seaway.

Two methods of securing watertightness where a Z bar passes through a deck are illustrated on p. 55.

Pillars. These are fitted to the beams to stiffen them after the beams have been faired, and at places where heavy weights have to be supported, and for local stiffening, *e.g.* see supports of barbette on p. 84, also supports under capstan flats and dynamo engines, etc. The pillars between the several decks are arranged in the same vertical line for mutual support.

Under the blast of heavy guns the deck plating is stiffened by fitting girders and pillars. Some pillars have to be hinged, so that they can be turned up out of the way when necessary, as when working the capstan by hand under the cable deck. The types of pillars used are shown on p. 56. Where pillars cannot be

worked, girders under the deck are substituted. All pillars are arranged to withstand tension as well as compression.

It should be borne in mind that the bulkheads, etc., also serve as pillars, and this must be considered when arranging the pillaring.

Wood Decks.

The wood covering is teak plank, generally 9 inches wide except on the cable deck, where wider planks are worked, with a finished thickness of 3 inches. The waterways are laid first, and follow the run of the ship's side, their width being from 15 to 18 inches. All the remaining planks are worked parallel to the middle line, and ended on the waterway as shown. The planks

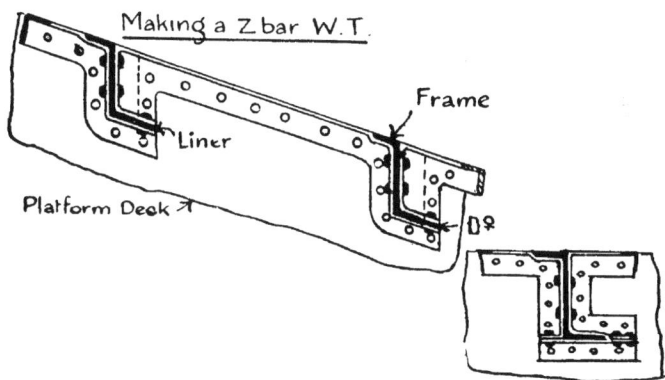

cannot be run off to a fine edge on the waterway, as their fastenings could not be placed near enough to the ends to obtain efficient caulking. To obtain the ending of a plank continue the edges until they intersect the waterway, and then square in from where the edge of plank intersects waterway, and a distance equal to $\frac{1}{3}$ the width of the plank, or slightly more than the width of a caulking iron, and join this point c with the other intersection a, as on p. 58.

The plank round barbettes is worked as shown on p. 58. It will be noticed that the planks in line with the centre of the barbette, and a number on each side of these, are ended against the barbette, but when the angle between the edge of plank and barbette becomes too acute for the fastening to be placed near the end, it is ended as shown.

The butts of the planks are placed between the beams, where

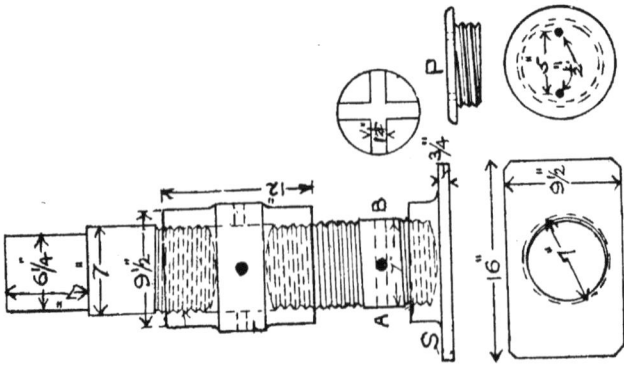

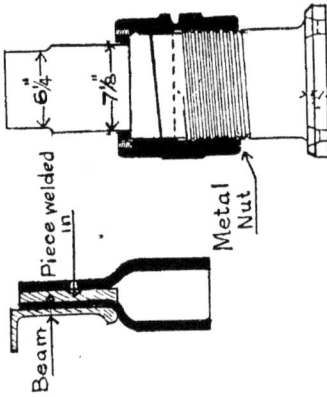

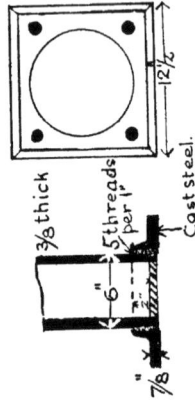

Pillars.

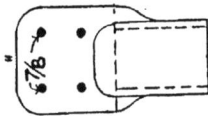

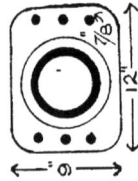

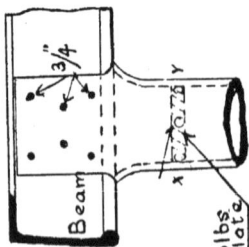

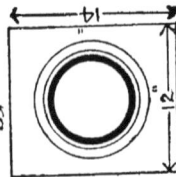

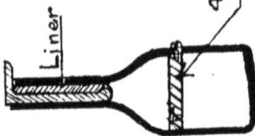

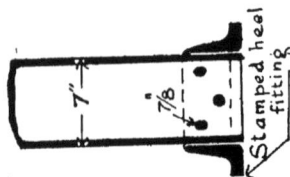

there is a steel deck under, and are properly shifted, so as to obtain four passing strakes between butts in the same beam space.

The fastenings are galvanized iron bolts, with square necks to prevent them turning in the wood when the nuts are being hove up, and are not generally tapped through the steel plating. Where, however, the point of the bolt cannot be got at, a special bolt (*b*) without a square neck and with a slot in the head is used, which is tapped into the plating. This also applies where planking has to be worked on a very thick deck, as the fastenings are not right through the deck.

In the case of wood flats in spaces below the water line, the bolt is tapped through the plating, and has a nut on the point as well, to ensure watertightness should the nut come off.

The method of fastening the wood planks when there is no steel deck is shown in Fig. p. 58, and in such cases the butts must come on the beams.

The method of laying wood deck is similar to that described for sheathing.

The back of the corticine is covered with knotting, and along the edges galvanized strips 1 inch wide are worked, which are secured to the plating by metal screws (see p. 58), the plating under the corticine being first thoroughly cleaned and coated with an adhesive mixture of resin and tallow.

PLATING BEHIND ARMOUR.

This is worked either in one or two thicknesses ; in the former edge strips and butt straps being necessary, but in the latter, as on p. 59, the plates form edge strips and butt straps to each other. The frames are spaced 2 feet apart, and vary in type, in some ships Z or channel bars are used, and in others frame and reverse frame bars (see sections, pp. 60, 61, 62, 63), intermediate frames being worked between the ordinary framing immediately under the armour shelf.

The frames are secured to the decks at head and heel by means of flanged bracket plates, and the structure is stiffened by horizontal girders ; special care must be taken in working the plating behind armour to secure watertightness.

Fashion plates as shown on p. 64 are worked at the junction

Method of working plank against Waterway.

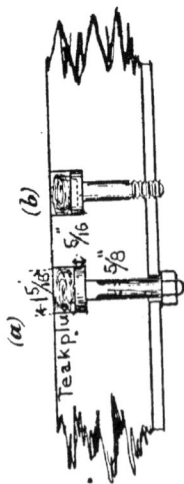

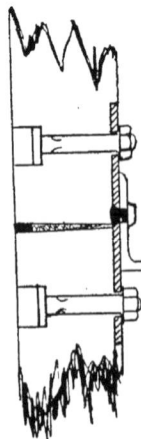

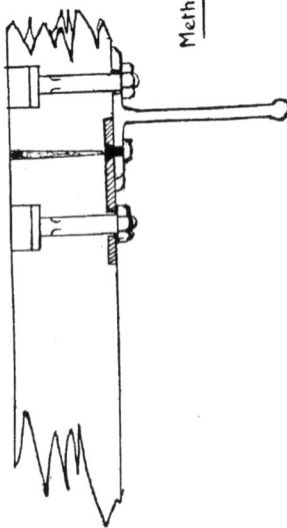

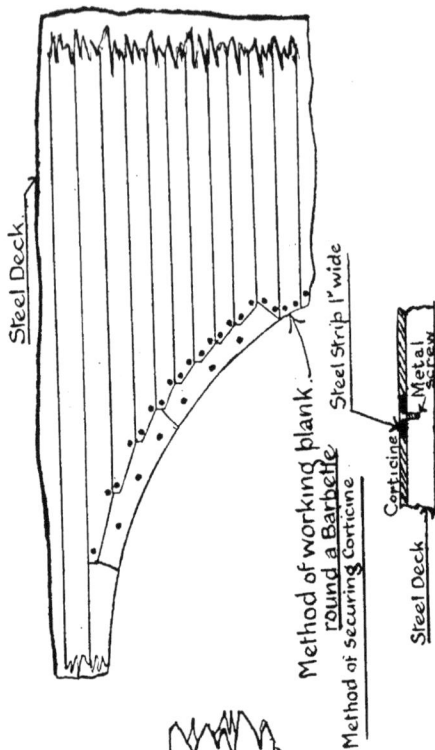

Steel Deck.

Method of working blank round a Barbette

Method of securing Corticine

Steel Strip 1" wide

Corticine

Metal Screw

Steel Deck.

(a)

(b)

Teak plug.

1 5/16"

5/8"

5/16"

of this plating with the shell plating. Where the armour shelf becomes narrow the connecting angle at lower edge of plating is worked on the inside.

Before the armour is erected the shelf is faired, bases being erected at intervals, levelled and sighted in, as shown on p. 65. The heights as A, B, C, etc., are measured and plotted on a curve, when it can be seen at what point the shelf is highest. The joint line must first be pitched at this point and sighted in by means of the horizontal battens on the section mould, holes being bored in the backing to indicate this line.

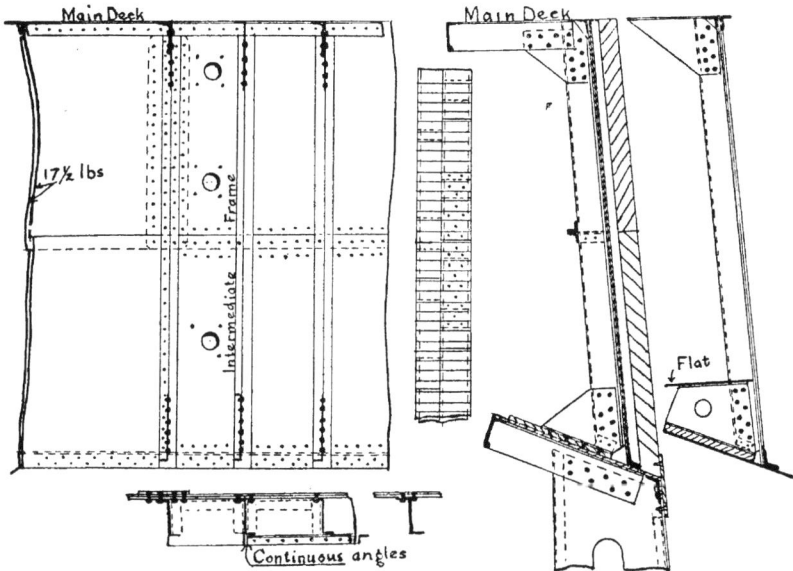

FRAMING BEHIND ARMOUR.

The position of the holes for the armour bolts is obtained by a mould made from the actual armour plates, which is correctly pitched at the joint line and fore and aft position of plate. Small holes are bored through the centre of the wooden plugs which represent the bolt holes, the large holes in the backing being cut out by means of a " ring " engine.

On p. 66 is shown the method of getting the plates into position.

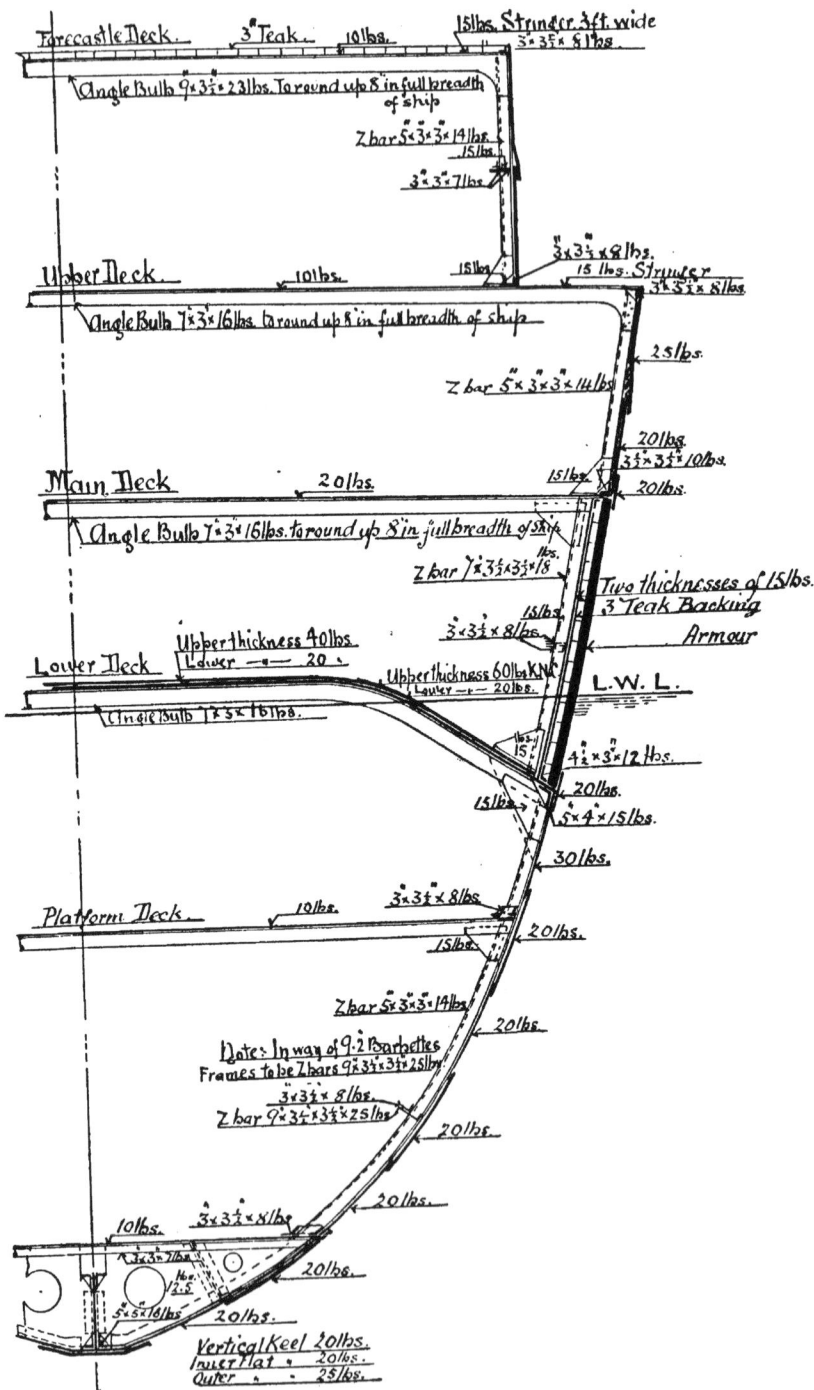

Forecastle Deck.   3" Teak.   10lbs.   15lbs. Stringer 3ft. wide
3 × 3½ × 8 lbs.

Angle Bulb 9 × 3½ × 23lbs. To round up 8 in full breadth
of ship

Z bar 5 × 3 × 3 × 14 lbs.
15 lbs.
3" × 3" × 7 lbs.

Upper Deck.   10lbs.   15lbs   3 × 3½ × 8 lbs.
15 lbs. Stringer
3 × 3½ × 8 lbs.

Angle Bulb 7 × 3 × 16 lbs. to round up 8 in full breadth of ship

25 lbs.

Z bar 5 × 3 × 3 × 14 lbs.

20 lbs.
3½ × 3½ × 10 lbs.

Main Deck.   20 lbs.   15 lbs   20 lbs.

Angle Bulb 7 × 3 × 16 lbs. to round up 8 in full breadth of ship

Z bar 7 × 3½ × 3½ × 18 lbs.

Two thicknesses of 15 lbs.
3" Teak Backing

15 lbs.
3 × 3½ × 8 lbs.   Armour

Lower Deck.   Upper thickness 40 lbs.
Lower   "   — 20   "
Upper thickness 60 lbs K.N.   L.W.L.
Lower   "   — 20 lbs.

Angle Bulb 7 × 3 × 16 lbs.
15   4½ × 3 × 12 lbs.
20 lbs.
15 lbs   5 × 4 × 15 lbs.

30 lbs.

Platform Deck.   10 lbs.   3 × 3½ × 8 lbs.
15 lbs   20 lbs.

Z bar 5 × 3 × 3 × 14 lbs.
20 lbs.
Note: In way of 9·2 Barbettes
Frames to be Z bars 9 × 3½ × 3½ × 25 lbs.
3 × 3½ × 8 lbs.
Z bar 9 × 3½ × 3½ × 25 lbs.
20 lbs.

20 lbs.

10 lbs.   3 × 3½ × 8 lbs.
20 lbs.

20 lbs.

Ho.
2.5
5 × 5 × 16 lbs.   20 lbs.

Vertical Keel 20 lbs.
Inner Flat   "   20 lbs.
Outer   "   "   25 lbs.

SECTION FORWARD SHOWING CONSTRUCTION OUTSIDE DOUBLE BOTTOM.

$7 \cdot 7$ lbs.

$10 \cdot c$"

to it.

15lbs. Stringer

Z Bar 5 × 3 × 3 × 14 lbs.

lbs.
e7 60 K.N.C.

Crown Plating 10.

$9 \cdot 2$" Ma

bs.

k Coaming a

$\dfrac{7 \times 3 \times 16 lbs.}{10 lbs.}$

$7 \frac{1}{2}$ × $3\frac{1}{2}$ × 25 lbs.
(*Intercostal*)

SECTION AT CENTRE OF F^{O}RE 9·2 INS. BARBETTE.

Erecting Stem Casting.

A mould is made on the mould loft floor to the outline of stem, and on it are marked the ends of the casting, the positions

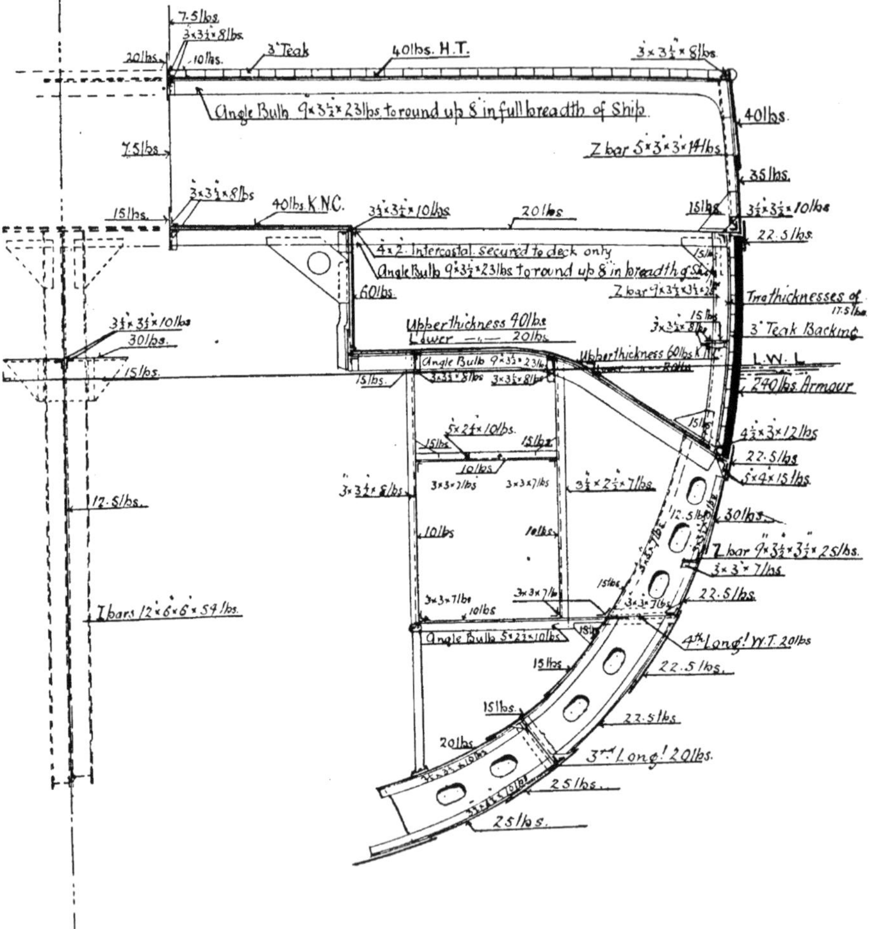

Section through Engine Room.

of decks, load water line, and fore perpendicular. A short batten is fixed to the upper end of mould to clear the deck lug, since it may be found that the casting will be longer than the pattern if the shrinkage has not been exactly allowed for. From a point A (see p. 67) a line is drawn at right angles to the L.W.L., crossing the mould at the point B, and another line is drawn at an inclination of $\frac{5}{8}$ inch to a foot to AB, crossing the mould at the point C. Since the L.W.L. of the ship on the blocks is at

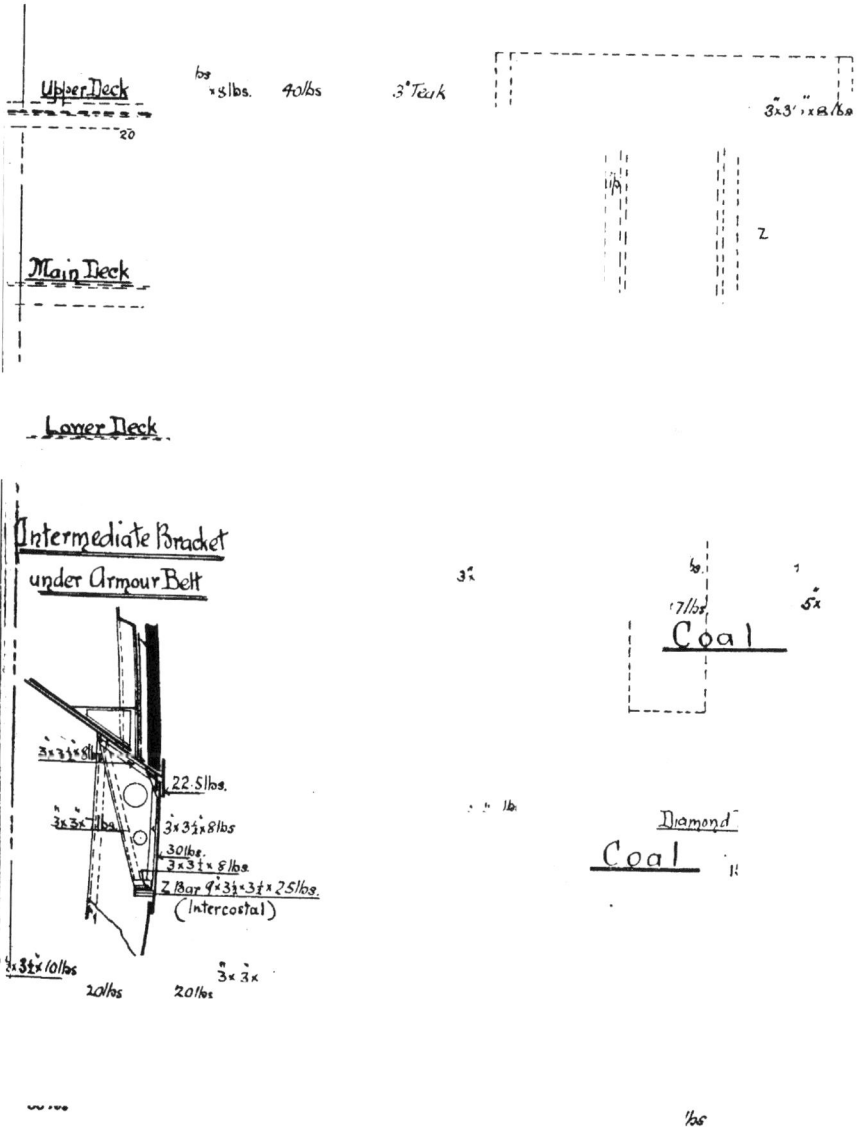

Upper Deck ×8lbs. 40lbs .3"Teak 3×3'×8lbs

20

Main Deck

Lower Deck

Intermediate Bracket
under Armour Belt

3×3¼×8lbs

.22.5lbs.

3×3¼×8lbs

3×3×lbs

.30lbs.

3×3¼×8lbs.

Z Bar 9×3¼×3¼×25lbs.
(Intercostal)

×3¼×10lbs

3×3×

20lbs

20lbs

3½

7lbs

5×

Coal

Diamond

Coal

lbs

MIDSHIP SECTION.

a slope of $\frac{5}{8}$ inch to a foot, it is clear that the line AC must be vertical when the casting is at its correct slope on the blocks.

The mould is now used for marking off the casting, and is placed round the latter as it lies on the ground, and by means of section moulds one face of the mould can be brought to the middle line of the casting, which can then be marked on the outside of the casting. This middle line has next to be marked'

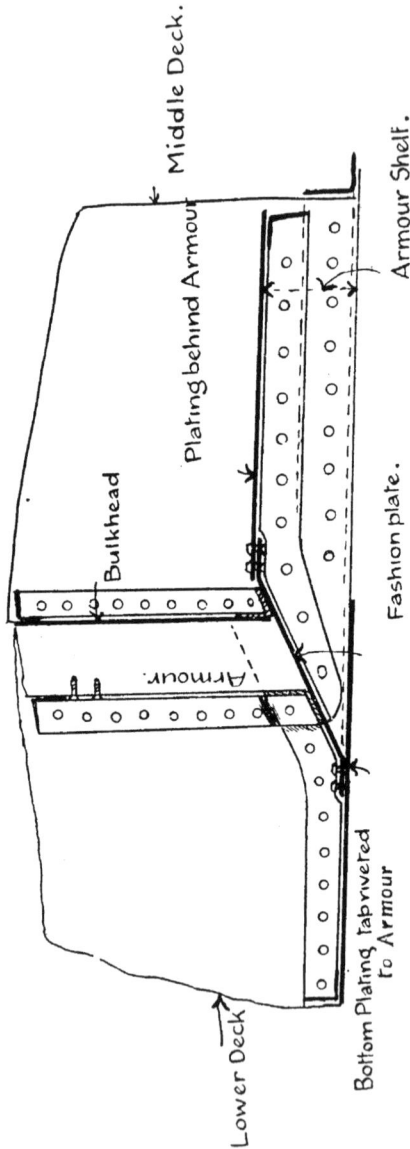

METHOD OF WORKING PLATING AT CORNER OF CITADEL.

on the inside of the casting, which is done by placing a batten across the arms of the casting at intervals, and squaring down the middle points, the marking being completed by transferring the point C and the fore perpendicular to the casting. Before

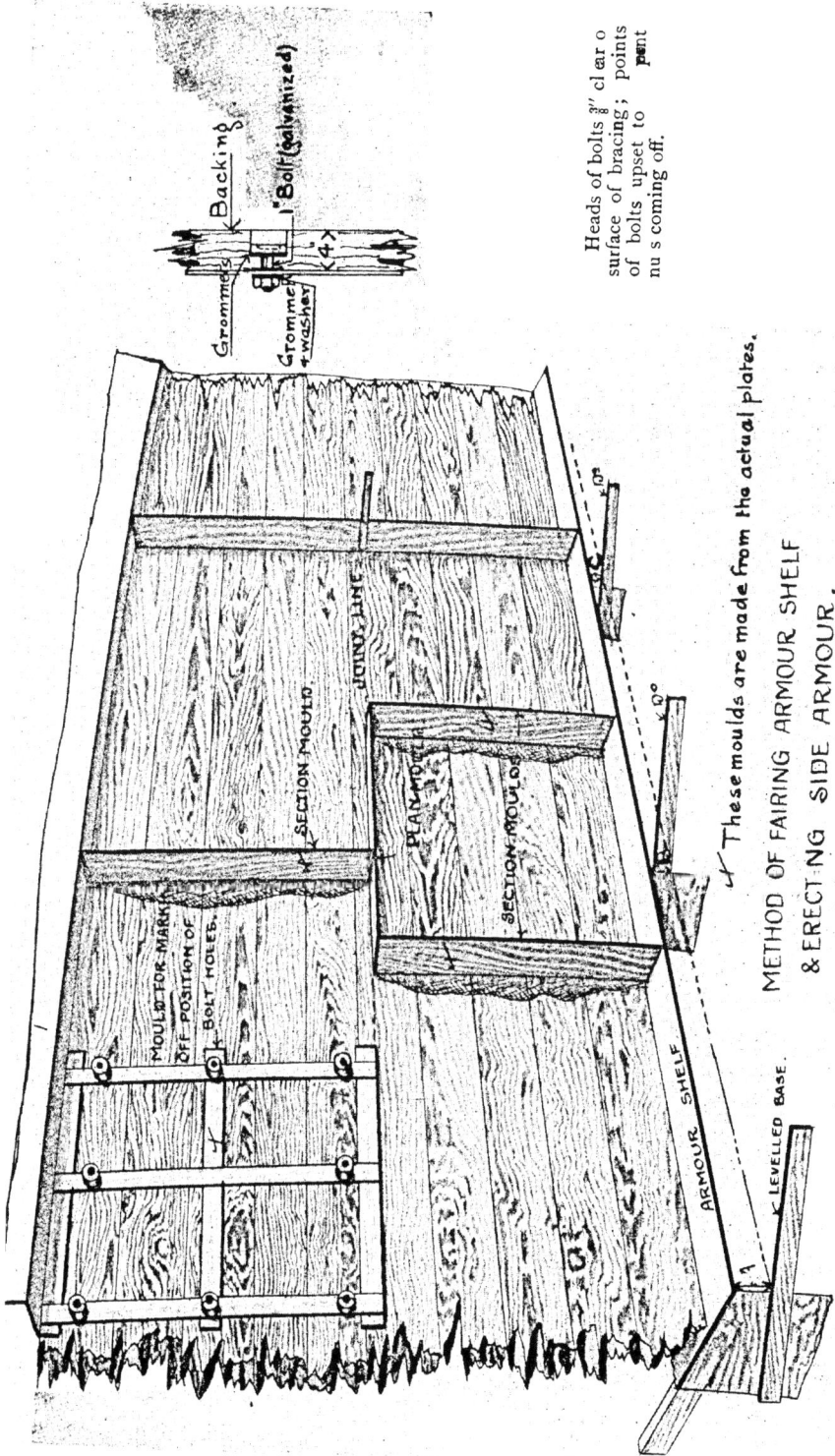

Heads of bolts ⅞" clear o
surface of bracing; points
of bolts upset to pent
nu s coming off.

Backing

Bolt (galvanized)

Grommet

Grommet
+ washer

< 4 >

These moulds are made from the actual plates.

SECTION MOULD

JOINT LINE

PLAN MOULD

SECTION MOULD

MOULD TO MARK
OFF POSITION OF
BOLT HOLES.

ARMOUR SHELF

LEVELLED BASE.

METHOD OF FAIRING ARMOUR SHELF
& ERECTING SIDE ARMOUR.

removing the mould it is noted how the point A stands with respect to the lug at the top of the casting, and it is convenient to attach a batten to the lug to correctly give the position of this point.

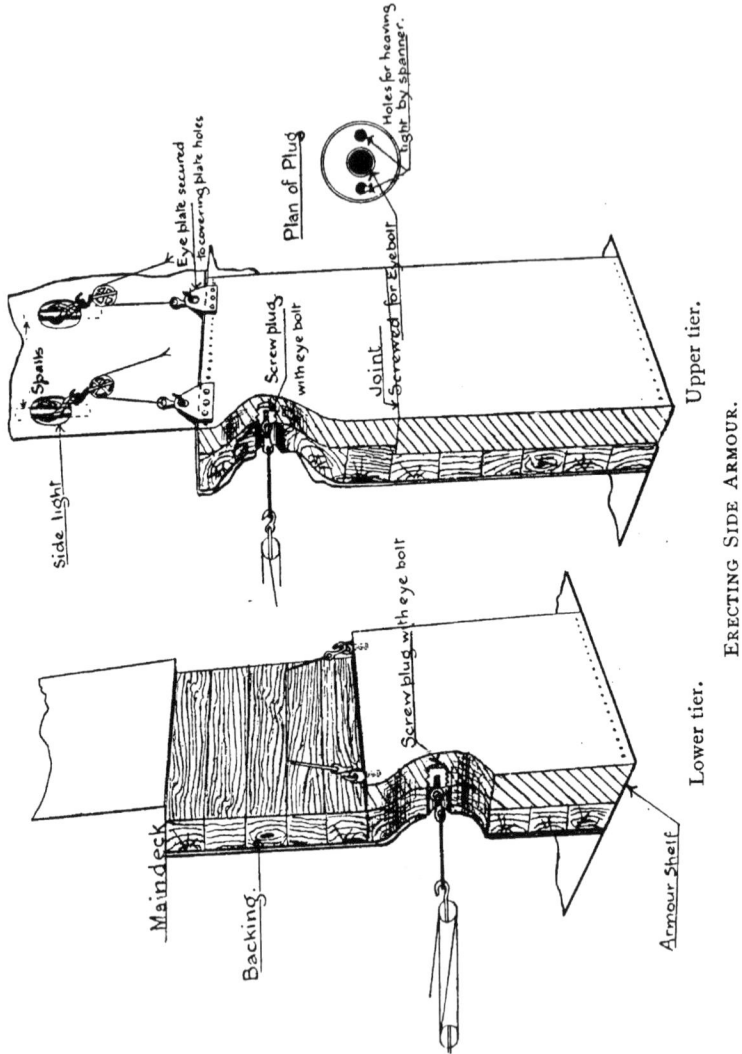

ERECTING SIDE ARMOUR.

The mould is removed and the casting hoisted into position, and to ensure its being correctly erected the following conditions have to be fulfilled :—

(1) It must be at its correct slope. This is checked by dropping a plumb line from the point A, the casting being

adjusted until the bob coincides with the point C. This will also ensure its being in a vertical plane.

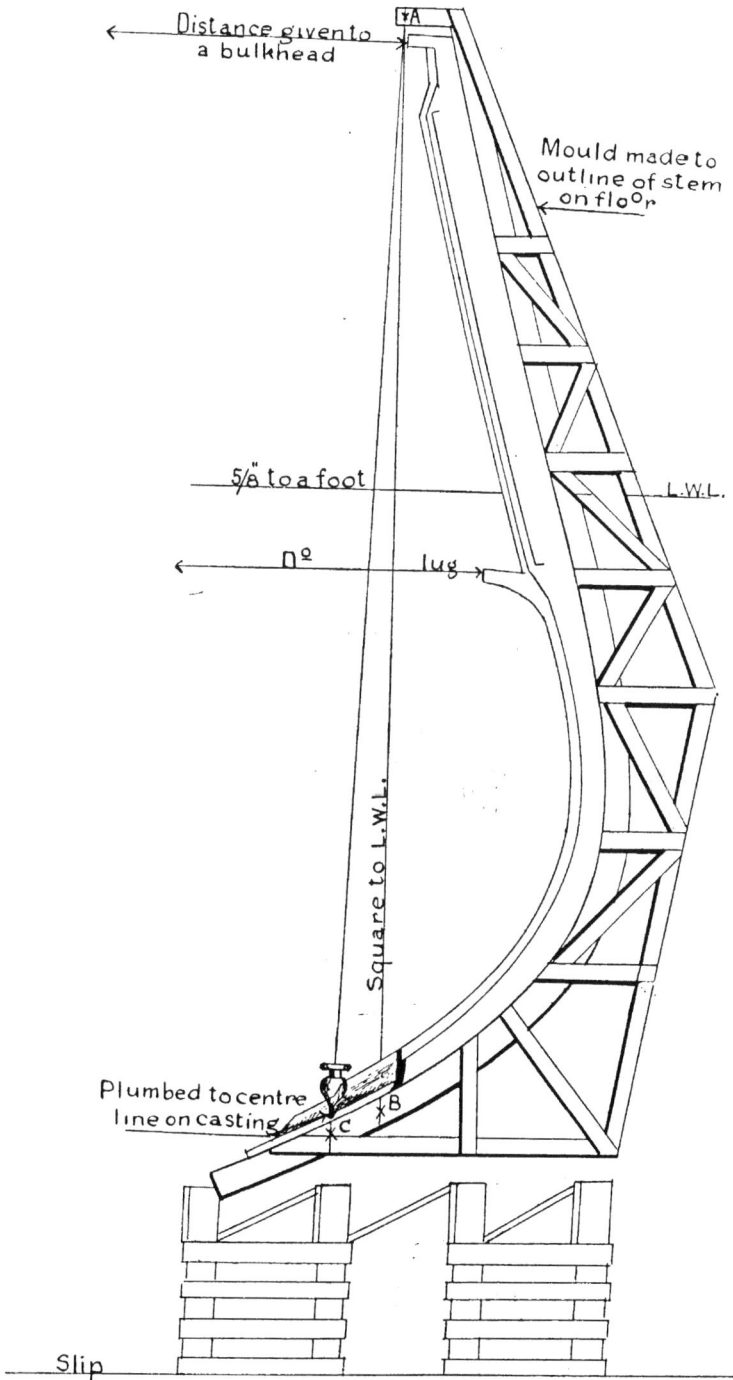

Distance given to a bulkhead

Mould made to outline of stem on floor

⅝" to a foot

D° lug

L.W.L.

Square to L.W.L.

Plumbed to centre line on casting

A

B

C

Slip

(2) It must be in the middle line plane of the ship. This is checked by dropping plumb lines from the middle line of the outside of the casting to the middle line of keel marked on the blocks.

(3) It must occupy its correct fore and aft position. This is ensured by giving measurements from the deck lugs and lower end of casting and fore perpendicular to a bulkhead.

The mould also serves as a check on the form of the casting, as it shows if the correct amounts have been allowed for shrinkage in making the pattern, and thus provides useful information when making future patterns.

SIGHTING IN CENTRE LINE ON LARGE CASTINGS.

A method of doing this is illustrated below. The casting is laid on chocks on the ground and parallel battens fixed at the

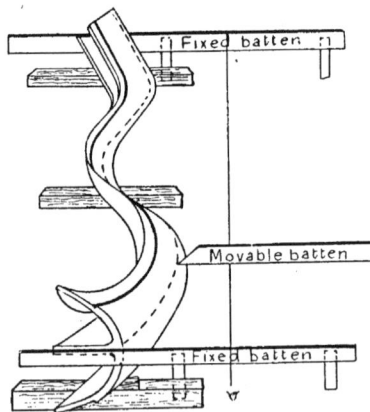

Sighting in centre line

extremities and made to outwind. Points on the centre line are obtained by means of a movable batten, the lower edge of which is sighted in with the upper edges of the end battens.

SHAFT BRACKETS.

Pages 69, 70, 71, show details of shaft brackets, methods of attaching to structure, and special framing in the vicinity to receive them.

Erecting Shaft Brackets. Moulds are made as described on p. 72, at each end of the boss of the casting, the upper end of the upper arm giving the direction of the palm plate, and the lower end of lower arm the connection to the stern casting. Battens

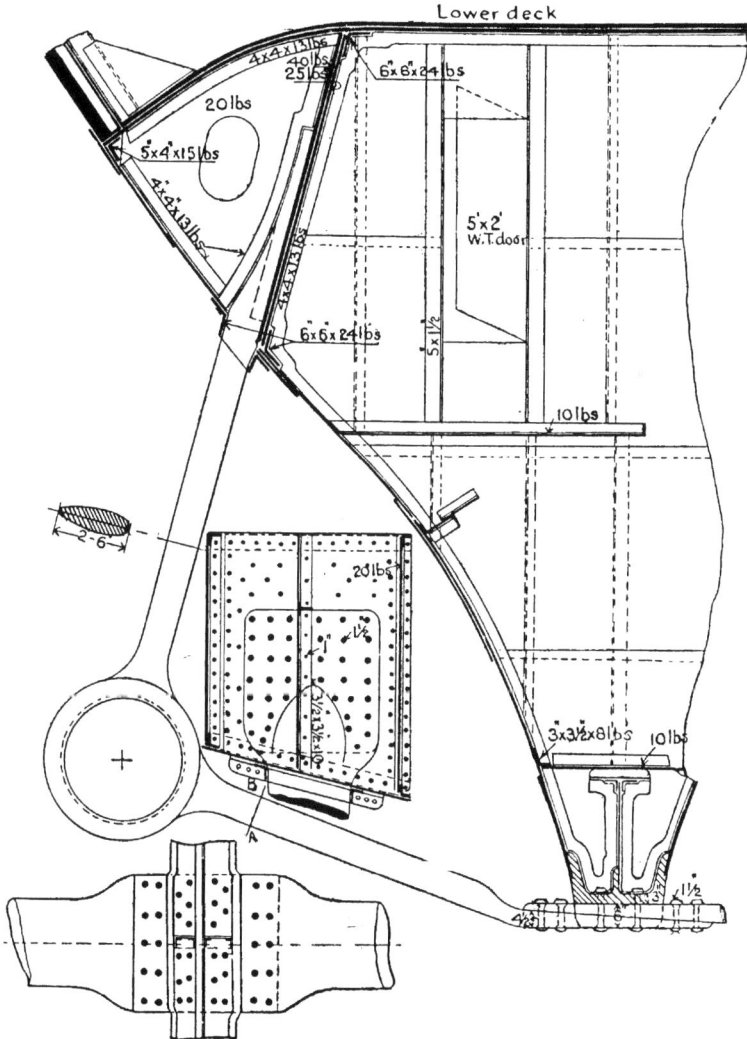

are placed across the arms parallel to the middle line of the body, and these battens will be vertical when moulds are erected at ship.

The moulds are erected at the ship as follows: The distance of the centre of the shaft out from the middle line and up from keel is given for each mould, which enables the centres to be pitched as in Fig. p. 72, and then the moulds are placed with the battens vertical and secured. The casting before being got into position is checked by means of these moulds, one being erected at each end of it, and the two battens made to outwind, with the arms of the moulds in the same plane as the arms of the casting. By

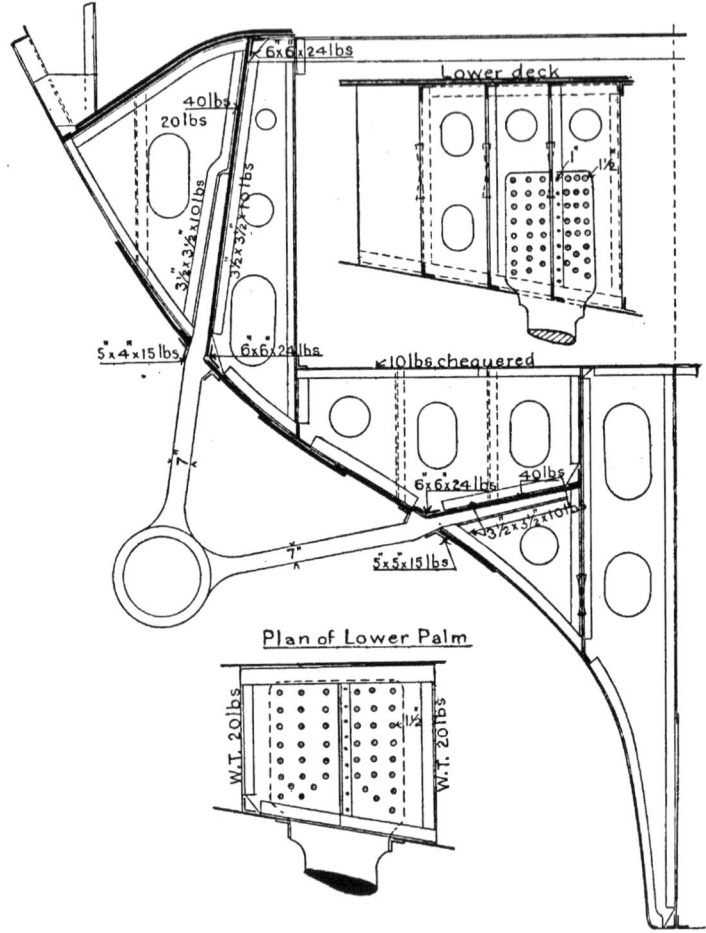

Lower deck

Plan of Lower Palm

means of a straight edge batten bearing against the upper arms it can be seen at once if the upper palm is twisted.

The lower palms can be tested in a like manner, section moulds being made to the stern casting to give the shape to which the lower palms must be machined.

Any machining of the palms found necessary having been done, the castings are erected, being correctly pitched by means of the moulds.

Another method is to erect the casting with the centre of boss in the same vertical plane as the centre of shaft. The distance

between the centre of shaft and centre of boss is measured, and spilings taken for any machining which may be necessary to the lower palms. The plate to receive the upper palms is then fitted in place, and takes account of any twist there may be in the arms of the brackets.

The centre line of shaft has to be sighted through the entire length of rooms, shaft passage, etc., to enable the engine or turbine seatings to be lined off, and the holes for the boss casting to be cut in the bulkheads and framing. The method of doing this is illustrated by Fig. A, p. 73, and consists in fixing up two points

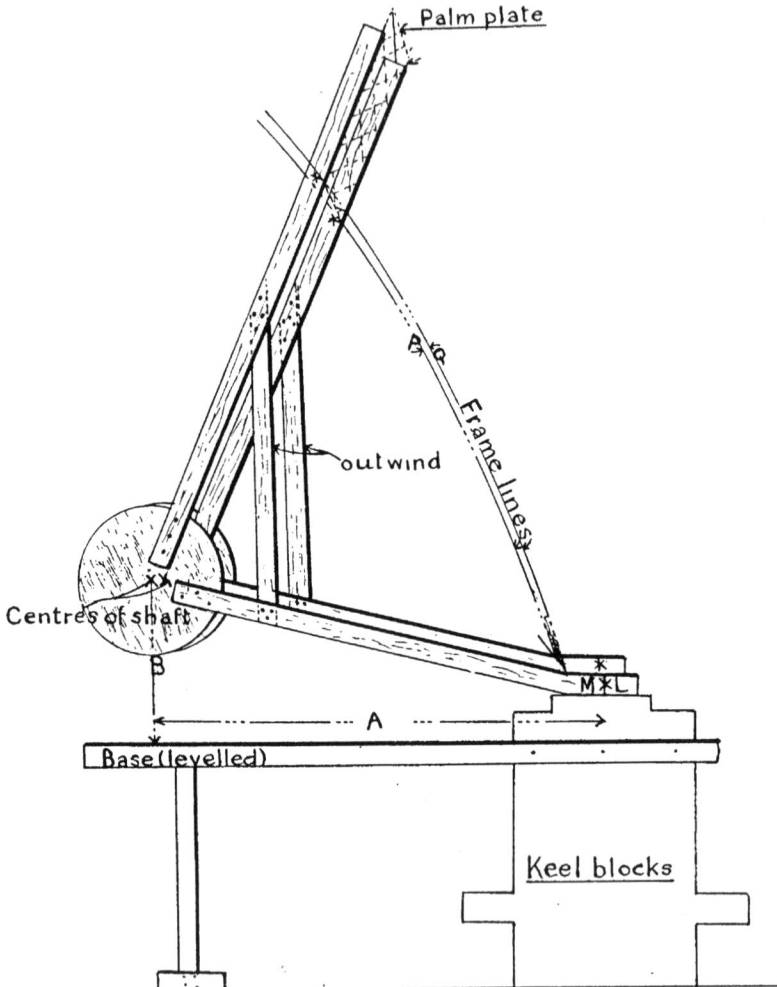

in the centre line of shaft, one at the forward end of engine room and the other just abaft the shaft bracket, by spilings from L.W.L. or keel, and middle line of ship, and placing plates with small holes in them at these positions. It will be necessary to have holes in the bulkheads and frames to permit of sighting, and it is usual to mark these holes when lining off on the scrieve

board, and punching them out about 6 inches in diameter, the full
opening for the boss casting and shipbuilder's tube being cut after
the frames and bulkheads have been erected and the centre of
shaft sighted in (see Fig. B, below). The centres of these holes

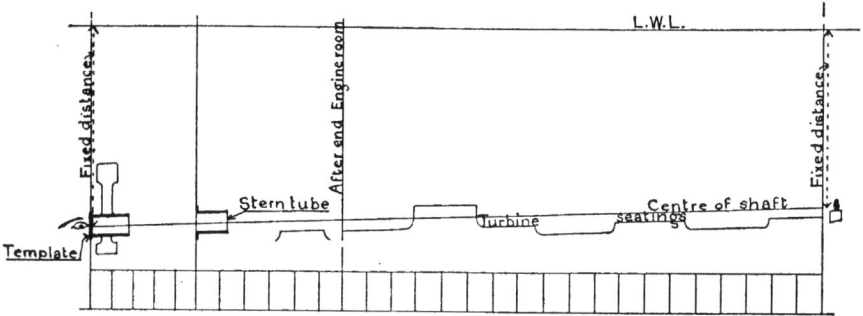

FIG. A.

are found thus: A batten is placed across the hole and its edge
sighted in with the spot of light; the edges of the batten are
marked by on the plating, and the batten is then placed vertical,

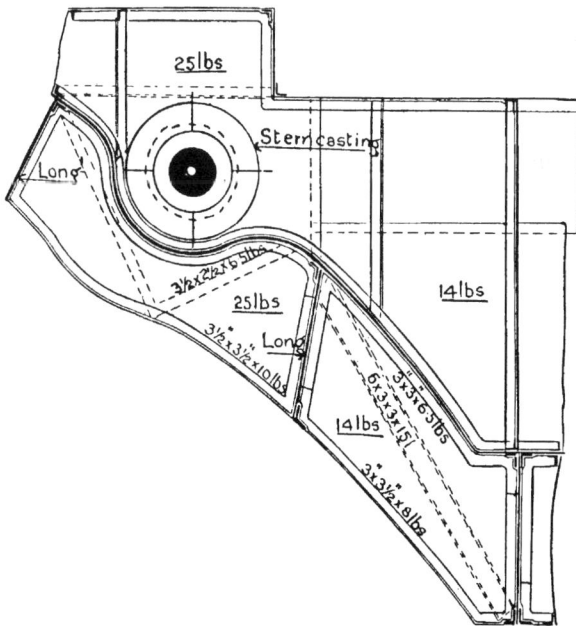

FIG. B.

sighted in, and marked by again; the correct centre of shaft
will be given by the intersection of these two lines.

The centre of shaft is readily obtained by a series of sights similar to that illustrated below. It consists of a frame, A, attached to the structure, in which are two sliding frames capable of movement in directions at right angles to each other by means of thumb screws. There is a circular disc with a slit through the centre capable of being turned through a right angle attached to one of the sliding frames. A sight is obtained by adjusting the

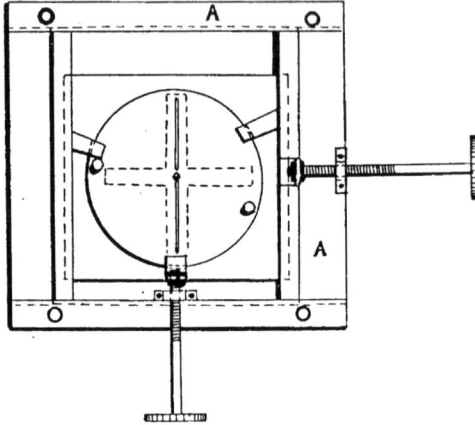

frames until the light is seen through the slit, and then turning the disc through a right angle and adjusting until the light is visible. The centre of the disc is the sight required.

RUDDERS.

There are two types of rudder, viz. ordinary rudder, in which the whole of the area is abaft the axis, and balanced rudder, in which a portion is on the fore side of the axis.

A rudder consists of a cast-steel frame, the spaces between the arms being filled in with fir, and the sides plated over. The arms and centre portion of casting are of H section, to save weight, and bossed out in way of the lifting holes. The rudder head is cast in one with the frame.

The methods of suspension vary, in some cases the pintle being at the lowest point of the rudder, and the whole weight taken there (Fig. A, p. 75), in others the pintle being about the middle of the rudder, and fitting into a lug on the stern casting (Fig. B, p. 75), while in another type the whole weight is taken inboard, there being no pintles (pp. 76, 77).

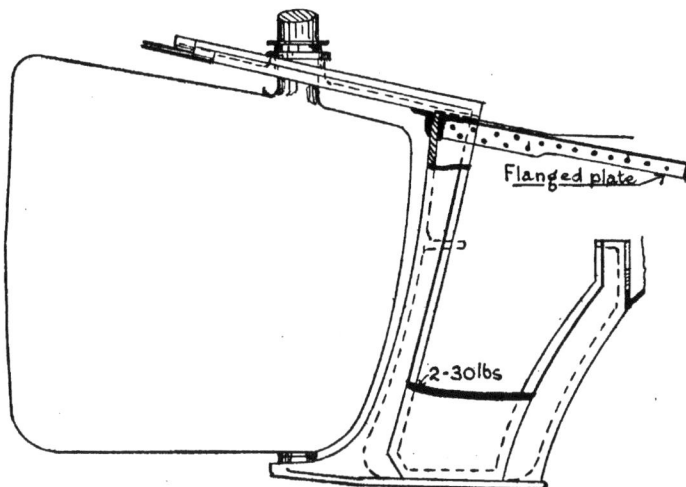

Flanged plate

2-30lbs

Fig. A.

The rudder below has the arm formed as shown, to enable it to be shipped and unshipped. The rudder has to be placed at

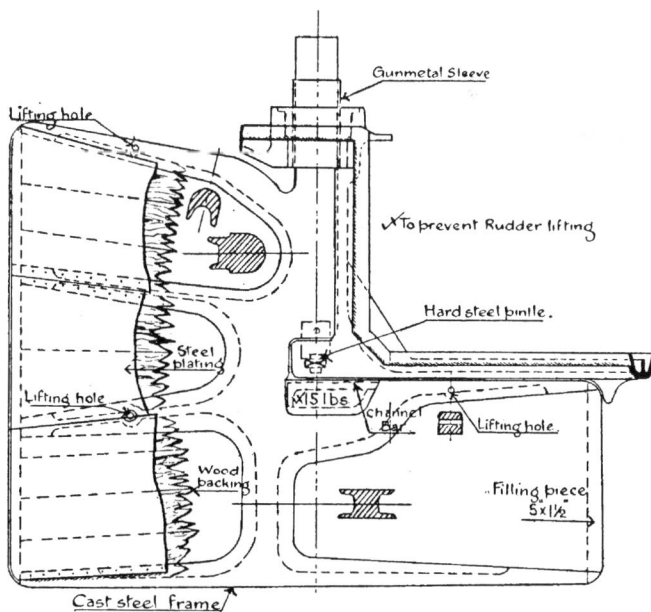

Gunmetal Sleeve

Lifting hole

To prevent Rudder lifting

Hard steel pintle.

Steel plating

Lifting hole

15 lbs

Channel Bar

Lifting hole

Wood backing

Filling piece 5 x 1½

Cast steel frame

Fig. B.

right angles to the middle line for this operation, the plate shown serving as a locking plate to prevent rudder unshipping.

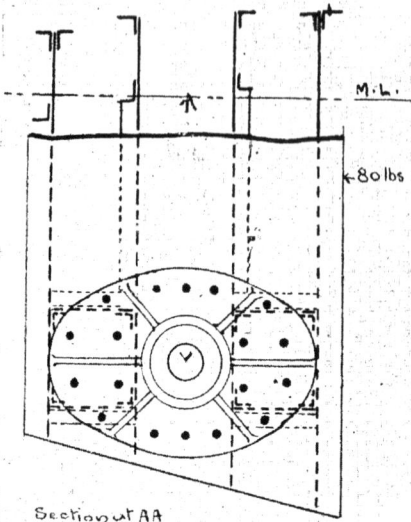

M.L.

←80 lbs

Phosphor **Bronze**

80 lbs
bs

Rudder hea

Section at AA,

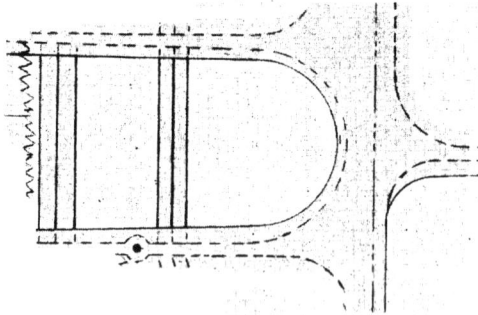

A,

To enable the rudder to be shipped an eyebolt is screwed into the head, or an eyeplate bolted to the head of the rudder.

The lowest point of the rudder is kept some inches above the straight line of keel to prevent damage. As shipping and unship-

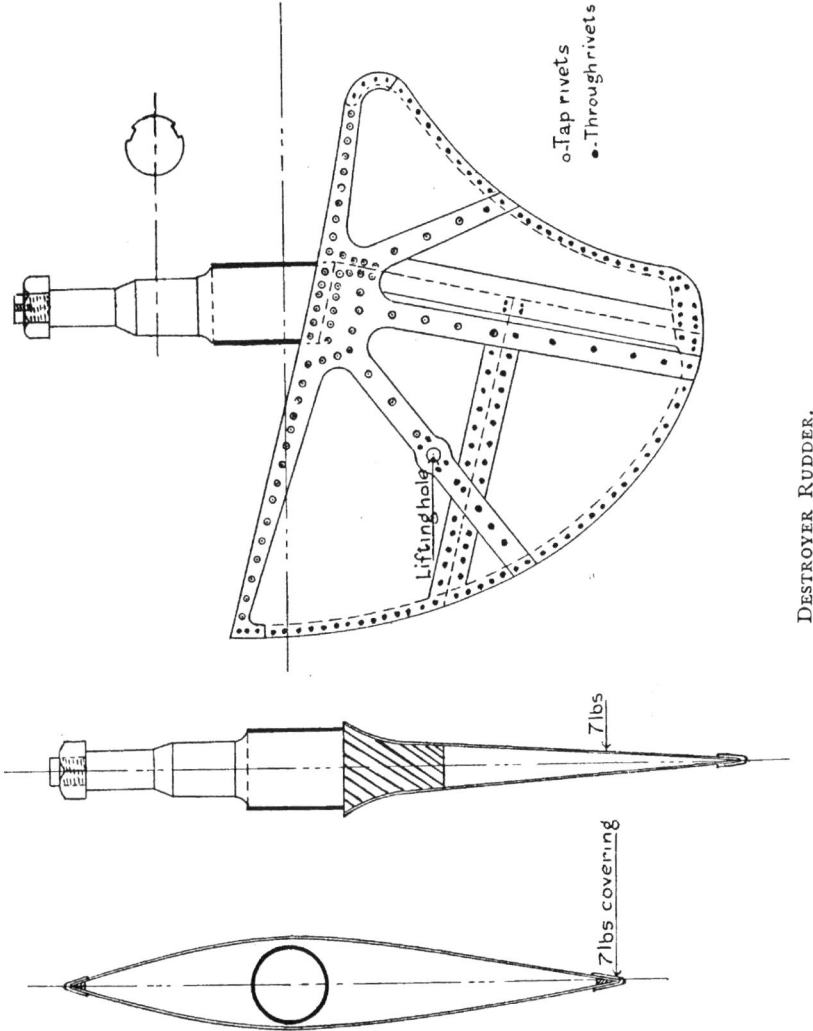

DESTROYER RUDDER.

ing have to be done in dock, it is necessary to make certain this is possible, especially in the case of rudders with great length of rudder head.

ENGINE AND BOILER SEATINGS.

Special framing is fitted under the engines and boilers, existing framing being incorporated in the arrangements as much as possible. The longitudinal framing of engine seatings is continuous, and the transverse therefore intercostal (details for turbine and reciprocating engines being shown on pp. 78, 79, respectively).

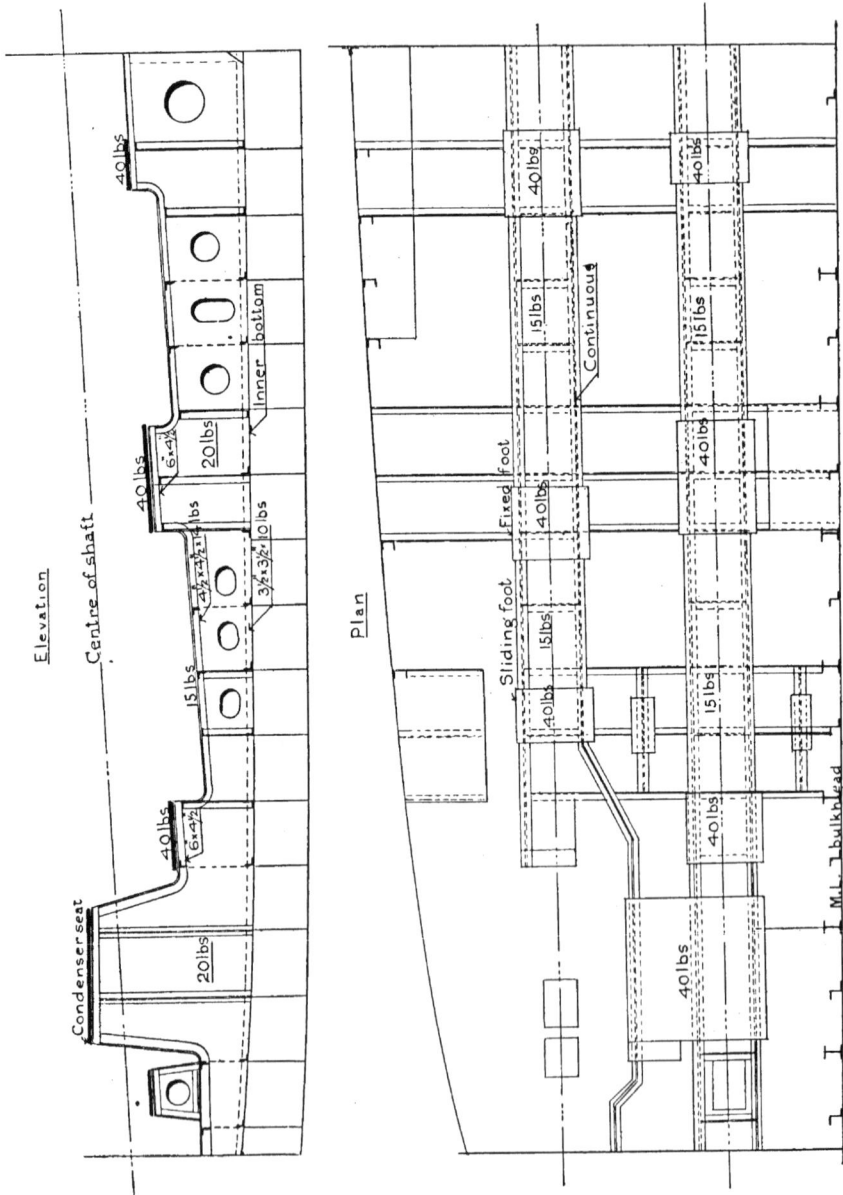

Holes for access to all parts of this framing and manholes for the double bottom spaces immediately under it must be provided.

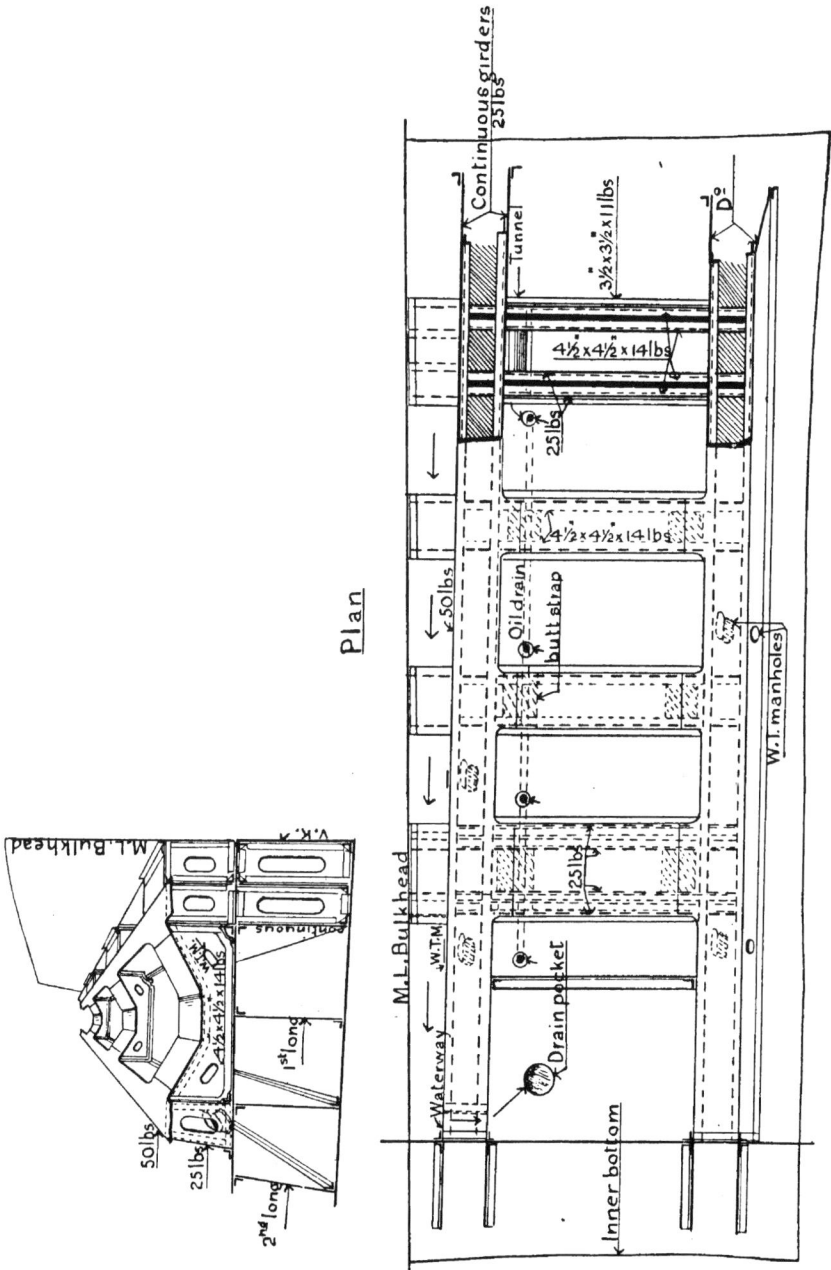

In reciprocating engines the crank pits are watertight, to confine the oil and water to these spaces (see above).

To permit of expansion of the turbines some of the feet are arranged to slide under wedge-shaped pieces, which are bolted to girders as shown below.

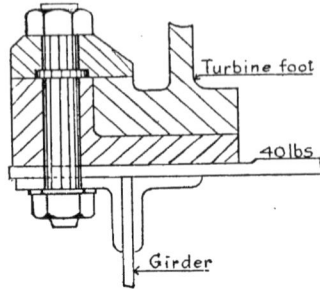

The construction of boiler bearers depends on the type of boiler fitted, and sketches of these for cylindrical and water tube types

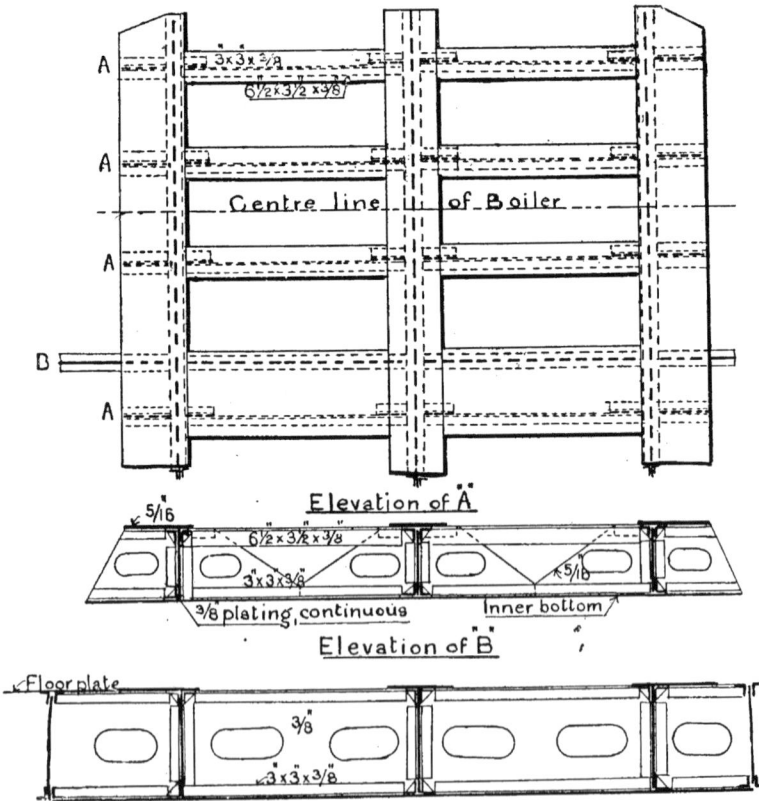

are given on pp. 80, 81, 82. On pp. 80, 81 are shown seatings for water tube boilers for Babcock and Yarrow types, respectively.

The boiler seatings can be worked as soon as the inner bottom is riveted off and those for the turbines after the centre line of shaft has been sighted in, so that all angles connecting this framing to

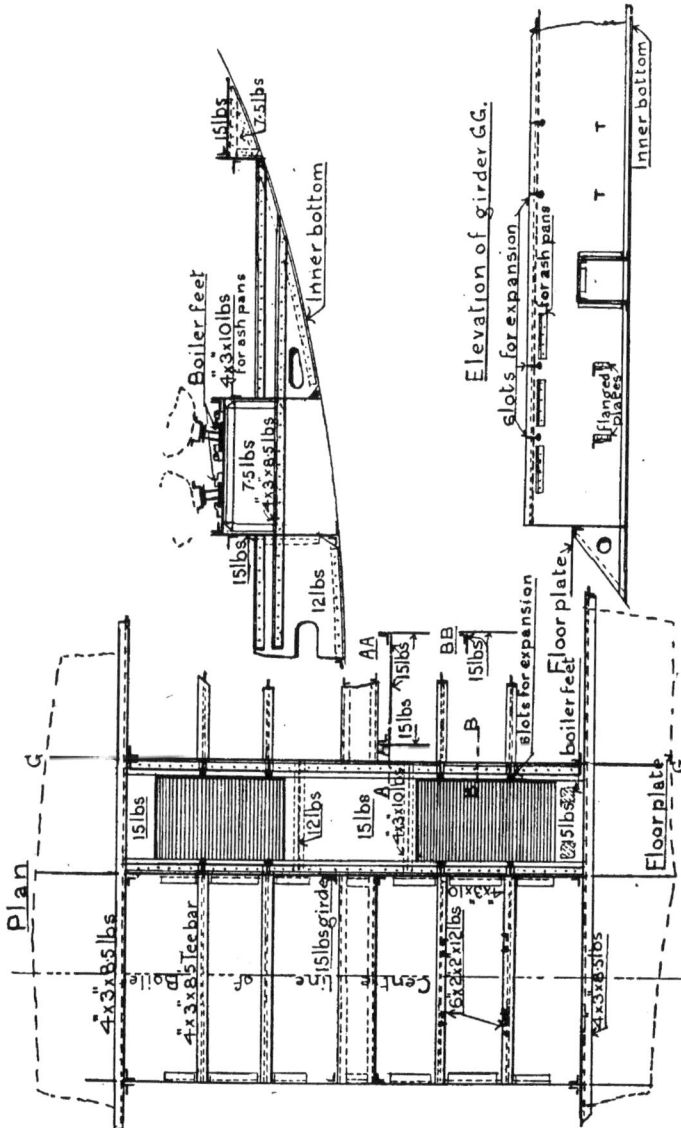

inner bottom are riveted off before water-testing is commenced. In ships intended for ramming, chocks have to be fitted to the boilers, as on p. 82, which also shows the rolling chocks. Steel wire stays are fitted for this purpose to water tube boilers.

G

In both cases allowance must be made for expansion of the boilers, this being done for cylindrical boilers by elongating the holes in the rolling chocks, and for the water tube type by sliding feet on the boilers (see p. 81).

BARBETTES.

In arranging the details of barbettes the requirements are to provide a structure to which to secure the armour and to take the weight of the revolving parts. In both cases the lower edges of the plates rest directly on the deck, so that no shearing forces

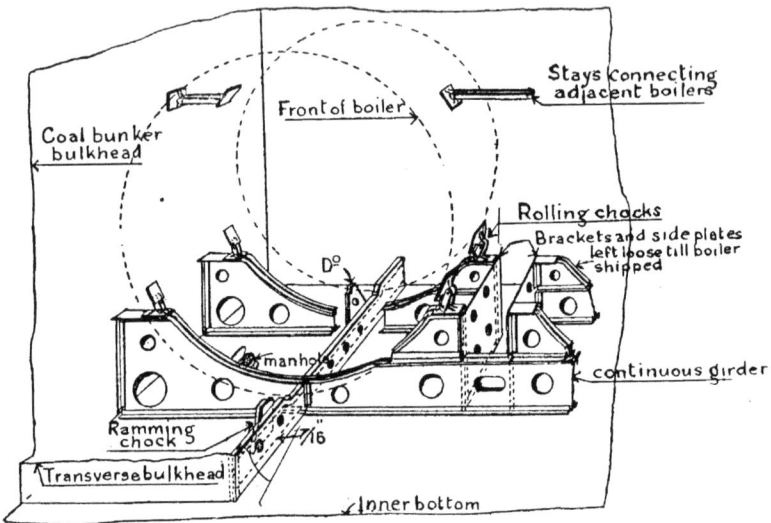

shall come on the rivets connecting the plating to the deck. Referring to Fig. p. 83, the structure of two typical barbettes is given, in one case backing being fitted behind the armour, and in the other the backing is omitted. The ring support is connected to the plating behind armour by elastic platforms, as indicated, which are intended to prevent damage to the ring support should the armour be struck and plating behind it forced inwards. The training rack seating has holes cut in it at intervals just above the rack to enable the riveting to be done.

The framing below a barbette has to be arranged so that the whole of the weight shall be distributed over the ship's structure as much as possible. In carrying this out, use is made of all bulkheads, and additional frames, beams, etc., worked as necessary, the frames in bottoms also being of the lightened plate type, suitably stiffened. The section of the barbette given on p. 84 shows how it is supported.

There are certain dimensions which have to be carefully adhered to, viz. the distance between the top of the barbette armour and the roller path, the diameter of the inside edge of the roller path seating, and the diameter of the training rack seating.

After the structure has been riveted off, the upper edge of roller path seat, the training rack seat, and the under side of roller path seat are machined, the latter to give a slight clearance to the clips which project from the turntable under this plate. A further important point is that the rivets through the roller path seatings

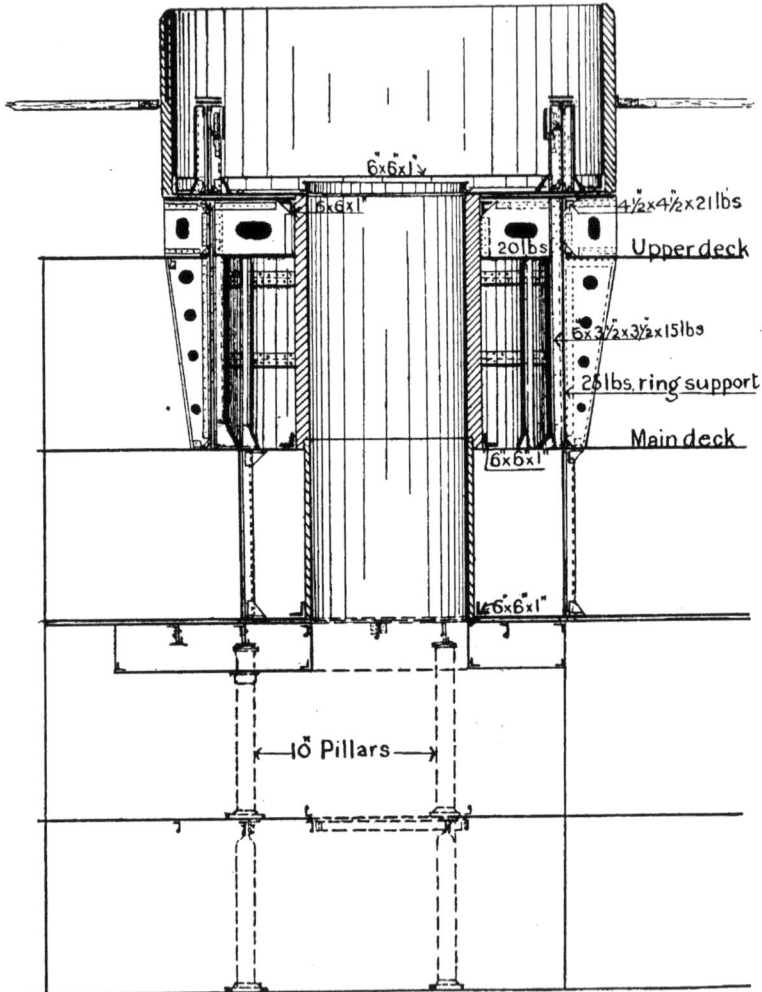

must be arranged to clear the bolt holes for the roller path. To check the correctness of the erection of the barbette armour plates, a line is given on their upper edge with the radius to which it should come, this line having been marked on the plates at the contractor's works.

In both cases the armour plates are secured to the plating by means of armour bolts, details of which are given on p. 83. Fig. H, p. 83, shows a bolt with a countersunk head fitted to a socket with a plug over the head, the latter being fixed to the plating by endless taps.

This method of fastening the socket is adopted to prevent risk of injury which would occur if ordinary taps were used, the heads of which would fly off should the armour be struck.

In the other form of bolt (Fig. F) there are no parts which would be likely to fly off if the barbette were struck. It should be noticed that an air space $\frac{1}{2}$ inch or so is arranged between the end of the bolt and the hole in the armour to act as a buffer, and further that the diameter of the bolt is reduced between the head of the bolt and the screwed portion. The reason for this is that greater resilience is obtained in this way. The latter form of armour bolt is also used for side armour.

It will be noticed that in the second case the plates are keyed together at the butts, the form of key being shown at Fig. E, p. 83, the depth of keyway reaching to top of plating behind armour. In the first case the butts are not connected.

Where the barbette armour forms a part of the screen bulkhead the framing and armour are worked as shown on p. 86.

The plating behind armour and for the ring supports is laid off on the mould loft floor, and the moulds made there, as described on pp. 240–248.

The plates are worked vertically, with edge strips and butt straps where of a single thickness, but form edge strips and butt straps to one another when in two thicknesses. All the holes for straps, stiffeners, and angles are formed before the plates are rolled, and the structure checked before riveting off, as no adjustments can be made after.

The centre line of barbette has first to be obtained, and this is done by taking measurements from the drawings giving its approximate position with respect to bulkheads in the vicinity, at the various decks. Plates with small holes in them are placed at the various decks and the line sighted through. The framework shown is erected, and the hole in the plate at the top of it is obtained. This hole is to take the pivot of the swinging arm on which the radius of the barbette is marked. The plates are removed and a wire stretched as in Fig. p. 88. A final check is put

on this centre line by means of a plumb bob. The centre line being
at right angles to the L.W.L. and the distance being measured,

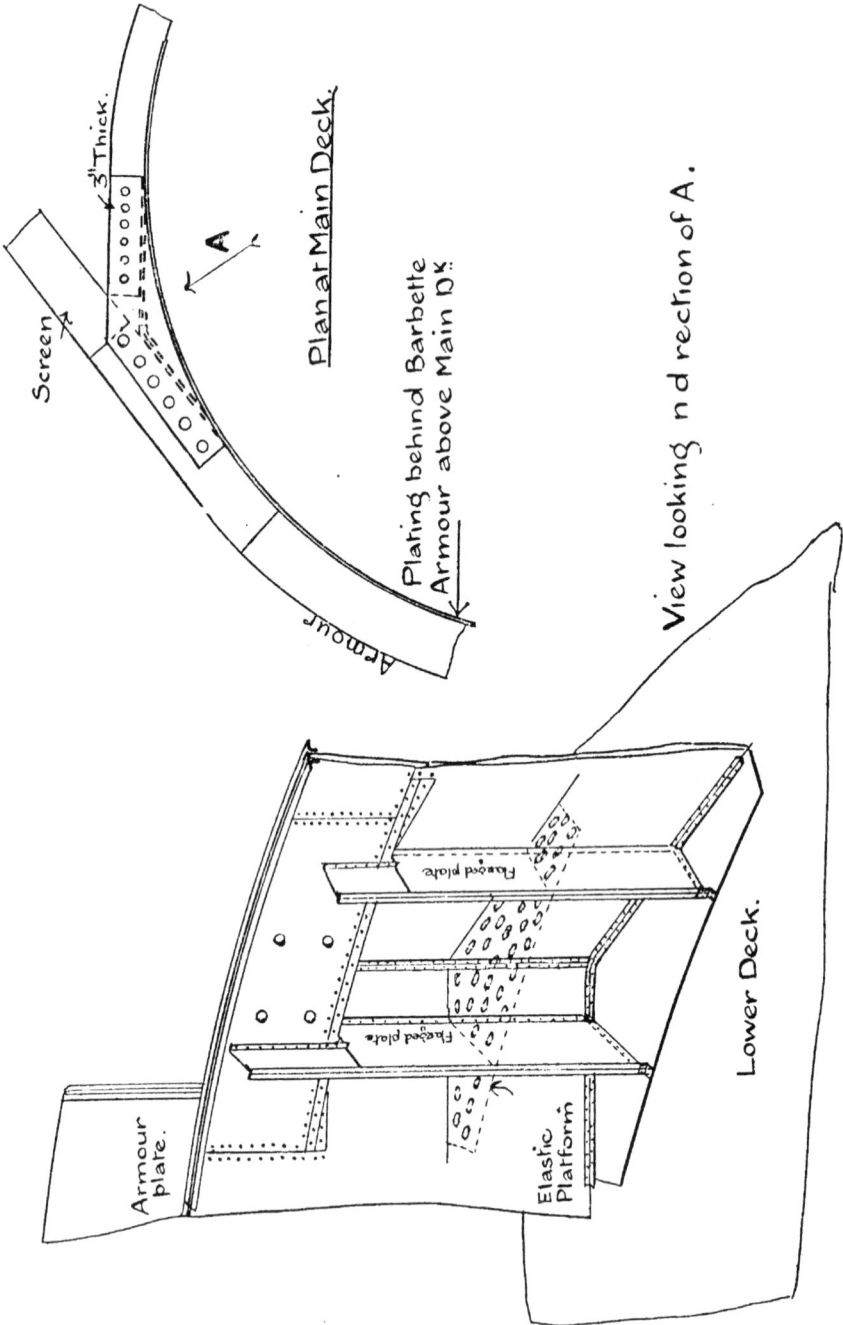

the amount the plumb bob should be from the wire can be easily
calculated.

The centre line is produced sufficiently high enough to obtain the upper centre for the trammel. An angle bulb erected on brackets as in figure below provides this centre.

To make any adjustments cat screws and chains are used, with wedges at the deck to assist, the deck plating being always cut short of the barbette, and pieces to make up fitted after the armour is in place.

While the ship is on the slip, steel straight edges are placed on the bulkheads, the one on the transverse bulkhead being level and that on a fore and aft bulkhead parallel to the L.W.L., to enable the list and trim to be found at any time after the ship is launched.

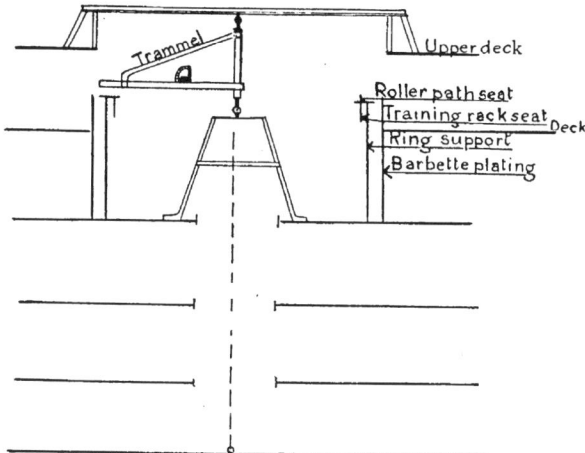

Templates are made to the sections of the roller path and training rack seatings, showing the holes for the rivets and securing bolts of the roller path and rack. In demanding these plates allowance must be made for machining, in the case of the roller path both on the inner edge and on the thickness, and on the training rack seats for thickness only. If these plates are first punched roughly to form, care should be taken not to encroach on the allowance for machining, as owing to the conical shape of punched holes the lower edge will be cut under.

After the barbette structure is complete the surfaces previously mentioned have to be machined. They are first trammelled, the trammel being fixed up as illustrated. On each occasion of doing this it is necessary to check the trammel by means of a clinometer, to see if it has been moved. The arm of the trammel is first placed fore and aft, and the inclination of the L.W.L. found by placing the clinometer on the fore and aft steel straight edges previously mentioned. The trammel arm is now placed across the ship, and made to agree with the list of the ship measured by means of the clinometer.

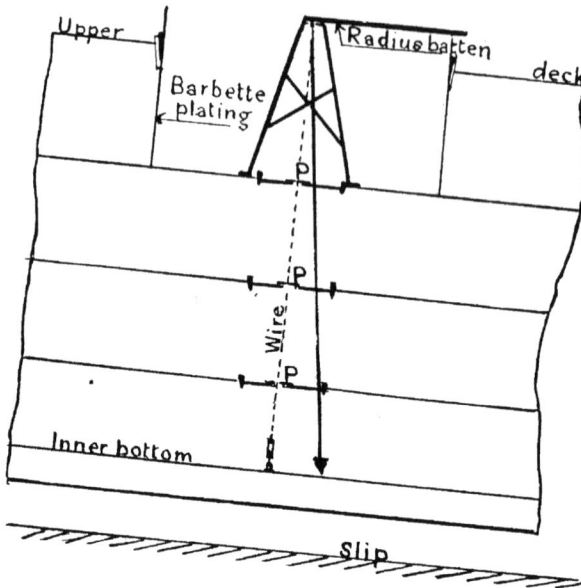

FUNNEL HATCHES.

The large openings made in the decks for funnel hatches weaken the deck, and compensation must be provided. This is done by means of deep coamings about 2 feet in depth connected to the deck by angles. It is also necessary to make up for the loss of protection, and this is done by fitting "armour gratings," and since these take away from 20 to 30 per cent. of the effective area of the opening, the latter must be made large enough to obtain the necessary area excluding the grating.

Means must also be provided to permit of the expansion of the fore and aft and cross bearers, which is done by slotting the holes in them through which the bolt fastenings pass, and cutting the plates $\frac{1}{2}$ inch to $\frac{3}{4}$ inch short at each end. The U pieces permit of the lateral expansion of the cross bearers should the expansion of the bearers be greater than the amounts arranged for. Certain of the gratings are hinged for access, and to admit of these being lifted readily their weight should not exceed about 70 lbs. Any hinged gratings which exceed this weight are split up as necessary, so that the several parts shall not exceed the limit given.

The casings are bolted to the coamings, as riveting cannot be done, the space being so confined, and along the lower edge of the outer coamings 2-inch holes 7 inches apart are cut to provide a means for the escape of hot air from the boiler rooms, sliding louvres being fitted for closing these openings when necessary.

There are two methods of constructing armour gratings, in one of which the divisional plates are tenoned into the end plates, and the ends hammered over as shown on p. 90, and the other in which the ends of the divisional plates are flanged and riveted to the end plates. The former method is preferable, as there is a considerable saving of weight, while at the same time they are as easily constructed. The gratings rest on horizontal bars riveted to the cross bearers, the horizontal flanges of the angles being about $1\frac{1}{2}$ inches. The length of the grating is less than the distance between bearers to allow for their expansion. The gratings are prevented from unshipping, through the motion of the ship, by clips made of pieces of bent plate as shown.

The gratings are placed with divisional plates fore and aft whenever possible, and the spacing of $2\frac{1}{2}$ inches in the clear is maintained by distance pieces of gas tubing through the centre of which passes a $\frac{7}{8}$-inch bolt.

The depths of the plates of armour gratings varies with the thickness of the deck, and is as follows—

Deck.	Bar.	
Thickness.	Thickness.	Depth.
1 inches	$\frac{1}{4}$ inch	5 inches
$1\frac{1}{2}$,,	$\frac{3}{8}$,,	6 ,,
2 ,,	$\frac{1}{2}$,,	7 ,,
$2\frac{1}{2}$ } 3 } ,,	$\frac{5}{8}$,,	8 ,,
4 ,,	$\frac{3}{4}$,,	9 ,,

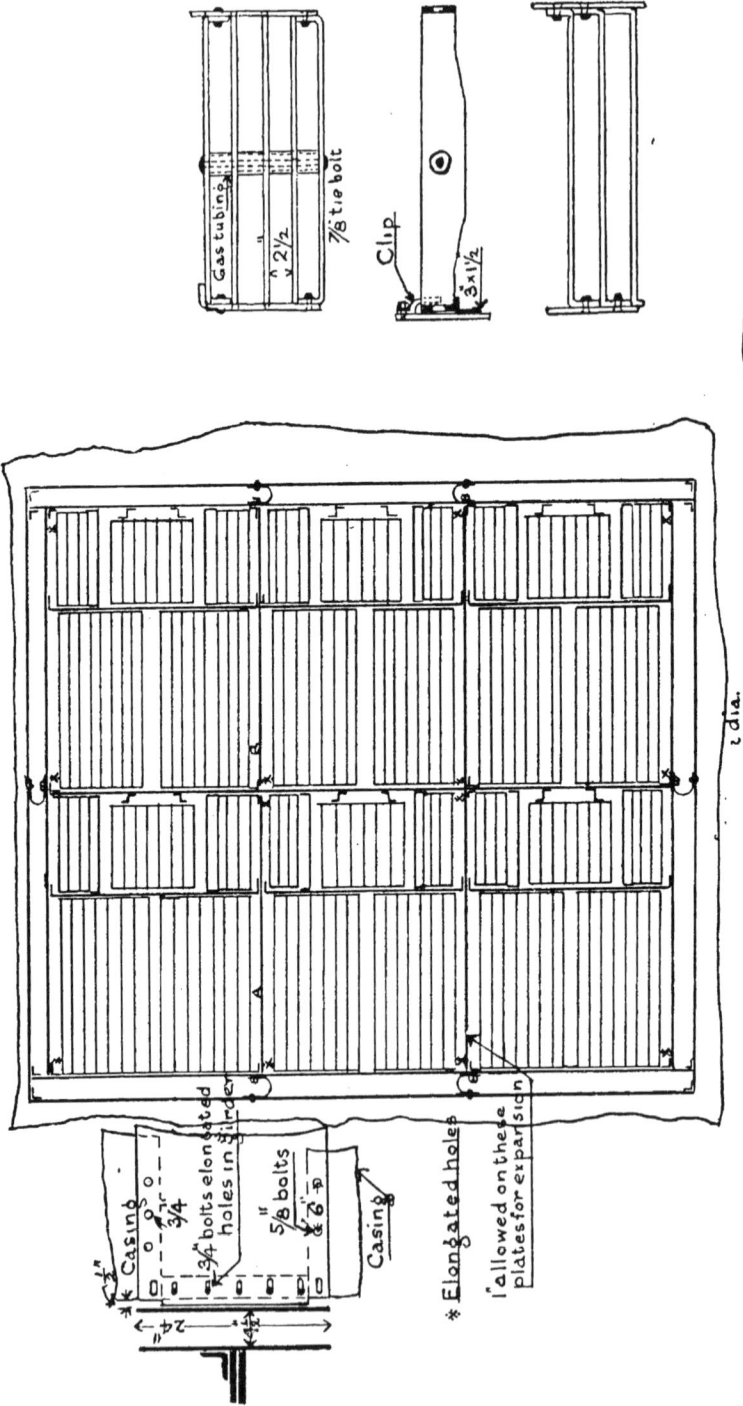

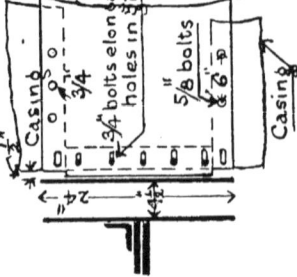

The full depth of the bar is maintained over ¼ the length of the middle portion of the bars of the grating, tapering to ⅝ the depth at the ends.

The upper surface of the gratings is at the level of the top of the deck.

Generally the openings required for shipping machinery are larger than the funnel hatches, and as the beams must be cut to the larger opening short pieces of beam have to be fitted between the outer coamings and end of existing beams before the deck can be completed, the two being connected by double riveted straps and by double angles to the coaming (see Fig. C on p. 43).

Between the funnels watertight bulkheads, continuations of those below, are worked intercostally between the decks.

The bulkheads surrounding the funnel coamings are lagged to prevent the adjacent spaces becoming overheated.

Launching Arrangements.

The launching arrangements commence with the hauling up of the bilgeways, which are secured to prevent them floating when the tide rises, with the risk of upsetting shores, etc.

The cradle in which the ship is to rest consists of the bilgeways, built up of balks of timber bolted and dowelled together (and are solid for the whole length of the ship), the slices or wedges, and either the poppet board and poppets, which are employed at the ends of the ship, or the stopping-up, fitted at the midship portion. The groundways are built of yellow pine with teak or oak cross bearers, except where they are secured to the land ties, where oak is used, and are solid right forward, right aft, and where the fore poppet will come when the stern lifts. The riband is of oak or teak, secured to the groundways by coach screws and teak dowels. The sliding plank which forms a smooth surface on top of the groundways is of teak, about 6 inches thick, and secured to the groundways by rag-pointed nails. The butts of the sliding plank and riband are not cut square, but sloped down the slip and the edges rounded off to prevent splintering as the bilgeways slide down (see Fig. p. 93).

When the bilgeways have been hauled up in place, and before making moulds for the poppets, the dog shores are got in place,

and section moulds made at convenient intervals to give the
shape of the stopping-up and poppets, allowance being made
for the thickness of the grease, about $\frac{1}{2}$ inch, and slices, and, in the
case of the poppets, the poppet board.   The stopping-up is made

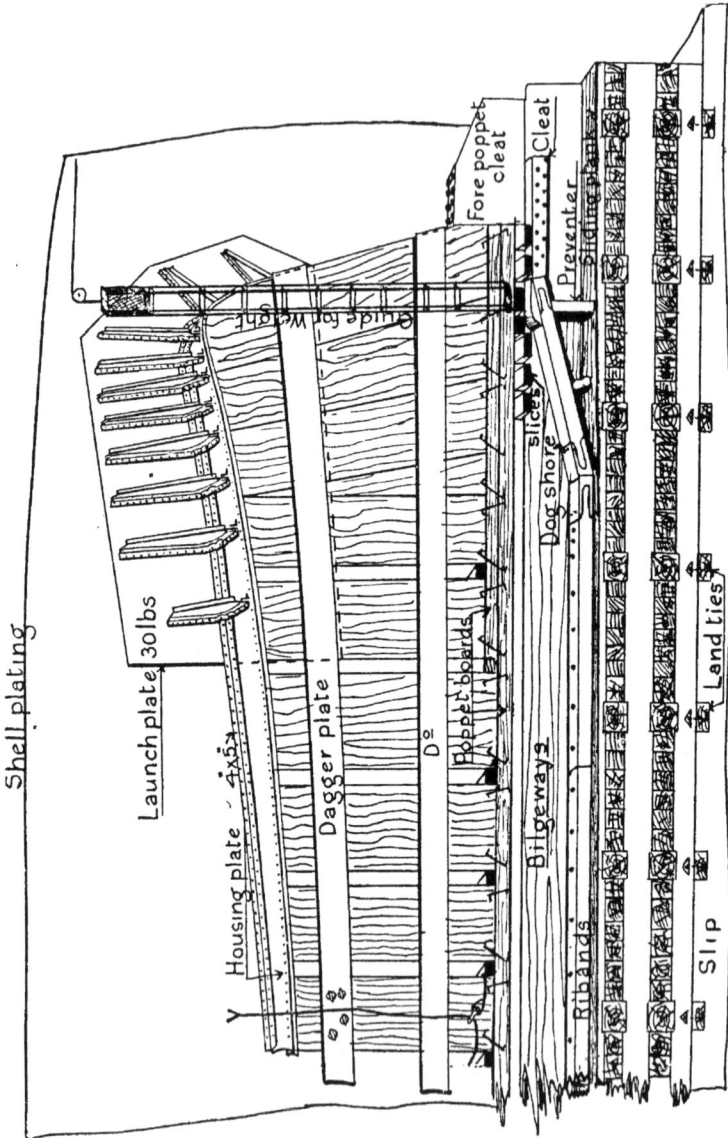

in as long lengths as can be conveniently handled.   A housing
angle is secured to the bottom plating to keep it from sliding outwards, except right amidships, where the bottom is flat enough
to render this unnecessary.   At the extreme end of the forward

and after poppets a housing plate is fixed perpendicular to the

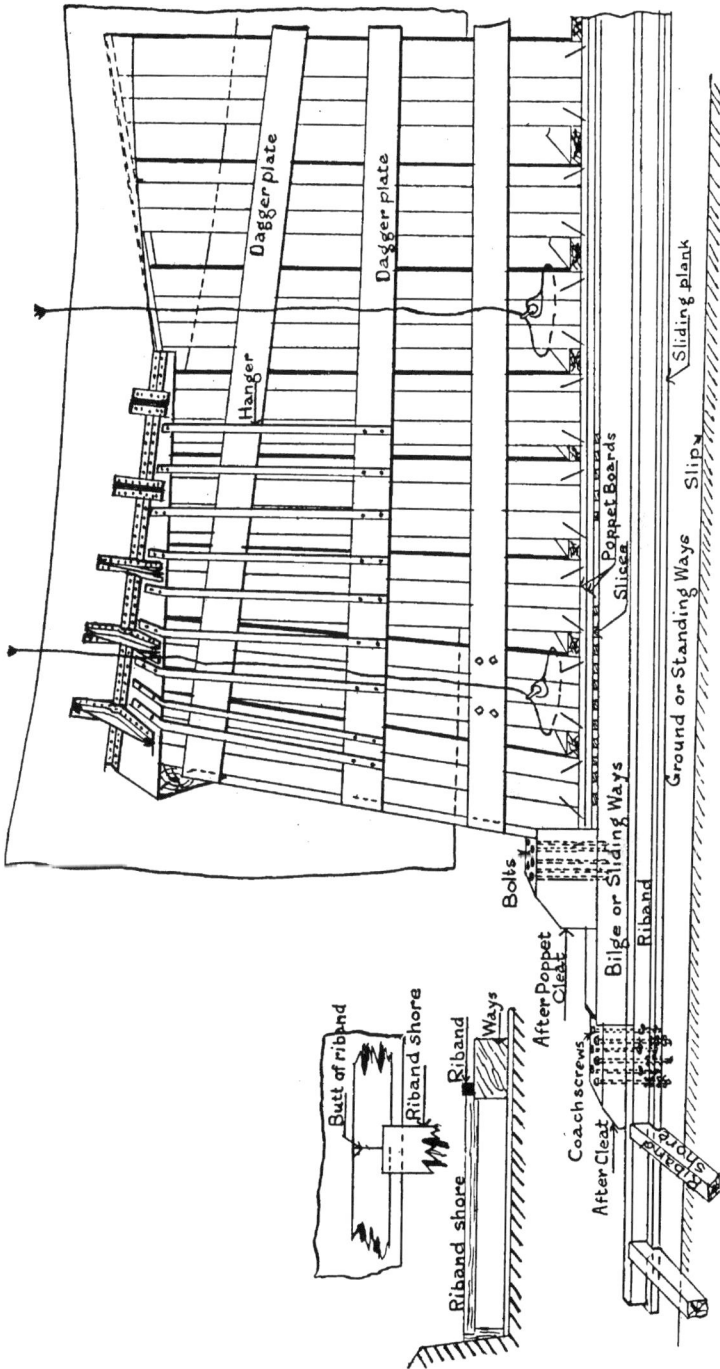

ship's side and stiffened by brackets (see pp. 92, 93, 94). In the case of the forward poppet, on which a great strain comes when the stern lifts, and where the side is not inclined greatly

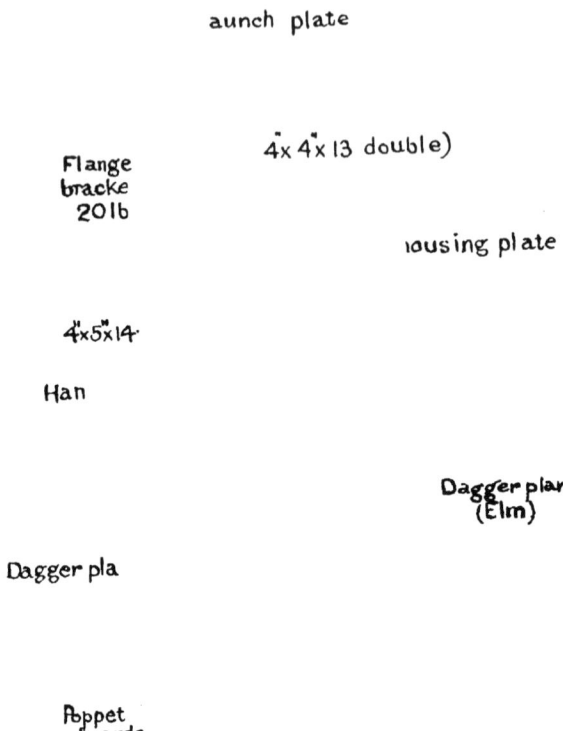

aunch plate

Flange
bracke
20 Ib

$4 \times 4 \times 13$ double)

ıousing plate

$4 \times 5 \times 14$

Han

Dagger plav
(Elm)

Dagger pla

Poppet

S

METHOD OF SUPPORTING CRADLE WHILE GREASING WAYS.

to the vertical, there is obviously a strong tendency to shear the rivets securing the angles and brackets to the ship's side.

Thus very closely spaced rivets are required on the angles to the brackets, etc., and there would be great difficulty in placing the brackets to keep clear of the riveting in connection with the ship's structure. For this reason the brackets and angles securing the housing plate are riveted to a poppet or launching plate, which is riveted to the ship's side. The forward and after poppets are prevented from spreading forward and aft respectively, under the weight of the ship, by the fore and aft poppet cleats, secured to the bilgeways. In order to distribute the strain brought on the structure of the ship in the wake of the fore poppets when the stern lifts, internal shoring with balks of timber called internal fortification is fitted, as shown on p. 96. The same is done in way of the after poppets, where the weight of overhang is taken, and in the machinery spaces. The poppets, which are built up with the grain vertical, are kept the requisite distance apart by means of distance pieces placed between the poppets at the top and bottom as shown. To give the poppets further rigidity two or three strips of plating called dagger plates are placed along them on the outboard side, and a dagger plank fitted on the inboard side, partly to serve the same purpose and partly to facilitate supporting the poppets when the bilgeways are turned out. The dagger plates and plank are secured to the poppets by eyebolts, and the poppets secured to the poppet board by dogs. To prevent the poppets moving laterally they are tenoned into the poppet board, and the latter projects into a V-shaped groove in the cleats as shown. The bilgeways are next turned out in order that the groundways may be coated with grease. For this purpose the riband has to be removed, and steps taken to ensure that the bilgeways when turned in again shall occupy exactly the same position as before. This is done by fitting stops to the bilgeshores, groundways, etc., as may be convenient, down the slip. One of these stops secured to a bilge shore is shown. Means have also to be taken to support the poppets and stopping-up, which have been kept in position by packing pieces (to make up the thickness of grease and slices) on the bilgeways. The method of supporting the cradle is shown in Fig. p. 94. The stopping-up is supported by shores wedged up into a cleat or housing as shown, and by strips of plating called hangers outside when possible. In addition to these supports, as

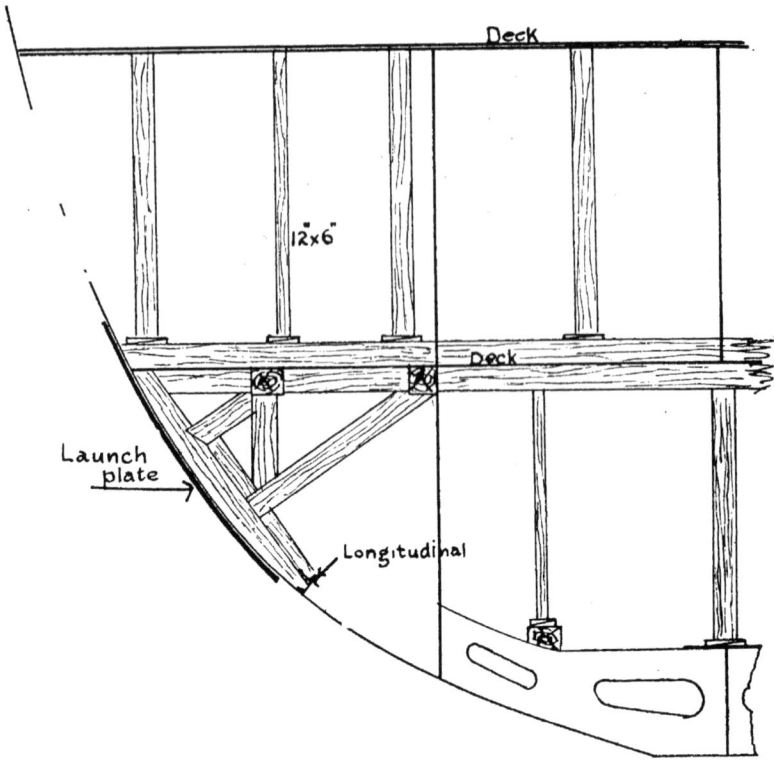

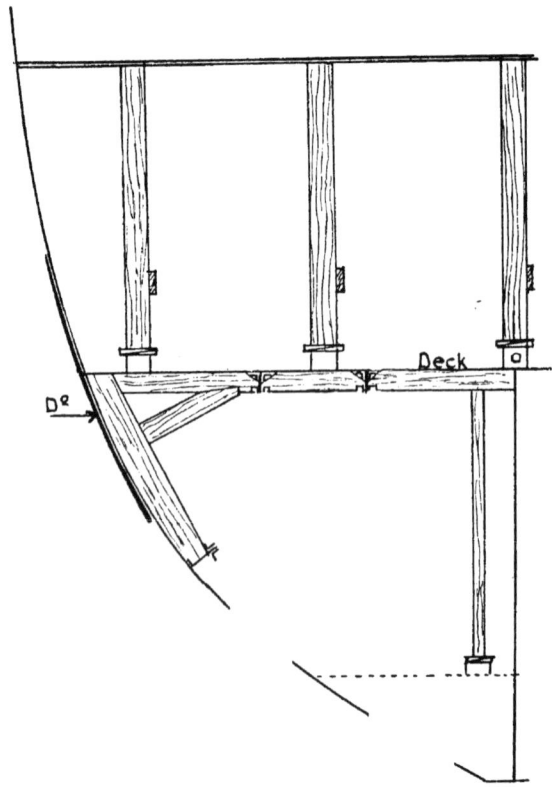

the bilgeways are turned out, short struts about 3 inches square in section are placed at intervals, and wedged up between the poppet board or stopping-up, as the case may be, and the ground-ways. The ends of these shores are chamfered off to facilitate removal. The bilgeways are turned out on to temporary trestles, erected at intervals of about 16 feet. If there is sufficient space between the top of the bilgeways and the ship's bottom for them to be "turned out" inboard instead of outboard, as in Fig. on p. 94, this is advantageous, as it will not be necessary to remove the ribands and riband shores, and hangers will not be required to temporarily support the stopping-up, as the latter can be shored from the ribands. Thus the holes in bottom for hangers will be saved as well. The depth of stopping-up determines whether this can be done. The bilgeways are turned out for their entire length, and a stop placed on every second or third trestle to prevent them

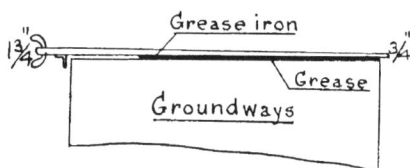

returning. The groundways are then cleaned and dried by burn-ing shavings on them in preparation for the grease. Battens of the same thickness as the grease are required to be placed along the position of the inside of the riband, and on the inside of the groundways. The tallow is smeared over as evenly as possible to the required thickness, and then scraped off level with the tops of the grease battens.

This coating of tallow is covered with slum, a mixture of tallow (2 parts) and train oil (1 part), to a depth of about $\frac{1}{4}$ inch. The surface of the slum is serrated in order to prevent the train oil which is put on last of all from running off. The bottom of the bilgeways is also coated with tallow and slum. The bilgeways are now turned in, and the outer grease battens removed to permit of this. To prevent the bilgeways from squeezing out the grease as they slide in, grease irons are placed on the groundways at intervals of about 20 feet, short lengths being cut out of the inner grease batten to allow of their insertion. The grease irons are 7 feet long and 6 inches wide, a piece of angle iron being

H

riveted on the end to prevent the bilgeways carrying the irons with them. The riband can now be fixed in position, and iron keys, about 12 in number on each side, are placed between the riband and bilgeways to ensure the necessary clearance, varying from $\frac{1}{2}$ inch at the fore end of the bilgeways to $1\frac{3}{4}$ inches at the after end. Pieces of batten are secured to the top of the riband to butt against the bilgeways, recesses being cut for the keys, to prevent dirt, etc. finding its way between riband and ways. Directly the bilgeways are turned in, a number of slices are placed in position and driven in to take the weight of the poppets and stopping-up, after which the supports can be removed. The riband shores reaching from the sides of the slip are now placed in position, one at each butt of the riband, and usually an intermediate one to each length of riband. Near the

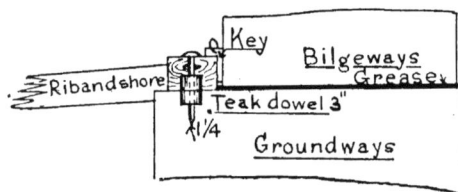

lower end of the slip the riband shores are spaced closer, as in the event of the ship being caught by a strong wind or tide and turning before she leaves the ways there would be a great side strain on these ribands. Auxiliary dogshores or stop cleats, about 7 feet long, are then fitted at the after end of bilgeways, as shown on p. 93, to assist in keeping the bilgeways in position previous to the launch. To prevent the bilgeways from moving towards each other as the ship is launched, spread shores are fitted between them, the ends resting in chocks.

Dogshores.—These are put in directly the bilgeways have been turned in. They are balks 12 inches square and 9 feet long of hard wood, generally African oak, which fits between the riband on the groundways and the dogcleat on the bilgeways, and thus prevents the cradle from sliding, the dogshores being kept in position by a trigger and a preventer. Iron shoes are fitted to the ends of the dogshore, riband, and dogcleat, as shown, to prevent the ends crushing. The slides for the falling weights are placed in position about this time, being secured to the poppets and to a bracket on the ship's side. The ship is now ready for *setting up*, which is the process of driving in the slices, which have been placed in position. These slices, or slivers, are long wedges, placed in pairs for the whole length of the ship, between the bilgeways and the stopping-up or poppet board, as the case may be. The ship is set up starting from the after end. Mauls and rams are used for driving the slices in. The rams are manned with from 4 to 8 men, and as the ship is set up, the slices immediately before and after those rammed are manned with mauls to keep them tight. The process of setting-up is usually started on the day before the launch, the after end being done at low tide, and is completed on the morning of the launch, being carried out simultaneously on both sides. The shores are all manned and set up at the same time, as they may become loose and fall.

On the day of the launch and after the setting up is finished the bolts through fore and after poppets are driven. Two such bolts are driven in fore and after poppets, their length being such that they stop short about 9 inches of the bottom of bilgeways. The process of setting up is finished off on the day of the launch, and the extent of setting up necessary to enable sufficient keel blocks to be removed varies a great deal with the conditions of the launch. It is usual to leave a number of keel blocks under the keel right forward, and to let the ship trip them as she goes down the slip. Their function is to assist the dogshores to hold the ship when all the other supports have been removed, and for this reason it is not required to set the ship up to any great extent forward, and rams are not used there. After setting up, all bilge shores and the skeg shore under the keel aft are removed, and the work of removing the remainder of the blocks, starting aft, is commenced. The removal of the blocks continues until the ship begins to "creep," *i.e.* shows signs of movement. The slices are roped together to haul them aboard after the launch, and those in way of falling weights cut to allow them to strike the dogshores. The stop cleats at the after end of the bilgeways are removed as the tide rises, and placed in a conspicuous position at the head of the slip. The keys are removed and placed on a board at the head of the slip. The weights which are to fall on the dogshores and thus release them have been previously put in position at the top of their respective slides, and

an endless rope is taken from either weight over a pulley at the top
of the slide, and passed round the bow of the ship to the other
weight, over the ornamental block on which it is to be severed.
The weights are further prevented from falling by shores, one
under each weight, and kept in place till the last minute. The
rope supporting the weights is hung with the weights at its ends
for a few days previous to placing it in position, in order to take up
all the stretch in it and thus ensure that the weights have the
greatest drop possible in the slide. Creep battens are fitted to
each bilgeway to indicate creep, and the amount is shown on a
dial. A tide indicator, giving water over end of ways, is placed at
the head of the slip, and is adjusted every quarter hour for an hour
or two previous to the launch. When everything is ready for the
launch and the tide indicator shows a sufficient height of tide, the

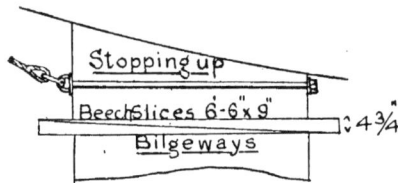

order is given, "Stand by to launch," the preventer, trigger and
the shores to the dog weights removed and placed at the head
of the slip, the rope supporting the weights is severed, and the
weights drop on the dogshores. If the ship refuses to move when
the dogshores are knocked away, hydraulic jacks are used. After
the launch, the bilgeways, stopping up, and poppets are hauled
away, steel ropes having been previously fixed to them, and
secured on the upper deck in readiness to be thrown to the tugs.

The ship is sighted both longitudinally and transversely, an
additional set of sights being set up over after A bracket to
fore end of shaft to obtain the breakage in way of propeller
shafting.

Tallow. It is of the greatest importance that the tallow used
in connection with the launch should be of good quality, and the
following are some of the tests carried out on it :—

(1) A sample is melted, and after being applied in a plastic
state to a clean wood surface to a thickness of $\frac{3}{8}$ of an inch, and
then exposed for 4 hours to a temperature of from 25 to 40

degrees F., should be free from cracks, and when cut with a knife should not adhere but give a clean cut.

(2) A second sample, applied to the groundways, and loaded to $2\frac{1}{2}$ tons per square foot, is allowed to remain about 12 hours, after which the tray is released, and allowed to slide over the tallow, and the latter should adhere firmly to the ways and not be altered in thickness by the action of the sliding weight.

The time of year at which the launch is to take place has an important bearing on the melting-point of the tallow to be used, this being about 118 degrees for summer use, and about 114 degrees for winter use. The " slum " applied to the surface of the tallow consists of scrapings of tallow from previous launches, mixed with train oil in the proportion of 2 to 1, and the mixture is then passed though a sieve.

The quantities used depend, of course, on the thickness it is proposed to apply to the ways, but for a large battleship it is approximately : tallow, 173 cwts. ; slum, 160 cwts.; train oil, 286 cwts. ; soft soap, 23 cwts.

The following table gives the results of some tests carried out on samples of tallow :—

Sample.	Temperature.	Crushing Strength.		
1	56	4	tons per sq. foot	
2	54	9·5	,,	,,
3	46	6·5	,,	,,
4	65	4·	,,	,,
5	62	5·5	,,	,,
6	54	8	,,	,,

Enumeration of points to be noted in connection with launch—

1. All underwater fittings screwed down or blank flanged.

2. Fairleads and bollards on board.

3. Outer and inner bottom watertight.

4. All manholes closed.

5. All holes in coal bunker and middle line bulkheads watertight.

6. Watertight doors in machinery spaces closed.

7. Side scuttles closed.

8. Temporary ladders, guards to hatchways.

9. All gear cleared from under ship, and keel blocks from after end of ship.

10. All holes made for temporary support of cradle whilst greasing ways, plugged.

11. Paint bare spots after removal of shores.

12. Put in steel wedges at dogshores after ship is set up.

13. Grease irons and keys all properly numbered so that they can be readily accounted for.

14. As keel blocks are removed numbers placed on board at fore end of slip.

15. Staffs showing tide over end of ways removed just before launching.

RIVETED WORK.

The most usual types of rivets used in shipbuilding are shown on p. 103, also the pitch and spacing for flush and lap-joints. 1, 7, 8 are used where through rivets cannot be worked; for example, the covering plates at top and bottom edges of armour, the thick plates attached to stem and stern posts, etc.; 7 is a special form used for connecting plates forming casemates, and on conning towers where ordinary rivets would be a source of danger should the heads fly off if the armour were struck; see also socket for armour bolt on p. 83. It is difficult in the case of three thicknesses of plating to ensure that the holes in each plate coincide, and if ordinary pan or snap head rivets were used, the holes in the plate not being concentric, the rivets would probably leak, owing to the holes not being properly filled; but by using rivets as 2, it can be ensured that the heads and points fill the holes. C.K. heads and points are also the best where the rivets may have to be caulked, as in some portions of the structure it is not possible to caulk the whole of a pan head, e.g. vertical keel angles 3 is used in ordinary watertight work, as outer bottom plating, 6 in non-watertight bracket plate frames, brackets to beam arms; 5 is similar to 3, but has a conical neck so as to fill up the conical hole in a punched plate. The reason that a punch gives a conical hole is that the die is larger than the punch (see Fig. p. 103), and the larger the punch the greater will be the difference between the size where punch enters and leaves the plate respectively. 4 is used for non-watertight work of minor importance, as cabin bulkheads, etc.

Flush butts   ## Lapped butts

Single chain riveting

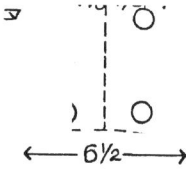

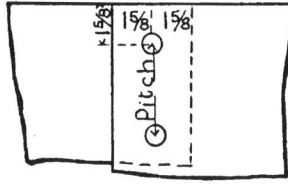

Double chain riveting

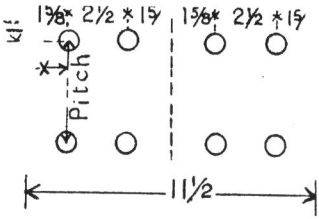

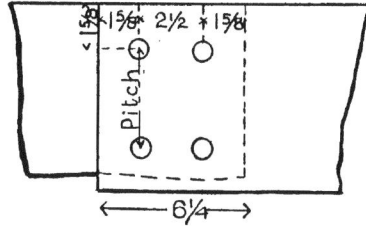

Treble chain riveting   ### Zig-zag riveting

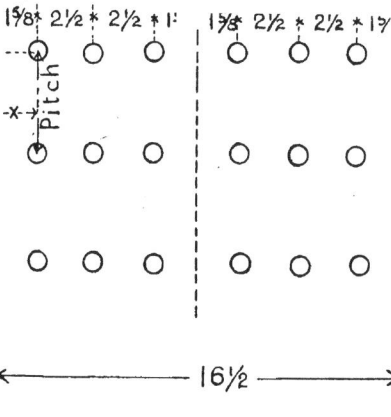

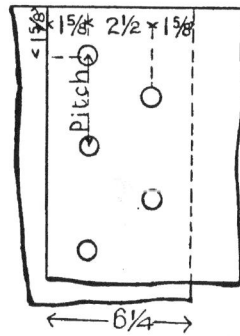

The distances marked x are 1¾ for H.T steel

1   2   3   4   5   6   7   8

In some cases, *e.g.* angle bars, the $2\frac{1}{2}$ diameter between the rows cannot be obtained with zig-zag riveting, and in such cases reeled riveting is resorted to. The rivets are kept the correct distance from the edge, and the pitch is measured diagonally from rivets in one row to those in the other.

It is of the greatest importance for watertightness to properly close the work by nut and screw bolts before commencing the riveting, otherwise collars may be formed between the plates, and also to ensure efficient caulking of the edges without having to resort to liners. The burrs caused by punching and drilling should also be chipped off.

Plates which have been drilled off in place should have the closing bolts eased back and the drillings removed from between the plates to ensure good contact.

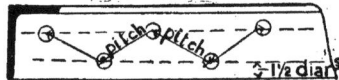

REELED RIVETING.

As each rivet is knocked down, the preceding ones are gone over again, as the last one dealt with will close up the work and tend to make the others slack.

In the case of rivets with countersink points, these should not be finished flush with the plating, but project slightly in the centre. This permits of the rivet points being caulked if necessary.

Rough hammered points should be finished with a good amount of metal outside the plating, as the holes in the plating being only slightly conical, there will be very little connecting the two plates if the point is finished flush.

Rivets are tested by hammering to see if they are loose, and any defective ones are caulked or removed and fresh rivets inserted.

	Pitch in diameter of rivet.	Remarks.
Watertight work	$4\frac{1}{2}$–5	
,, ,, ...	4 –$4\frac{1}{2}$	Butt straps
Oiltight work	$3\frac{1}{2}$–4	
Beams to deck plating	7 –8	
Frames to shell plating	7 –8	Ordinary frames
,, ,, ...	$4\frac{1}{2}$–5	Watertight ,,
Non-watertight work	7 –8	
Stiffeners	7 –9	12″ spacing in H bars

The caulking is done by means of the tools illustrated as under, which shows the various steps in edge and flush caulking.

The form of countersink in a plate will depend principally on the thickness of the plate. In thin plates it must be carried

Splitting Tools.   Making Tools.   Reeding   Spike
Flush.   Lap.   Flush.   Lap.   Tool.   Iron.

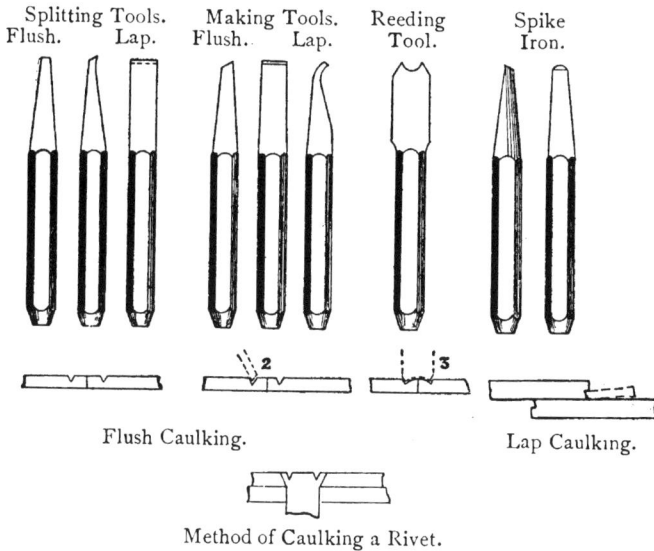

Flush Caulking.   Lap Caulking.

Method of Caulking a Rivet.

right through the plate and have a large angle. This will apply for plates up to about $\frac{3}{4}$ inch in thickness. The angle of the countersink will diminish as the thickness of plate increases, and in plates above $\frac{3}{4}$ inch it will not go right through the plates stopping short about $\frac{1}{10}$th.

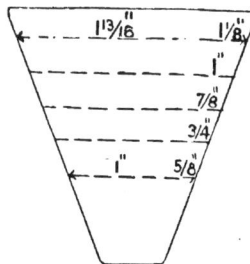

Fig. A.—Template for Countersink.

The template gives (Fig. A, above) countersinks for diameter of holes indicated. Thus for $1\frac{1}{8}$ inch diameter hole the diameter of countersink is $1\frac{13}{16}$ inch.

The size of hole formed in a plate will depend on nature of rivet, that is, whether a through or tap rivet. In the former case, the hole will be about $\frac{1}{16}$ inch larger than the rivet for ordinary rivets, while in the latter the hole will be $\frac{1}{16}$ inch less than the size of tap rivet to allow of the thread being cut.

The length of a rivet will depend on nature of finished point, that is, whether C.K., rough hammered, or snap. In the former, about $1\frac{1}{4}$ diameter of rivet, in addition to the thickness of plating, will be necessary to completely fill the hole and the countersink, while in the other cases a diameter will be sufficient. In the case of a tap rivet the length is generally $\frac{1}{8}$ inch more than the diameter.

It is essential that the form of countersink be kept the same, or some rivets will have to be chipped, while others will not fill the countersink and have to be taken out, causing considerable waste.

WATER-TESTING.

(See also Appendix, p. 324.)

A commencement is made with this as soon as possible, as it must be completed before the ship is launched. Compartments are surveyed to test riveting and caulking, and then filled with water, the necessary head of water being obtained by means of a stand pipe. Holes are drilled through the webs of beams to prevent air pockets being formed in the angle between the deck and the beams, and at the highest points of the compartment to allow of escape of air. To ensure there is no obstruction in the stand pipe, the air escape holes are left unplugged until the water flows freely through them, after which they are plugged with wooden plugs.

While the pressure is on, all rivets and caulking are examined and any distortion of the plating noted, especially where the crown of the compartment is of relatively thin plating. Any leaky rivet or defective caulking is made good, and should a flat show signs of undue distortion it is shored and additional stiffening or pillars worked as may be necessary, the compartment being retested until found satisfactory.

In the case of double-bottom compartments a few rivets are left out at the lowest point of the compartment for draining off the water, these holes being subsequently filled.

The pressure head for most compartments is about 5 feet above L.W.L., except for the foremost ones, on which the head is about

12 feet. In the case of feed tanks and hydraulic tanks these are tested to about 10 lbs. and 20 lbs. per square inch respectively.

Watertight doors are tested before fitting to a pressure equal to that to which they would be subjected if fitted in the ship. When testing these doors in the shop a small cock is fitted in the door to permit of the escape of air. This hole is subsequently plugged.

It is desirable that all fittings be in place before water-testing, *e.g.* in double-bottom compartments, distance pieces for sea cocks, blow-down valves, etc., bilge and docking keels or the angles connecting these to the shell riveted off, W.T. doors to bunkers, etc., stays in bunkers, hatch covers to decks, retesting is necessary unless this can be done, and in the case of fittings on shell and inner bottom, *e.g.* distance pieces, special arrangements have to be made for testing the efficiency of the work.

Compartments not filled with water have surrounding bulk-heads, etc., tested by hose, and all decks tested by flooding before any coverings as wood or corticine are worked. All fixed coaling shoots, scuppers, soil pipes, bunker ventilation pipes, air pipes to oil fuel compartments, supply and exhaust trunks to fans, and watertight ventilation trunks are tested by filling with water, and metal castings in connection with machinery underwater fittings, sea cocks, etc., to a pressure of 60 lbs. per square inch.

One of the large machinery spaces is usually filled to a head of 5 feet above L.W.L., and any deflection of the bulkheads bounding the space measured. This is done by fixing bases on the opposite side of the compartment tested between the lower deck and inner bottom, on which are placed at intervals a number of sliding pieces of wood with their points in contact with the bulkhead before the space is filled with water (see p. 108). When the pressure comes on the bulkhead these pointers are forced out, the distances being measured and curves plotted showing the shape of the bulkhead under pressure, and from this it can be determined whether any additional stiffening is necessary.

WATERTIGHT DOORS, MANHOLES, ETC.

Means of securing Watertightness. To permit of access along decks and through the watertight bulkheads, hinged watertight

doors as on p. 109 are fitted. These are generally of Mechan's
patent embossed type, and are provided with clips to keep them

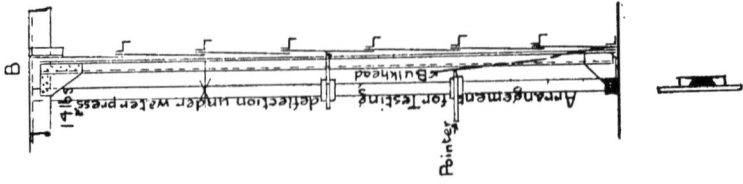

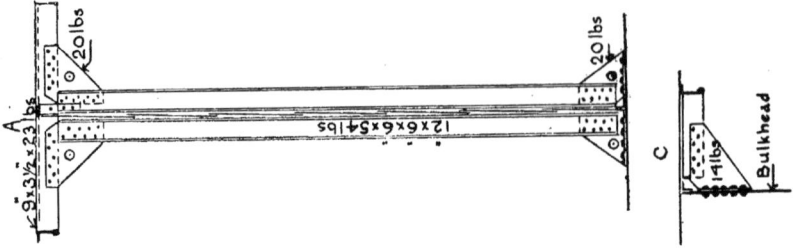

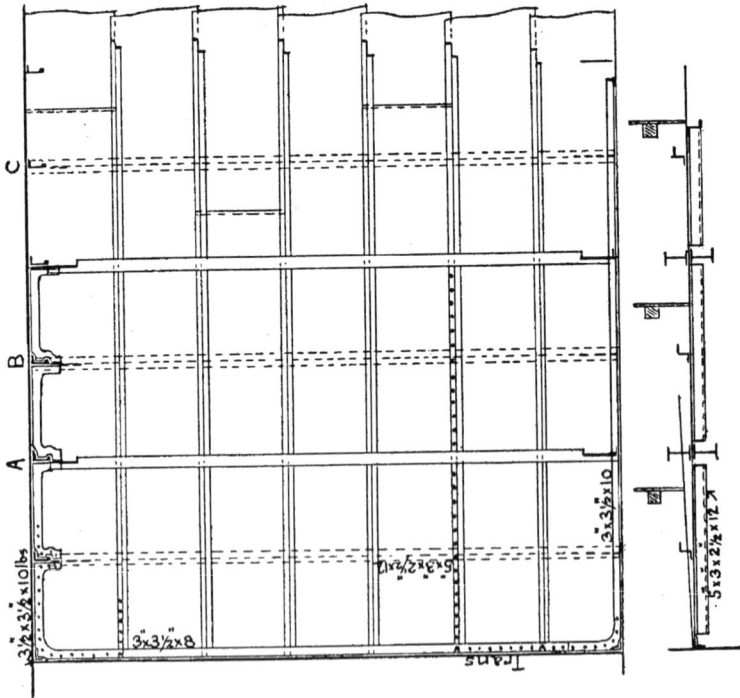

closed when necessary. The clip below the lower hinge should be
noticed, as it is of special form to permit of its clearing the hinge.

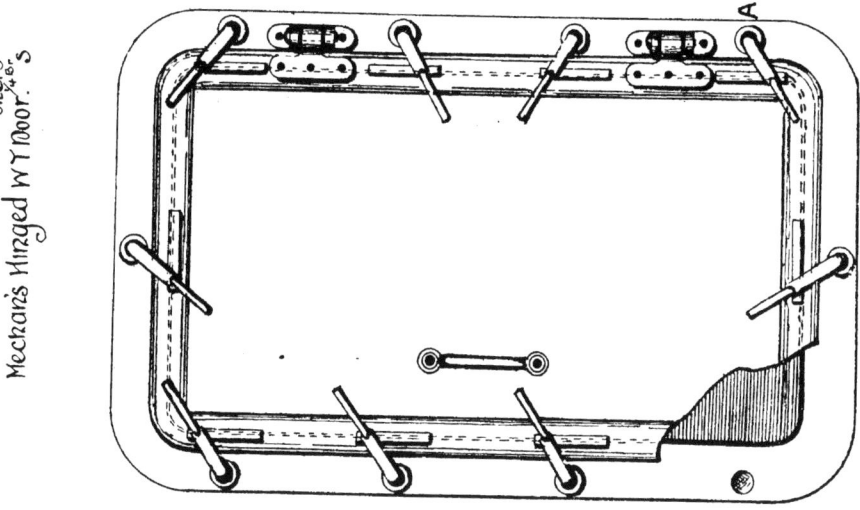

Mechan's Hinged WT Door. Steel 5
5 Steels

Sectio

Special clear

A

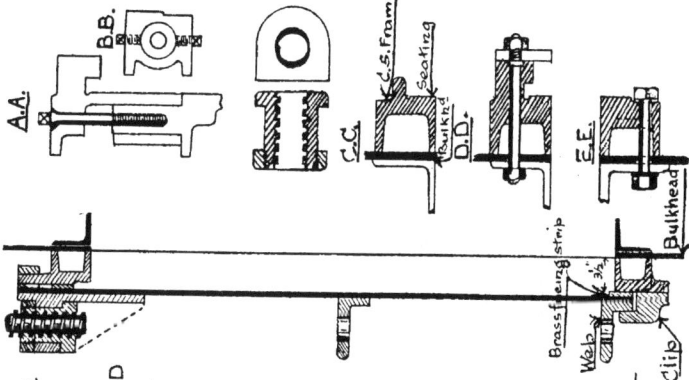

Sections

A.A.

B.B.

C.C.
C.S. Frame
Bulkhd
Seating
D.D.

E.E.
Bulkhead

Brass facing strip
Web
3½
Clib

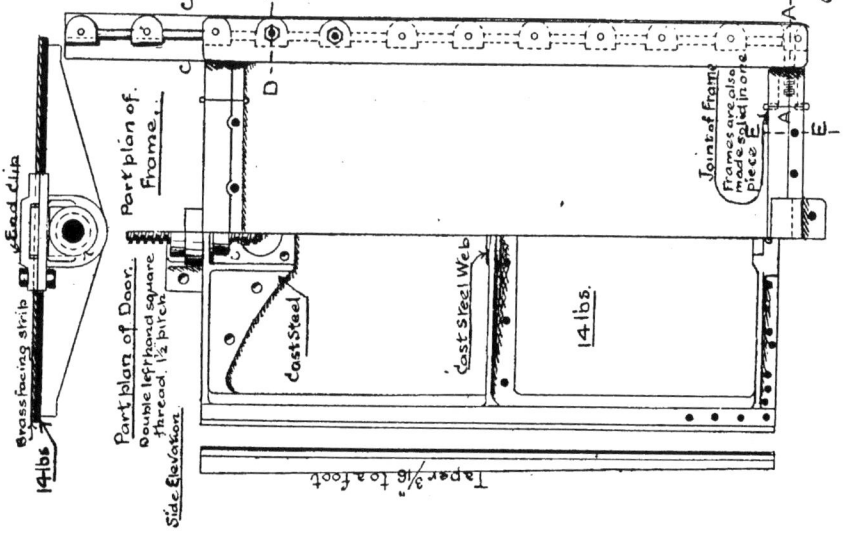

1012

Brass facing strip
End clip
14 lbs

Part plan of Door.
Double left hand square thread. ½ pitch

Part plan of Frame.

Side Elevation

Taper 3/16″ to a foot

Cast Steel

Cast Steel Web

14 lbs.

Joint of Frame
Frames are also made split in one piece

The holes in the hinge are elongated to allow of the door bearing evenly all round on the rubber. Vertical sliding watertight doors are fitted to the coal bunkers, these doors being worked by a chain in the stokehold, and also geared to the main deck. The details are given on p. 109.

For access to double-bottom compartments raised manholes are fitted on top of the inner bottom. Air escapes are fitted to the

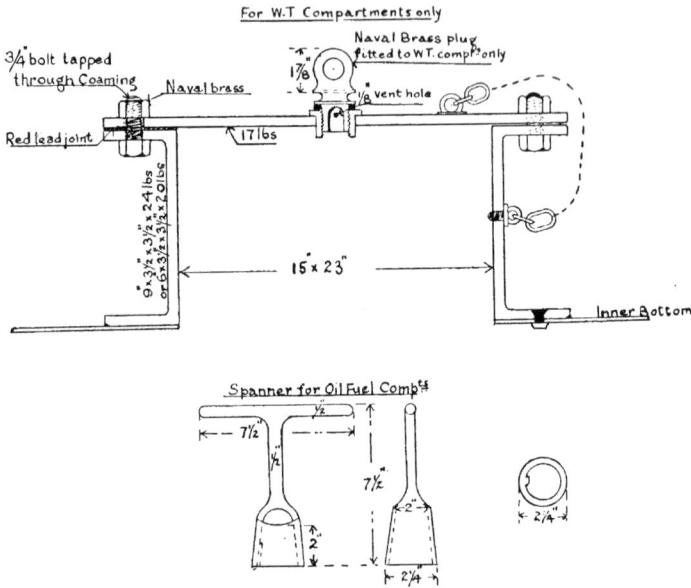

manhole covers, except those to oil fuel compartments, to permit of the escape of the air, so that the compartments may be completely filled. Details of these manholes are given above.

The space between the floor of gun house and top of barbette armour is made watertight, and a method of doing this is as illustrated.

APRON TO BARBETTE.

Side scuttles, skylights, etc., are made watertight as shown on sketches of those fittings. The method of passing an electric cable and also a fresh water pipe through a bulkhead and deck is given below.

Bulkhead pieces for pipes such as salt water mains are depicted below. Two methods of fitting these are shown, viz. A, in which

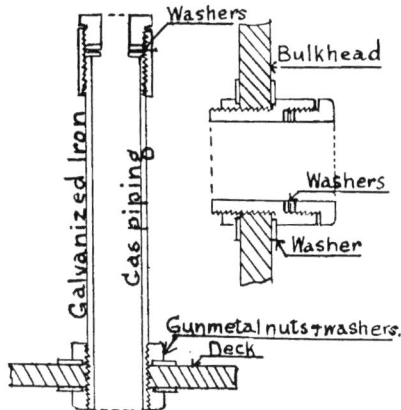

GLANDS FOR ELECTRIC CABLE.

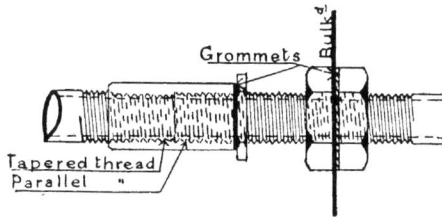

FRESH WATER PIPE PASSING THROUGH BULKHEAD.

the piece is fitted in place whole, and which require a hole in the bulkhead large enough for the flange to pass through, and B, in which one end of the pipe is screwed, the flange being fitted after

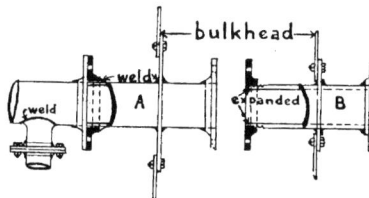

the piece has been put in place, requiring a hole in the bulkhead of the size of the pipe only. The end of the pipe is expanded after the screwed flange is fitted.

LIGHT AND VENTILATION.

Living spaces are ventilated by motor fans which obtain their air supply from downcast trunks. These trunks are provided with

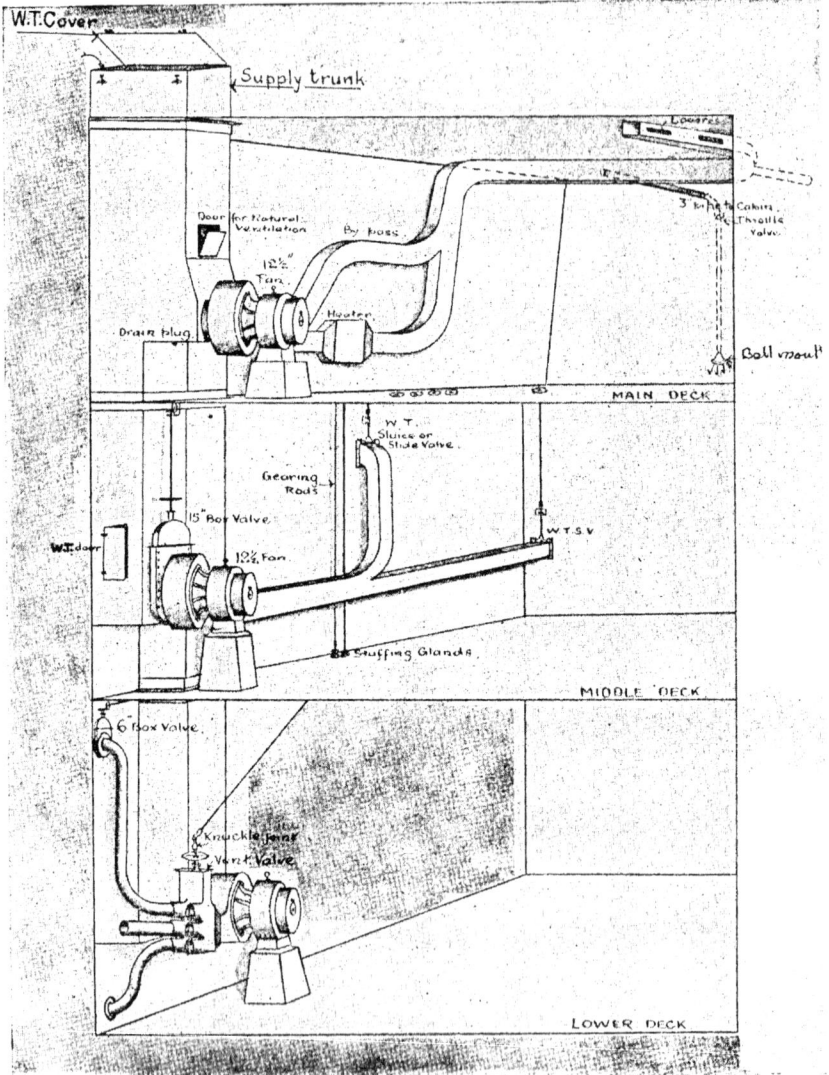

means of excluding water where necessary, as for example on the weather deck forward and aft, by fitting screw-down mushroom tops at the tops of the trunks. When these are closed, the air supply is obtained through the doors between decks. The trunks are also arranged to ventilate by natural means when the fans are not working.

The ship is divided up for ventilation purposes according to the watertight sub-division, *i.e.* all the compartments between any two consecutive main transverse bulkheads would be supplied with one or more fans within the spaces to do away with the necessity of piercing the watertight bulkheads. It is not always possible to adhere entirely to this, and in cases where the watertight bulk-

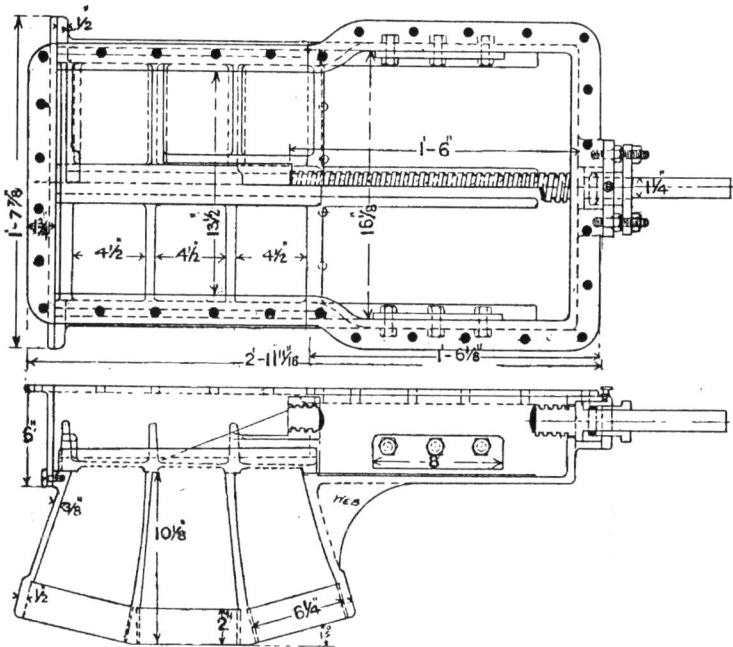

WATERTIGHT VENTILATION VALVE BOX.

heads have to be pierced, watertight slide valves are fitted at the bulkheads.

Watertight valve boxes from which 4 or 6 pipes can be taken off (see above) are fitted near the fan on its delivery side and the pipes led to the various compartments, those which pierce the decks having watertight valves at the decks to maintain their watertightness. Such fans as have not these watertight valve boxes have box valves between the supply trunk and the fan. All pipes to a height of about 5 feet above the main deck are made watertight.

All pipes and trunks entirely above this deck are not watertight, but if they pierce watertight bulkheads, slide valves are fitted at the bulkheads to ensure the watertightness of the latter.

A hot air supply is arranged for cabins and mess spaces, the air from the fan passing through a heater. The heater consists of coils of pipes enclosed in a steel tank through the interior of which steam circulates, the air passing over the outside of the pipes.

When cold air only is desired the air is directed from the heater through the by-pass trunk to avoid the loss of pressure which would occur if it were sent past the coils in the heater.

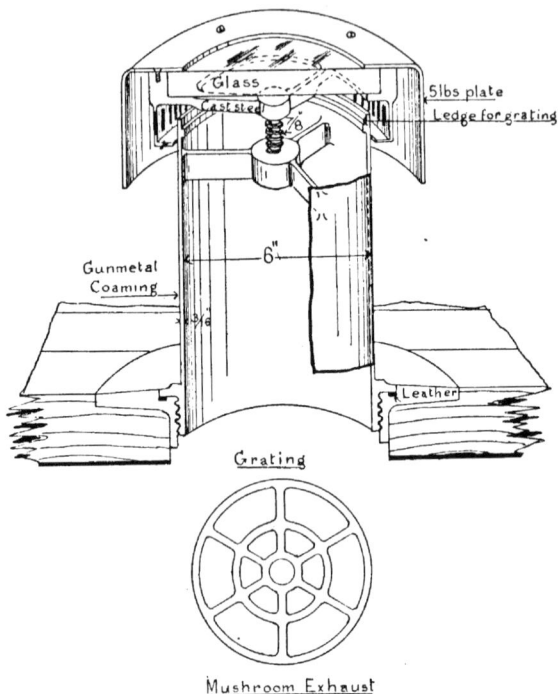

Mushroom Exhaust

The open ends of pipes and trunks are bell mouthed, and covered with wire netting.

The exhaust from living spaces is through the hatchways, side lights, skylights, etc.

In the case of certain spaces, exhaust will be through escape trunks.

The spirit room has special supply and exhaust trunks, distinct from the ordinary system. These trunks extend to the weather deck, and have mushroom tops at the upper end.

The sick bay has a branch supply led from a fan, the exhaust being through mushroom tops, this space not being fitted with jalousies as are cabins, to prevent air exhausting into adjacent living spaces.

The ventilation of the boiler rooms is carried out by means of steam driven fans. The engine rooms are ventilated by large electrically driven supply and exhaust fans, the trunks of fans being led to the weather deck and fitted with hoods, and so arranged as to be used for natural ventilation when the fans are not running.

All watertight box valves between the fan and trunk, valve boxes, watertight slide valves on bulkheads and under protective deck, are arranged to be worked at the valve and at the main deck, the remaining being worked at the valve only.

The heads and urinals have supplies from fans with sliding louvres in the pipes, the exhaust being through mushroom tops. As these spaces are generally at the sides of the ship, the side scuttles provide natural ventilation for them in addition.

The magazines and shell rooms have their own separate system of ventilation, as illustrated on p. 116. The details of box slide valve fitted in supply and exhaust pipes where they pierce the deck are shown on p. 117. These valves are tested in the shop to 10 lbs. per square inch.

The ventilation of coal bunkers is described on p. 118.

Certain watertight compartments need only be ventilated when opened up, and for such hose connections are provided on the adjacent trunks, to which hoses can be attached for ventilating them.

The between deck spaces are lighted electrically and by natural means through side scuttles, square side ports, skylights, etc., all of which must be capable of being made watertight. A sketch of side scuttle is given on p. 117. It will be seen to consist of three parts, viz. a steel casting riveted to the hull, a gun-metal ring to take the glass and indiarubber, and a cast steel deadlight, the deadlight hinge being so formed that the deadlight can be either closed directly on to the casting on ship's side, the glass being hinged back, or on to the back of the glass scuttle by shifting the deadlight into the front of the hinge. Under each scuttle is a copper drip pan, and outside above the opening in plating a rigol is fitted to prevent water from the weather deck running down the

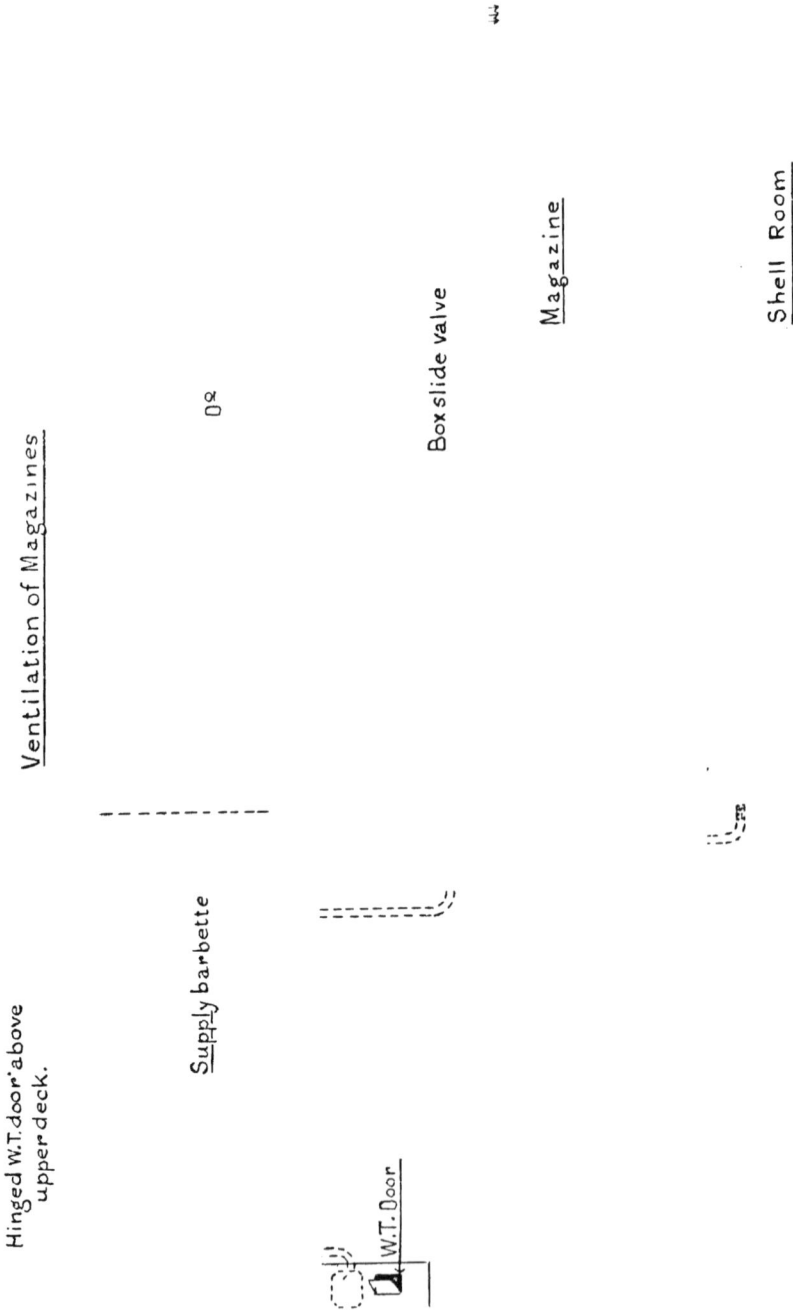

$\frac{3}{4}$

Shell Room

Magazine

Box slide valve

Ventilation of Magazines

02

Hinged W.T. door above upper deck.

Supply barbette

W.T. Door

side and through the opening. Side scuttles hinge on the fore side, those for the two sides of ship being of opposite handings.

Square Side Ports.—These are large ports for light and ventila-

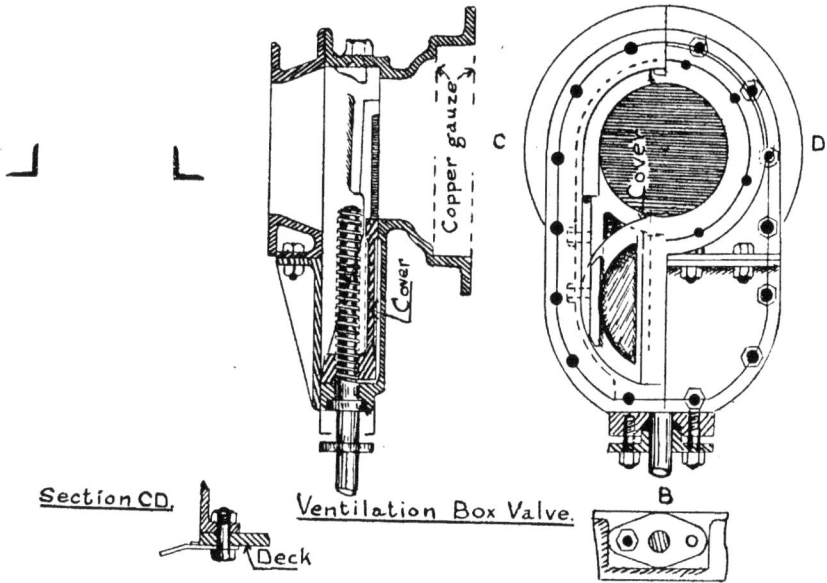

Section CD.

Ventilation Box Valve.

Deck

tion fitted in the side, without glass, and open outwards, and not inwards as in the case of side scuttles, being actuated by various

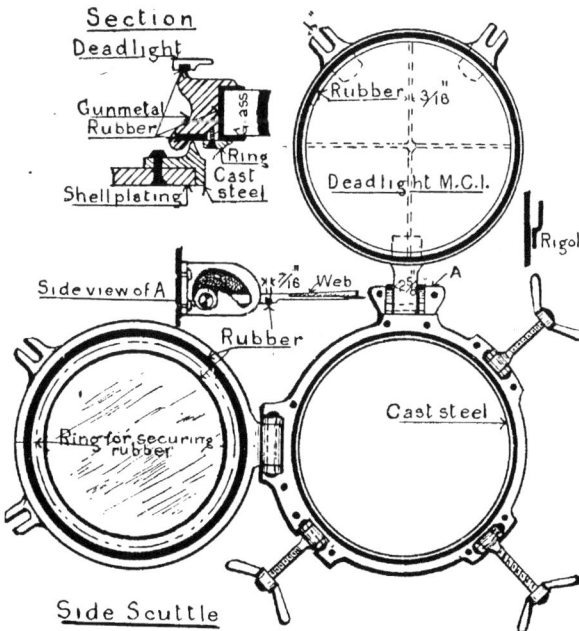

Section
Deadlight

Gunmetal
Rubber

Shellplating

Rubber 3/8"

Deadlight M.C.I.

Ring
Cast
steel

Rigol

Side view of A

Web

Rubber

Ring for securing rubber

Cast steel

Side Scuttle

means, as levers, worm, etc., as shown on p. 118   Watertightness is secured by means of rubber round the edge of the door, and a frame on the shell plating, the door being secured by clips and in addition by strongbacks to prevent undue strain coming on the clips when gun firing.

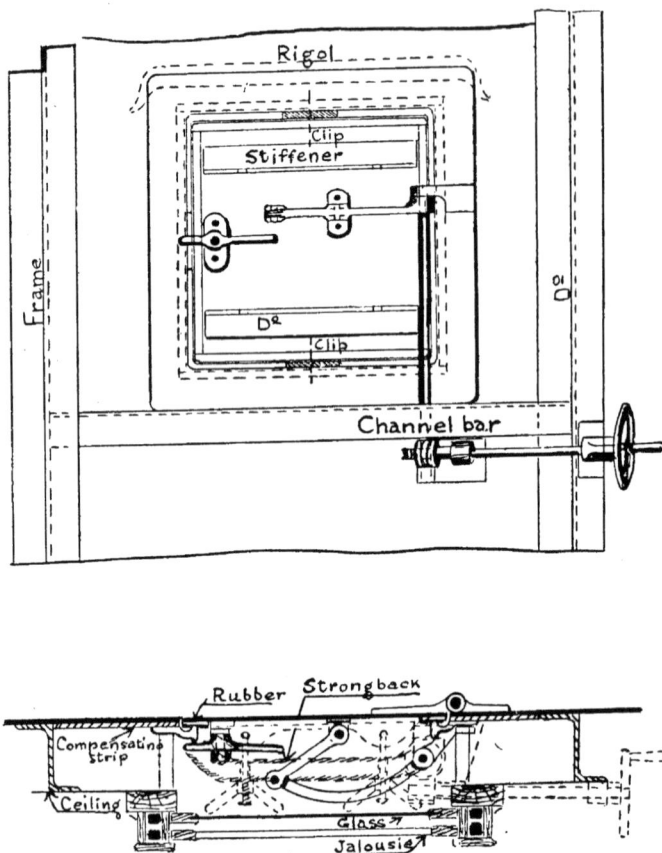

VENTILATION OF BUNKERS.

The bunkers are ventilated by natural means, pipes being led from opposite ends of each bunker, the exhausts being taken to funnel or other casings (p. 119). These pipes are quite distinct from those ventilating other parts of ship. To assist the circulation of air over the top of the coal, holes are cut in the beams. It is not necessary to cut these holes near the ends of the beams, as the triangular opening formed by bracket plate, ship's side, and deck serves this purpose. Temperature tubes are fitted to all bunkers, to enable the temperature in the mass of the coal to be obtained. These are steel tubes plugged at the lower

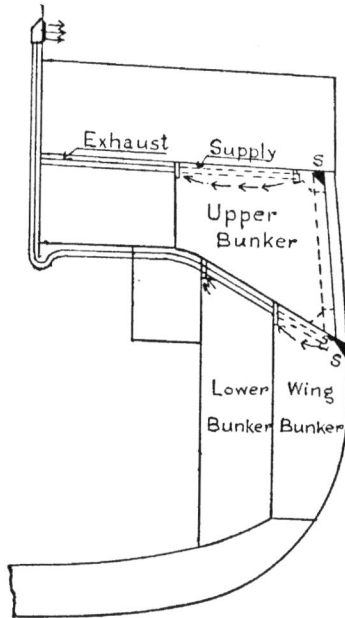

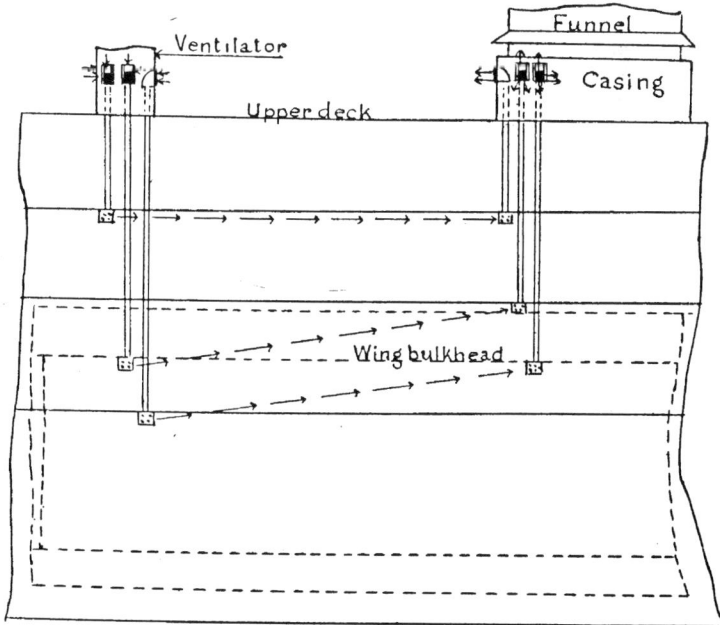

end and fitted with a screwed cap at the upper. Fig. C shows a pocket in bulkhead for temperature tubes, the depth of pocket being sufficient for insertion of thermometer. Figs. A and B show two methods of fitting screwed cap.

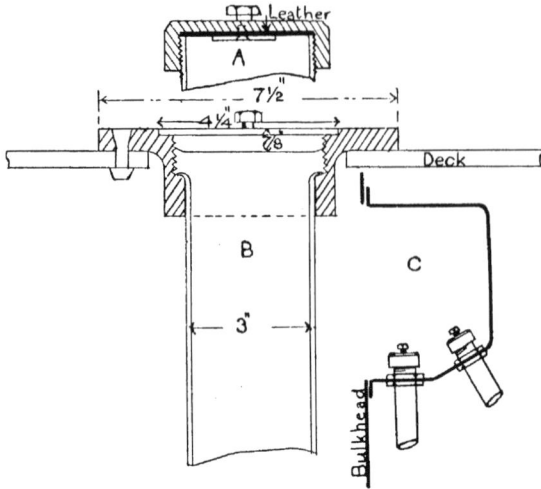

TEMPERATURE TUBES.

A locked sliding louvre is fitted at the upper end of each bunker ventilation pipe and a perforated plate over the lower end. A sketch of a louvre is shown below, which also shows the method of locking.

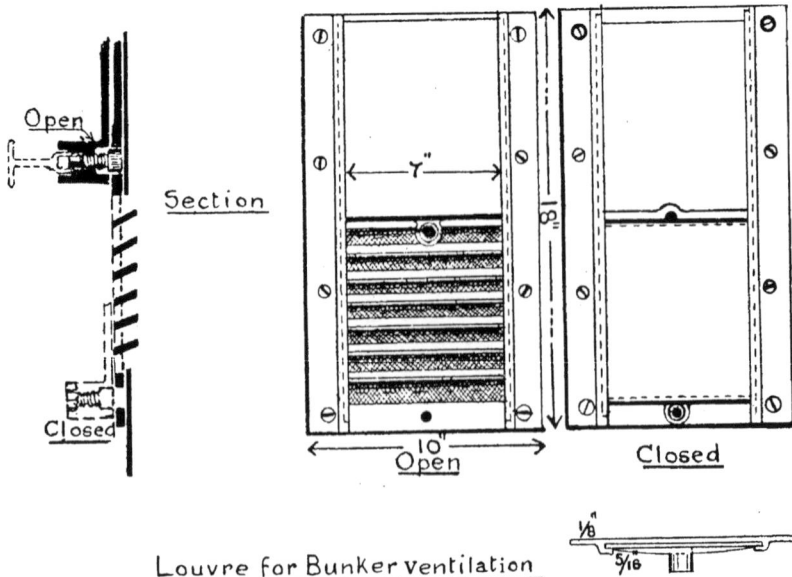

Louvre for Bunker ventilation

The ventilation pipes are galvanized and generally 5 inches or 6 inches in diameter, made watertight to a height of about 5 feet above the L.W.L. and tested by plugging the lower end, and filling with water to the height mentioned. For the remaining length of pipe airtightness only is necessary, and this is tested by the smoke test.

Magazine Cooling Arrangements. The drawing on p. 122 shows the arrangements fitted, and consists of a supply trunk leading down to a $12\frac{1}{2}$-inch Sirocco fan, to the outlet of which the cooling tank is attached, the latter being connected to the trunks running into the magazine. To *cool* the magazine, the throttle valve just above the fan is closed, and also the one in the exhaust trunk leading to the fresh air, so that the cold air simply circulates through the coolers and magazines. When *ventilating* only, the throttle above the fan is opened, and that in the exhaust turned to the position shown in the figure, so that fresh air is now drawn down the supply trunk by the fan and through the cooler into the magazine, the exhaust being taken to the open air. The cooling tank is a rectangular box with pipes fitted inside at right angles to the stream of air, the "brine" from the CO_2 machine circulating through the interior of the pipes.

It is of the greatest importance that the cooling system be insulated from the structure of the ship, as the temperature of the magazines should not rise above 70 degrees Fahr.

All trunks *outside* magazines are lagged with silicate of cotton covered with painted canvas or asbestos sheeting, and in cases where trunks have to be riveted to the ship's structure and are watertight, additional lagging is fitted inside the trunk.

The arrangement of trunks is such that the whole cooling effect can be concentrated on one magazine, either by means of the slide valves or by means of the hinged baffle plates.

The system is insulated by fitting teak chocks between the trunks and valves, and all bolt heads and nuts are also insulated.

The valves are geared so that they can be worked at a position adjacent to the magazines, but outside of them. The CO_2 machine is placed in a separate compartment from the rest of the cooling gear, and special ventilating arrangements are provided for these compartments. The exhaust trunks in these compartments are led close down to the deck to take off any gas which may be present in the compartment.

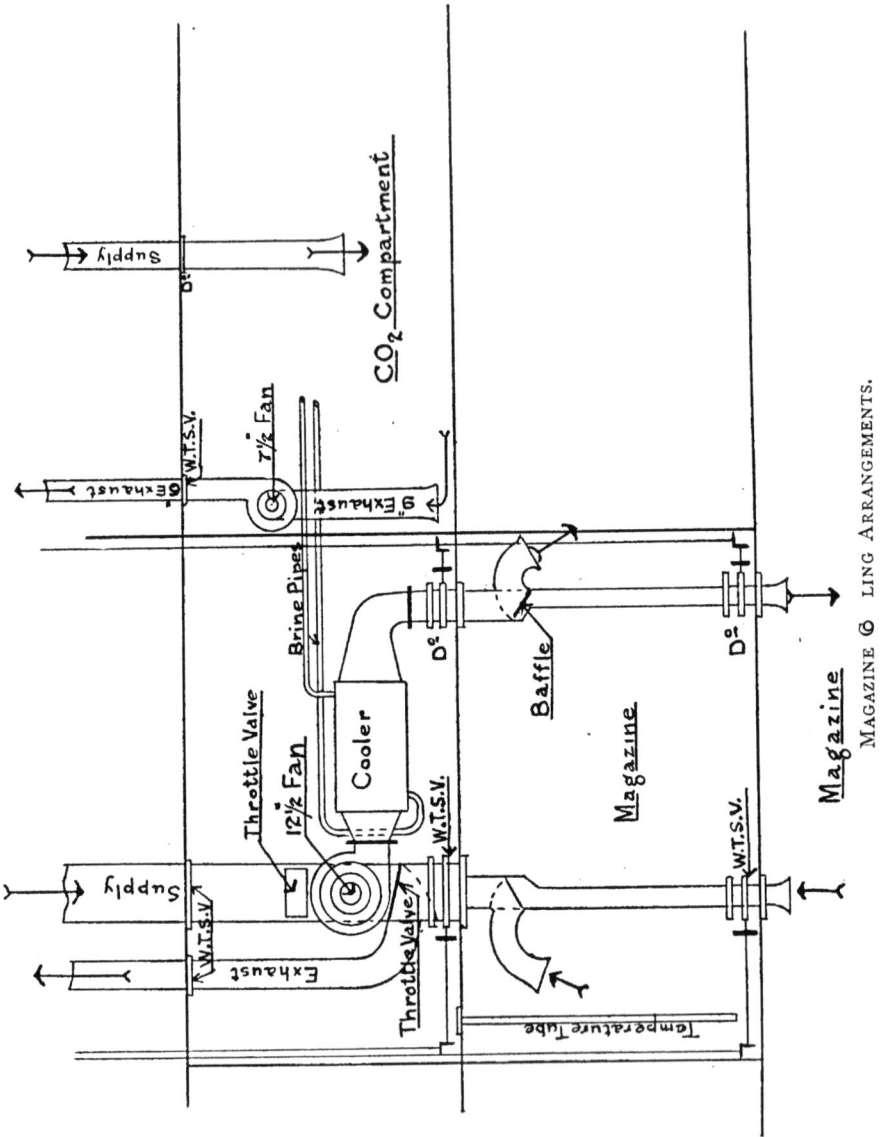

MAGAZINE @ LING ARRANGEMENTS.

PUMPING, FLOODING AND DRAINING.

The systems adopted have varied from time to time, and it will be convenient to describe them in detail.

(1) In the arrangement shown on pp. 123, 124 the pumping power consists of 9-inch Downton pumps, each of which is connected to

a 6-inch pipe running from end to end of the ship on top of the inner bottom, and called the main suction. Branches are taken from this into each double bottom space, both pumping and flooding taking place through these pipes. The pumps are placed in water-tight compartments at the top of the lower bunker, access being obtained from the ammunition passage through a W.T. door. The

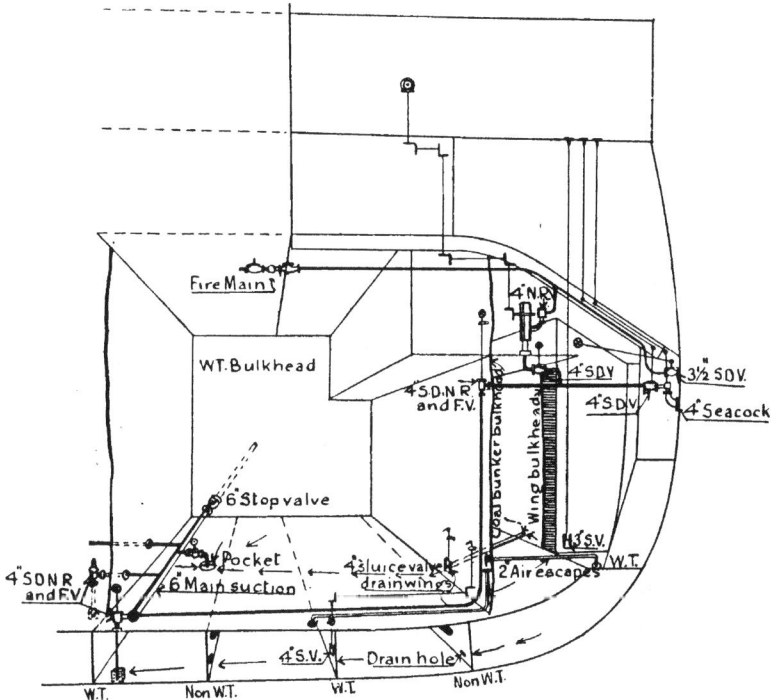

connections from each pump are—(1) tail pipe to sea cock and main suction, (2) delivery pipe to fire main and discharge overboard. The various connections are shown on the drawing. To prevent accidental flooding of the pump room should the pump be opened up for examination, there is a valve between the flooding pipe and the pump.

Drainage. The wings and coal bunkers are drained on to the inner bottom, in which pockets are formed to catch the water, the pockets being pumped out by branches from the main suction, and from the fire and bilge pumps. The double bottom spaces are drained from one to the other through drain holes cut in the non-

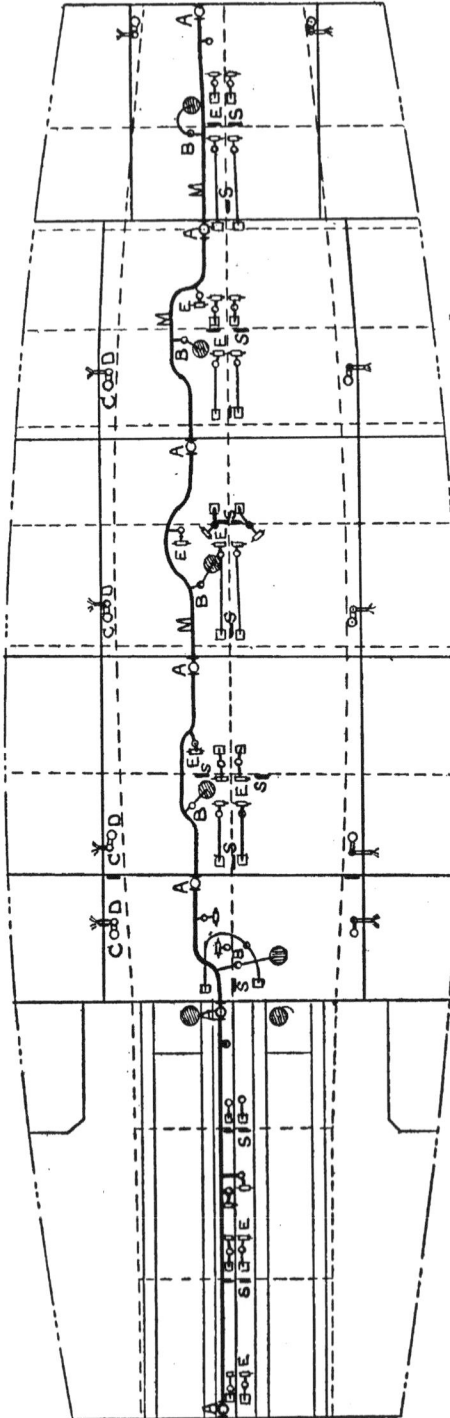

A—6" Stop valve
B—4" S.D.N.R. + F.V.
C—4" Seacock
D—4" F.V.
E—3½" S.D.N.R.V. with hose connection
S—5" Sluice valve
◉ — Pockets in Inner Bottom
M —6" Main Suction.

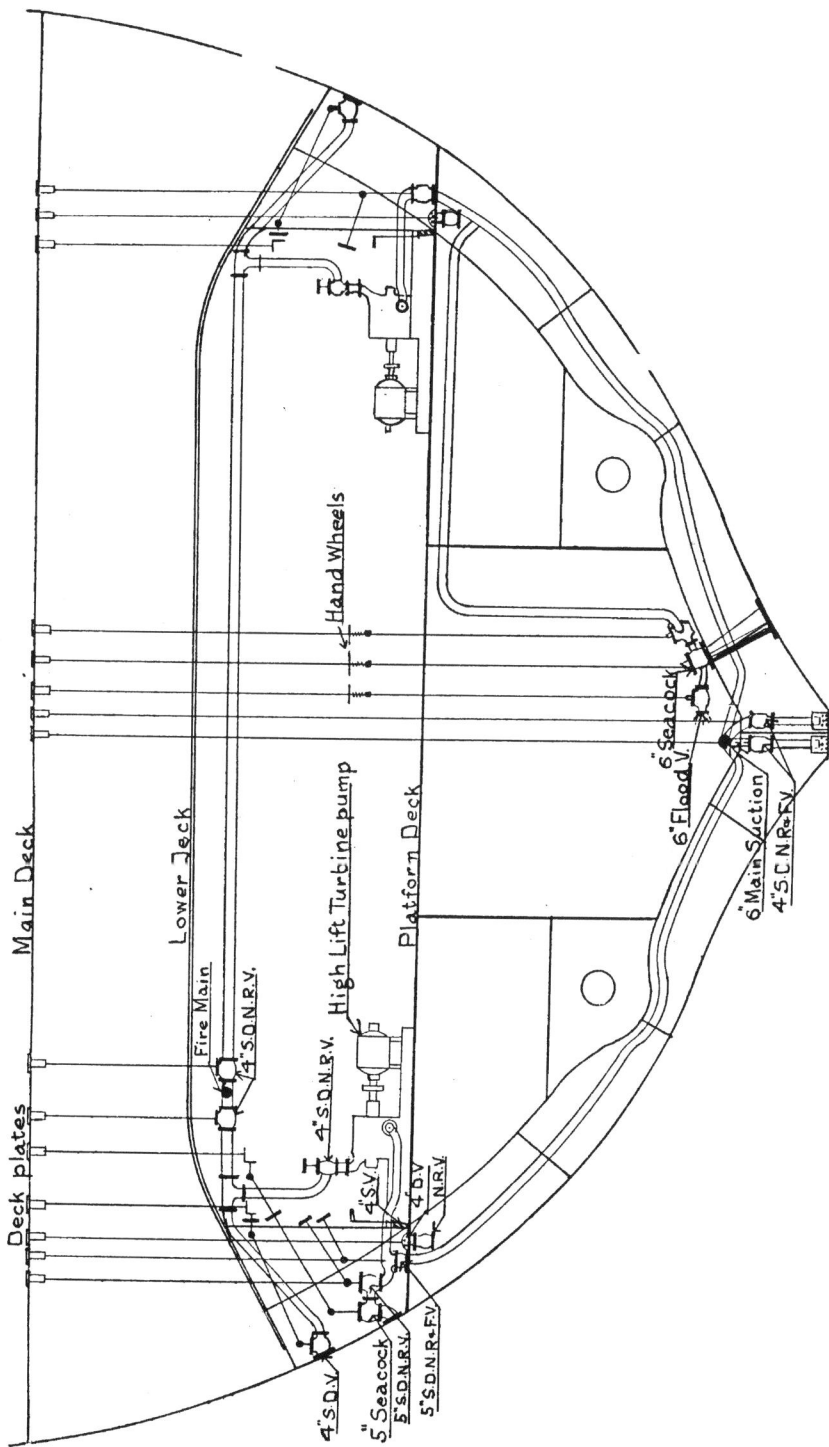

Main Deck

Deck plates

Lower Deck

Fire Main

4"S.D.N.R.V.

4"S.D.N.R.V. High Lift Turbine pump

Platform Deck

Hand Wheels

4"S.V.

4"D.V. N.R.V.

4"S.D.V.

5"Seacock

5"S.D.N.R.V.

5"S.D.N.R+F.V.

6"Seacock

6"Flood V.

6"Main Suction

4"S.C.N.R+F.V.

watertight longitudinals and sluice valves on the watertight longitudinals, as shown.

The gearing for actuating the various valves is shown on the sketch and the positions from which they are worked.

To allow of the air in double bottom spaces escaping when flooding, escape pipes are fitted at the top of each compartment, as shown (see also air escape holes at top of non-watertight longitudinals on pp. 12 and 15).

To deal with large quantities of water in the machinery spaces a main drain is fitted which drains all water entering machinery spaces to a large pocket on the inner bottom in engine room, to which is led a suction pipe from the circulating pump.

To prevent salt water getting into reserve feed tanks, there is a portable connection provided between main suction and tanks, which is only put in place when pumping out these spaces by ship's pumps.

(2) In this method the ship is subdivided for pumping, etc., purposes into self-contained portions which follow the watertight subdivision, the object aimed at being to avoid the piercing of watertight bulkheads as far as possible, there being no main suction pipe (p. 127).

For pumping out compartments before and abaft the machinery spaces, electrically driven centrifugal pumps are provided, capable of discharging from 30 tons to 50 tons per hour at a pressure of 60 lbs. per square inch ; smaller electric portable pumps are provided in case of a breakdown of the main pumps, these pumps are also arranged to be worked by hand.

Each pump has a suction pipe leading to a valve box from which branches are taken to the various compartments, and it is also connected by a rising main to the main service pipe running fore and aft the ship under the upper deck. Each pump is connected to a seacock to enable a supply of water to be delivered to the main service pipe, and also for flooding compartments through the valve box without passing through the pump. The suction pipes are led to the lowest part of the compartment in all cases.

To prevent risk of damage to the inner bottom through the forcing in of the outer bottom, due to grounding or any other cause, the portion of suction pipe between the inner and outer bottoms is bent. Residue stand pipes to oil fuel compartments

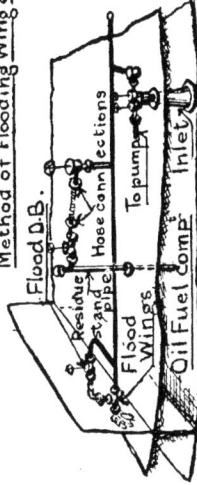

PUMPING AND FLOODING SECTION AT FORE PART.

which must be straight for sounding purposes have the lower end of the pipe weakened in section. In all cases the lower ends of pipes are kept about one inch clear of the outer bottom.

A discharge overboard is provided to each pump, through which the water pumped from the compartments is discharged into the sea below the water line.

The double bottom compartments in the wake of machinery spaces are largely devoted to storage of oil fuel, for which special filling arrangements are fitted which are illustrated below ; but

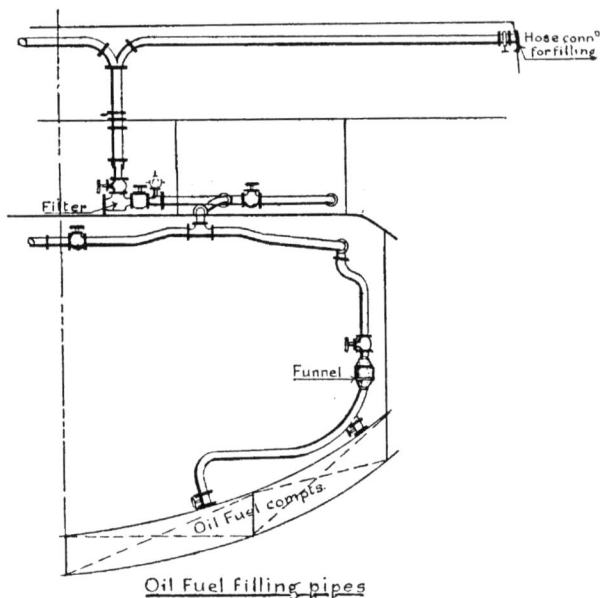

Oil Fuel filling pipes

to enable the compartments to be flooded with salt water, branch pipes are taken off the wing flood pipes with hose connections to residue stand pipes (see Fig. p. 127). The water is pumped out through a suction hose connecting the residue stand pipe to the fire and bilge pumps. The oil fuel residue is got rid of by means of portable electric pumps which can be connected up to the residue stand pipes.

The air escapes, which also provide ventilation for oil fuel compartments, are taken from the highest point of the compartment, and carried up above the weather deck (see p. 129).

The trimming tanks and double bottom compartments, other than those used for storage of oil fuel, have means provided for escape of air when flooding and of sounding, consisting of a combined air escape and sounding pipe, the upper end of the pipe having a screw cap. Air escapes are fitted in covers of

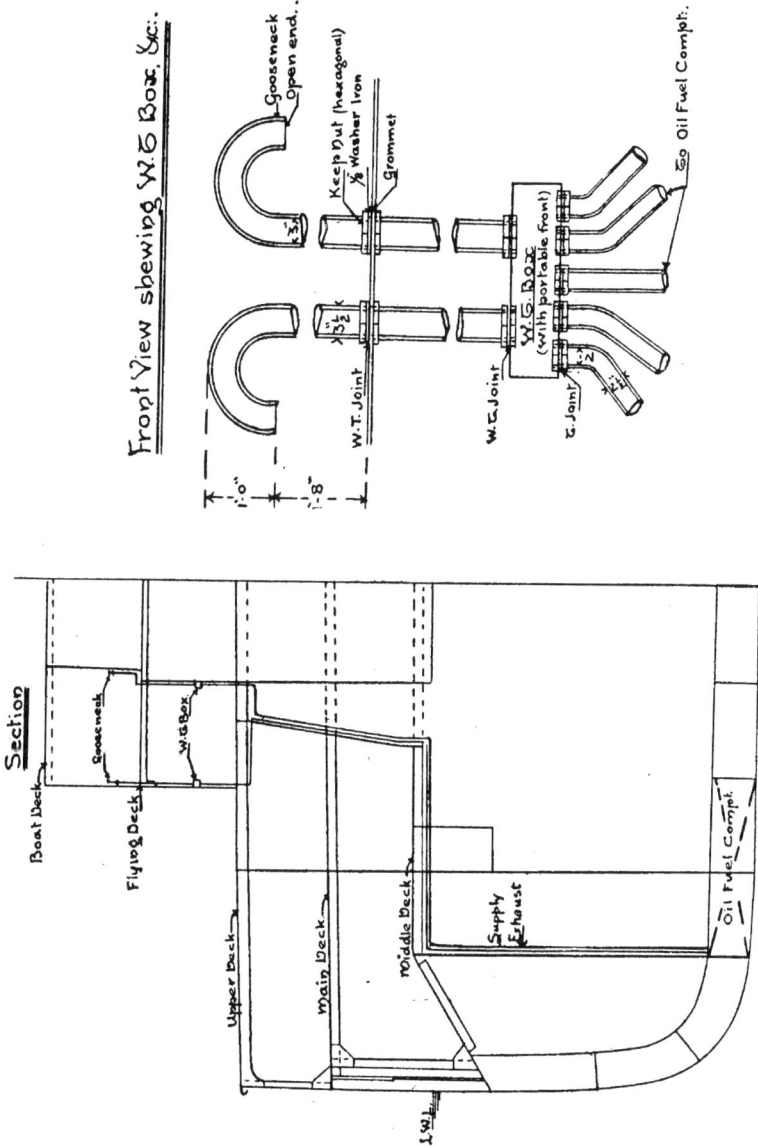

Front View shewing W.T. Box, &c.

Gooseneck open end.
Keep Nut (hexagonal)
⅜ Washer Iron
Grommet
W.T. Joint
W.T. Joint
T. Joint
W.T. Box (with portable front)
To Oil Fuel Compt.

Section

Boat Deck
Flying Deck
Gooseneck
W.T. Box
Upper Deck
Main Deck
Middle Deck
Supply
Exhaust
Oil Fuel Compt.
I. W. T.

METHOD OF VENTILATION TO OIL FUEL COMPARTMENTS.

Each pipe should be led up separately with gooseneck on weather deck or if a W.T. box is fitted the box should be at the upper deck level.

manholes to double bottom compartments (except those intended for storage of oil fuel), and to magazines, shell rooms, etc.

The magazines are flooded as shown on p. 130, the leads of pipes being as direct as possible. The wing spaces are flooded by means of branches off the inlets (see p. 127).

K

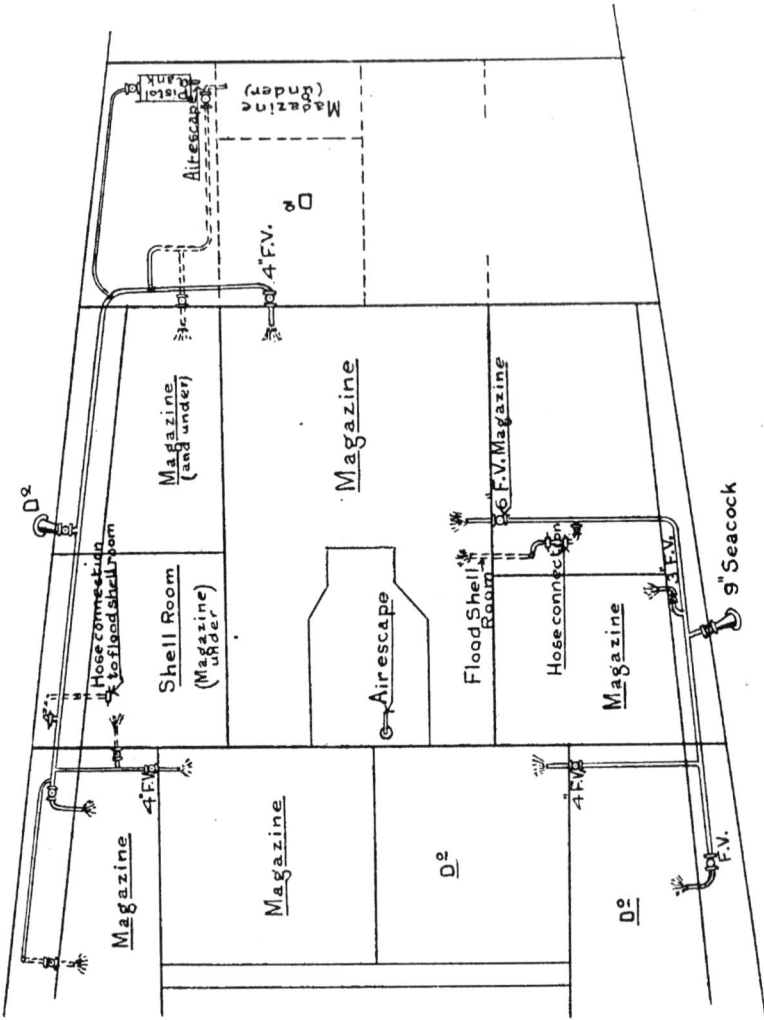

Flooding Magazines, etc.

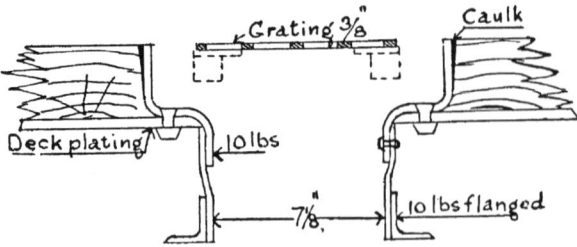

Section

Scuppers.

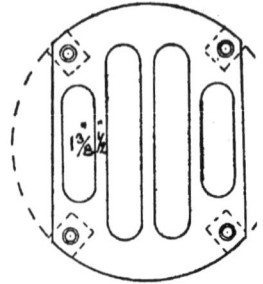

Plan

All flood pipes and those in which water is likely to lodge are of copper to avoid corrosion.

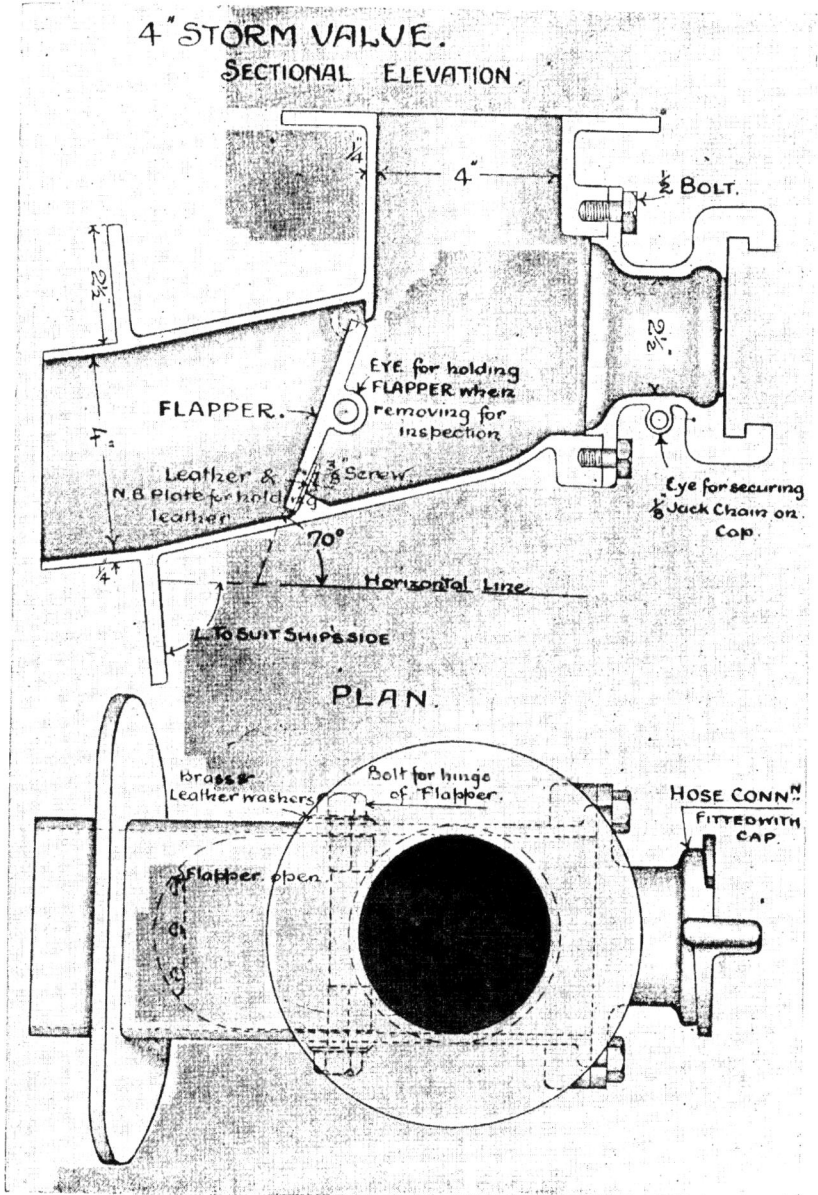

4" STORM VALVE.
SECTIONAL ELEVATION.

Drainage. Compartments generally are drained through drain valves into the bilges, whence the water can be pumped out.

The bridges, forecastle, etc., decks are drained on to the

upper deck, whence the water passes overboard through the scuppers, the lower end of scupper pipe having a storm valve fitted, as shown on p. 131, but without the hose connection.

The main deck is drained through scuppers or sluice valves, as shown on pp. 127 and below.

The middle deck is not usually drained, but places where water is likely to lodge are cemented over to protect the plating.

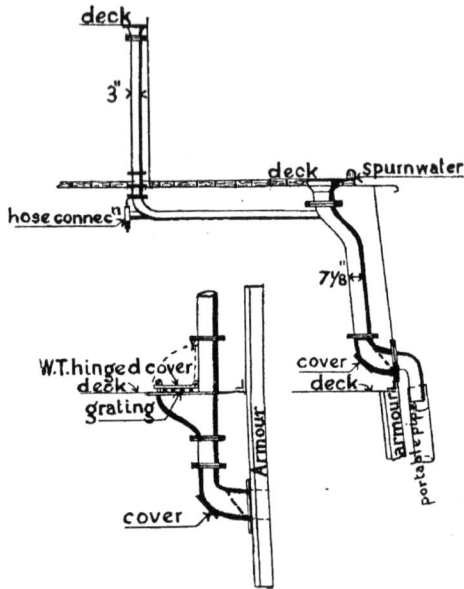

METHODS OF FITTING SCUPPERS.

The wing spaces drain from one to the other through sluice valves on the watertight bulkheads, thence through drain pipes to the pockets on the inner bottom (see p. 123).

Magazines are not drained, but hose connections are fitted for pumping out when necessary. In some cases drainage arrangements have been fitted to shell rooms.

Barbette floors are drained by means of pipes led direct to the bilges, a strainer being fitted at the orifice in the deck.

The coal bunkers drain through the bunker doors on to the inner bottom.

Compartments containing CO_2 machinery are drained to the bilge, the pipes being fitted with traps.

VALVES, ETC.

1. Foot Valve. This is a screw-down, non-return, and flood valve fitted at the foot of each branch to bottom compartments, a section through which is shown below. With a centrifugal pump a non-return valve is necessary at the open end to prevent the pipe emptying itself into the compartment should the pump stop. These pumps have to be "flooded" before they will draw.

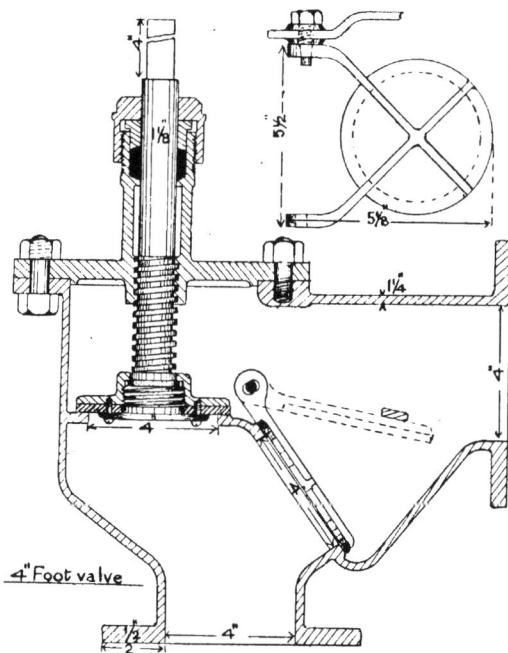

4" Foot valve

2. Flood Valve. This is a leather-seated valve for flooding magazines, etc., and is shown on p. 134. The valve is placed outside the compartment which it floods, and is fitted so that the pressure of the water in the pipe tends to close the valve and prevent leakage into the compartment.

3. Drain Valve. This is shown on p. 134, and consists of a screw-down valve, the seating being the deck itself, with a strainer surrounding it to prevent waste from getting into the valve. Beneath the deck a non-return valve is fitted to which the drain pipe leading into the bilges is attached: there is a cover to this for examination purposes. The non-return is necessary to prevent water entering the compartment should that below be flooded.

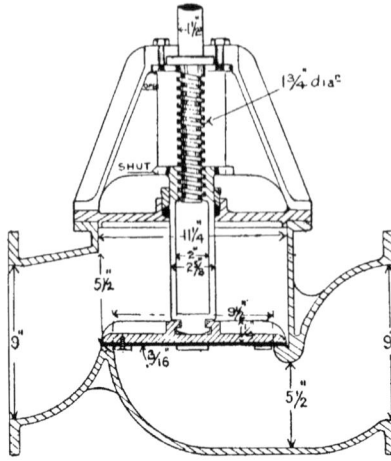

9″ FLOOD VALVE.

4. **Seacocks**. It is from these that the pumps draw their supply of salt water, and where they have to be fitted on the

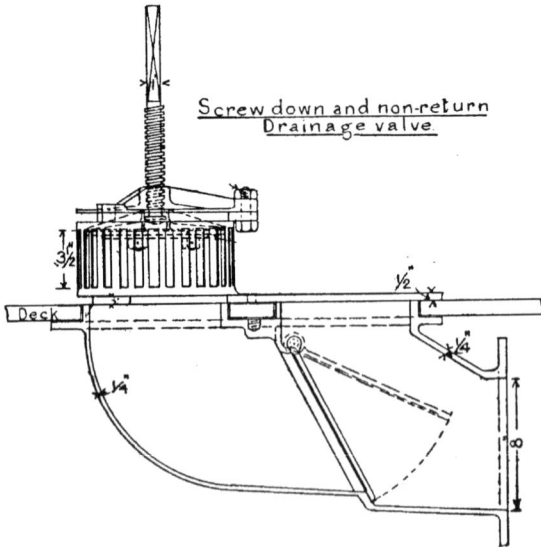

inner bottom a distance piece is provided between inner and outer bottoms, as shown on p. 135. This distance piece is conical in shape, the larger diameter being at the outer bottom to make up for the loss of area caused by the grating which is fitted to prevent weeds getting into the valve. The effective area through the grating depends on the size of the valve, as follows: for

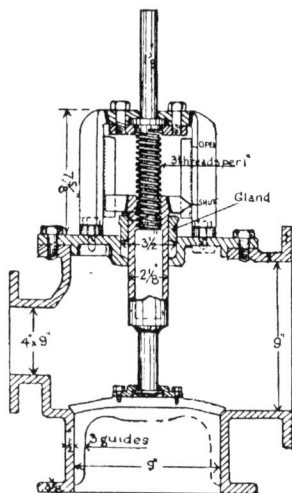

9″ SEACOCK.

valves up to and including 3 inches the clear area is not less than four times the area of the inlet valve; up to and including 6-inch valves, three times, and for valves larger than 6 inches, twice. The bars of the grating are fitted in a fore and aft

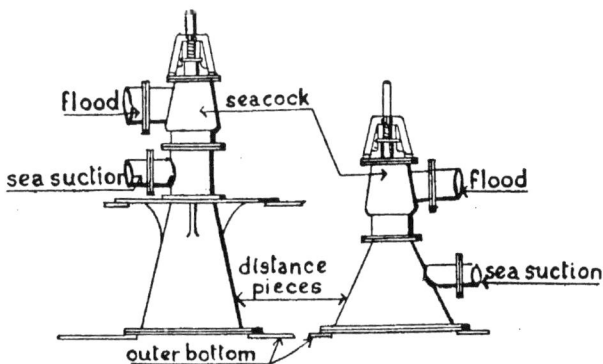

METHOD OF FITTING SEACOCKS.

direction, the orifices being not more than $\frac{3}{4}$ inch wide, except where weed boxes are provided, where space between bars of grating may be 2 inches.

For dry dock flooding a bonnet is provided which is attached
to the distance piece, and has a number of hose connections
to which hoses from hydrants on shore can be connected. As
the bolts for securing the grating have square shanks, to prevent

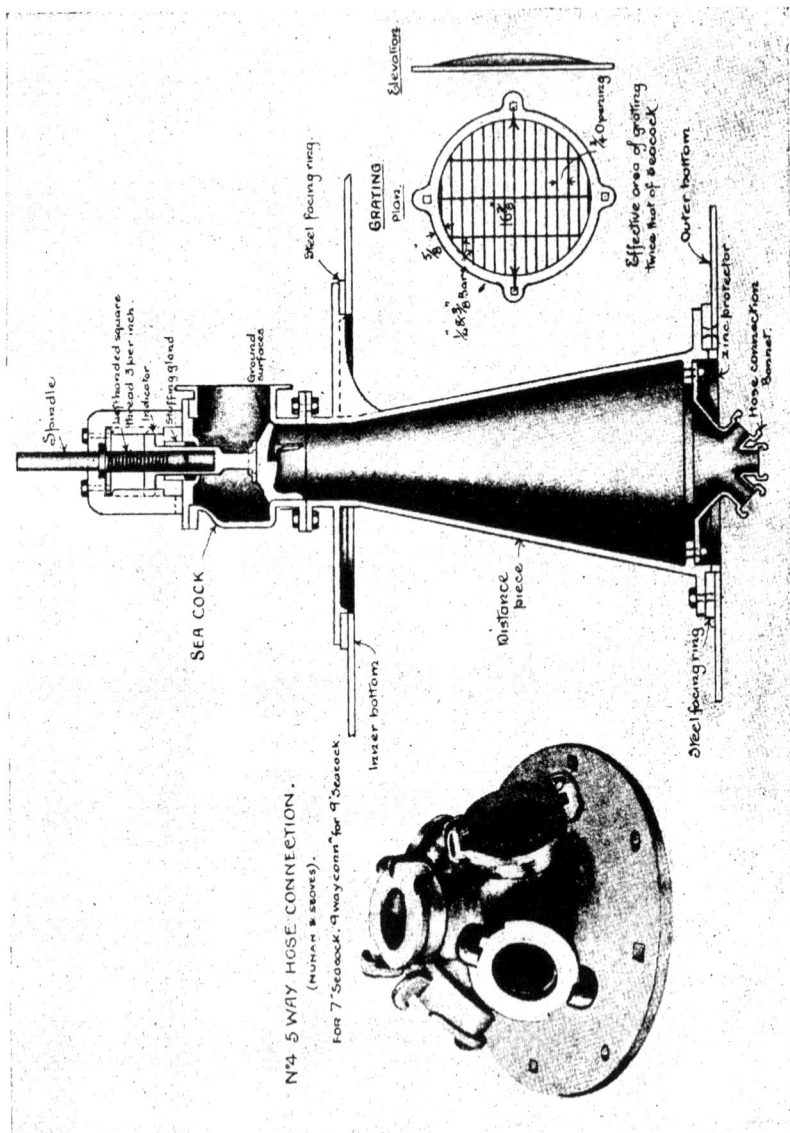

them unscrewing while the grating is in place, similar holes have
to be formed in the bonnet in addition to the circular holes for
the bolts securing the bonnet (p. 136).

Where dry flooding is not provided for, lugs only are required on the distance pieces for the attachment of gratings.

5. **Deck Plates.** These are fitted at the upper ends of rods actuating valves and watertight doors. They are generally placed on the main deck, and consist mainly of two cylinders, the outer one being secured to the deck and the inner revolving inside it (Fig. A, p. 138). The inner cylinder has helical grooves cut in it, in which slide the inclined faces formed on the nut. The nut is prevented from revolving by projections which work in grooves cut in the outer cylinder. Thus when the valve spindle is turned, the nut slides up and down without revolving, and in so doing causes the inner cylinder to revolve and indicate whether the valve is open or shut.

There are two types of this fitting, one for flood valves and the other for watertight doors and ordinary valves. The former, which is the one shown on p. 138, has a locking arrangement, consisting of a cap fitting on the upper end of the spindle and having a V-shaped projection on its side which fits in similar grooves in the inner cylinder, the tongue of the lock projecting into the annular space shown.

A screwed cover fits in the inner cylinder and has a slot in it, which is triangular in shape for flood valves and square for others. A box spanner is supplied for working the valves, the upper end of this spanner having a projection on it for removing the cover.

In some cases the deck plate has been dispensed with other than for flood valves, and a simpler fitting substituted. It consists of a spanner for working the valve, the upper end of the valve rod having a square for this purpose. The spanner is fitted to a band which can move freely between collars formed on the rod, and thus is always at hand. As this fitting requires to be placed against a bulkhead, there is no advantage in using it if a long lead of rods has to be arranged to enable this to be done.

6. **Locking Hand Wheel.** This fitting, with the one just described, permits of the valve being worked both from the deck or from an accessible position near the valve (Fig. B, p. 138). It consists of a hand wheel on the spigot of which is a sleeve which can be slid up and down. To connect the hand wheel to the valve spindle, the sleeve is raised and the pin inserted, the length of valve rod above being thereby disconnected, and it is not possible to arrange otherwise,

as the upper holes are covered by the sleeve. To connect up the rods so that the valve can be worked from the deck plate, the pin is removed from the lower position, the sleeve slid down, and the

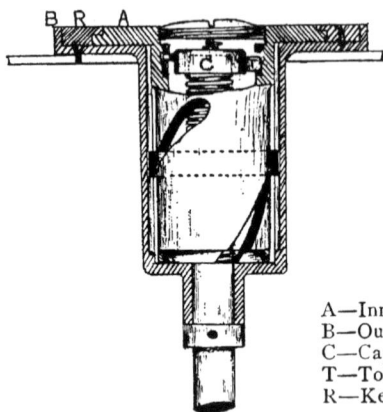

A—Inner revolving cylinder.
B—Outer fixed cylinder.
C—Cap.
T—Tongue.
R—Keep ring.

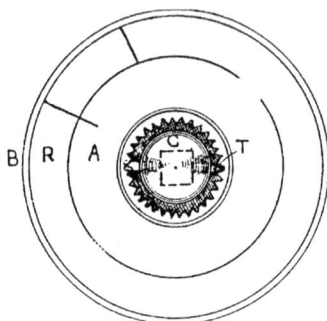

FIG. A.

pin inserted in the upper position. There is an indicator at the hand wheel to show whether the valve is open or closed, and when connecting up it is necessary to see that the deck plate and hand wheel agree.

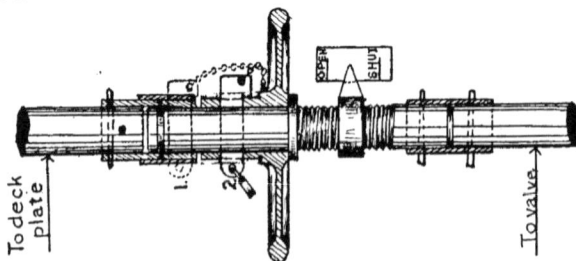

FIG. B.

7. **Universal Joints.** These are necessary owing to the indirect leads of the valve rods from valve to deck (p. 139). Details of

these fittings are shown in sketch, and a table of dimensions given on p. 319.

Universal Joint

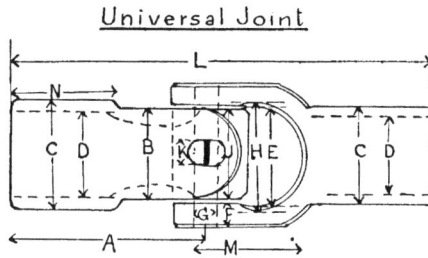

8. **Sluice Valves.** These are valves fitted on watertight bulk-heads to enable adjacent compartments to be drained one to the other. Zinc protectors are fitted to sluice valves where water is likely to lodge.

4" Sluice Valve

Elevation

Side view

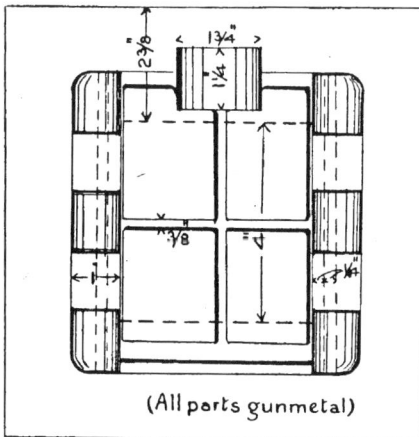

(All parts gunmetal)

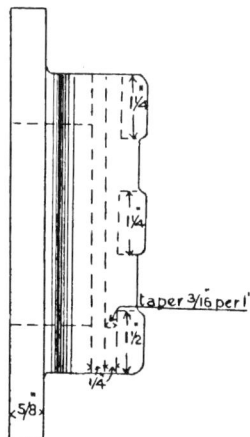

taper 3/16 per 1'

Plan

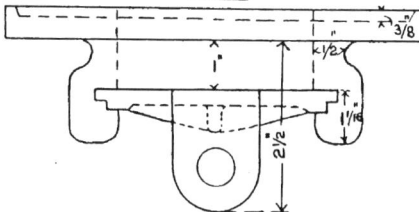

9. **Screw-down Valve with Hose Connections.** This is fitted to the rising mains for wash deck and fire purposes (see Fig. A, p. 140).

10. Double-faced Slide Valves. These are fitted to the valve chests, as shown on pp. 127 and 141. The valve chest on p. 127 should be

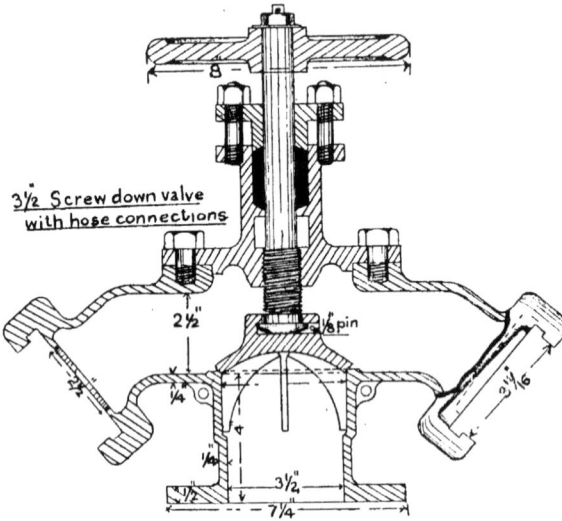

3½ Screw down valve
with hose connections

Fig. A.

compared with that on p. 141. In the latter, the double-faced slide valves are attached to the separate casting, and thus there is a

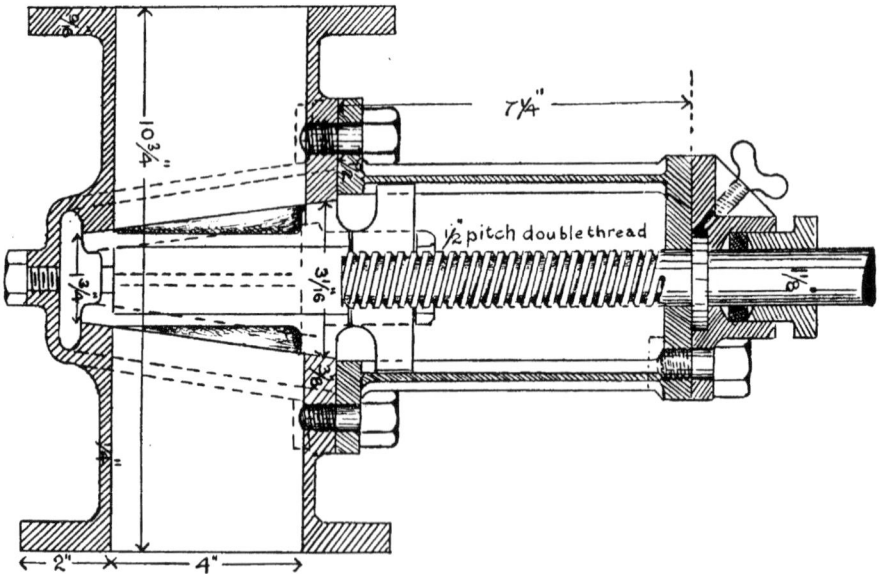

Fig. B.—Double-faced Slide Valve.

clear flow to the sea-cock, while in the former there is the obstruction of the valve spindles, the valves being of the screw-down type; further, the slide valve when open is quite clear of the flow of water, and no obstruction is offered when flooding or pumping, as is the case in the valve chest shown on p. 127.

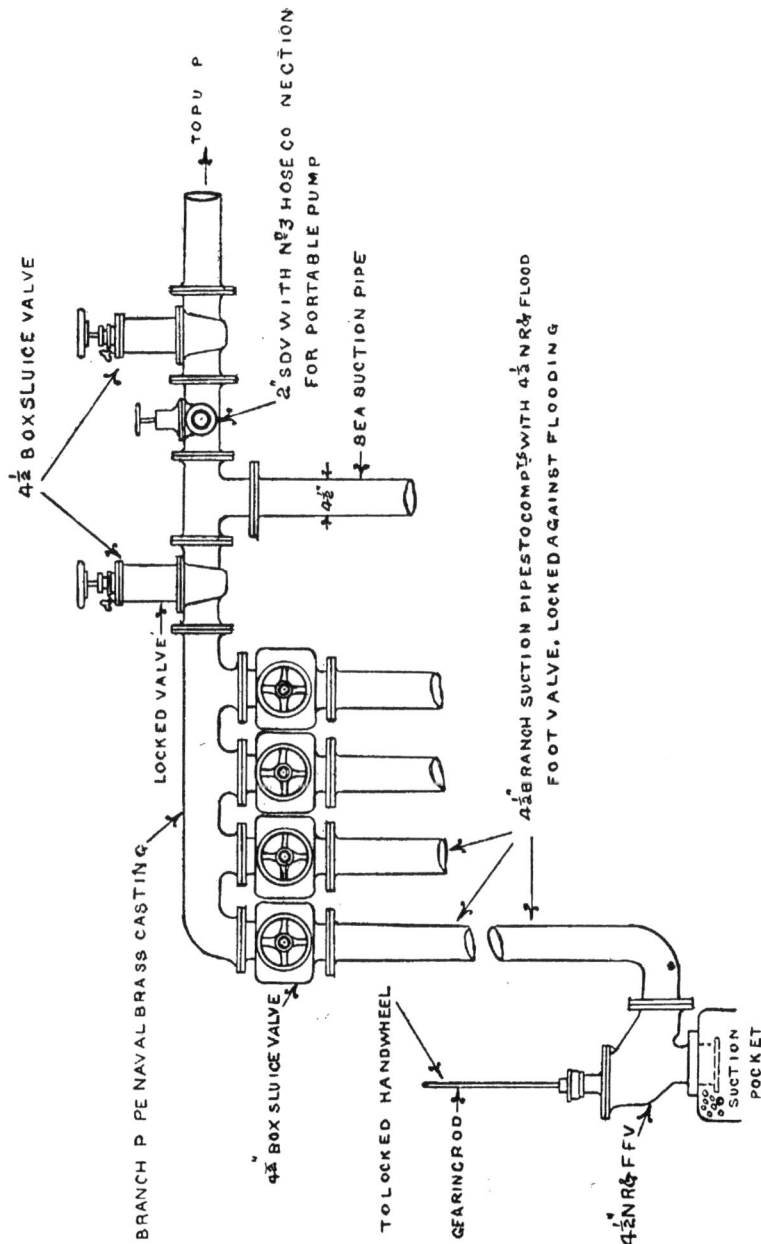

ARRANGEMENT OF VALVES AND SUCTION PIPES.

All valves are arranged to close with a clockwise motion of the hand wheel. Thus, in the case of a valve spindle in which the valve moves along the spindle, as the latter moves a left-handed thread is necessary, while in the case of a valve which moves up and down with the spindle, the thread will be right-handed. Most of the valves come under the former head, but an example of the latter is the drainage valve (p. 134). The valve spindle in this case has a square portion which works in a sleeve fitted to the lower end of the actuating rod.

FRESH WATER SERVICE.

The fresh water is stored in tanks built into the ship. These tanks are coated with rosbonite, which is melted and put on in a plastic state. The fresh water made by the ship's distillers is

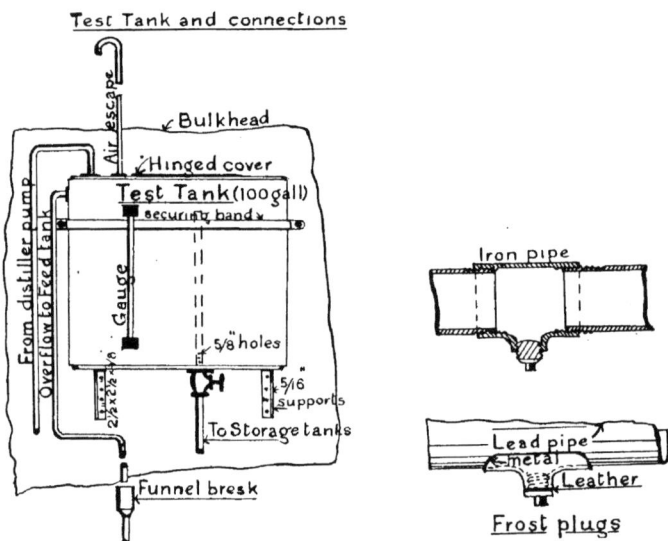

Test Tank and connections

Frost plugs

pumped through the test tank into the storage tanks. The test tank is a small tank usually placed in the engine hatch at main deck level to enable the medical officer to test the quality of the water, the connections being as shown.

The valve chests, suction and delivery, connect up the various pipes to tanks, ship's side, main delivery, and pumps, and permit of all possible requirements of supply and delivery of fresh water (see p. 143). The pumps are motor driven plunger pumps, with a capacity of 10 tons per hour.

Means must be provided for allowing of the escape of air, and also for ascertaining the depth of water in the tanks. This is done by means of the combined air escape and sounding tubes shown in the figure. Access to these tanks is obtained through watertight scuttles in the deck.

To prevent the possibility of damage to the system should the pumps be started with all the valves shut in the delivery pipes, spring relief valves, marked S. R. V. on p. 143, are fitted in the branches from the main delivery to each tank. When the pressure exceeds 60 lbs. per sq. inch, these valves open, and the water is simply returned to the storage tanks.

Frost plugs are fitted to prevent damage to the system through water freezing in the pipes.

The main delivery is a 2-inch pipe running the whole length of the ship, with branches to drinking tanks, baths, washplaces, bakery, galleys, sick-bay, etc. The drinking tanks are plain tanks with a funnel in the top for filling, fitted with a cover and an air escape and bib cock, and contain no filtering matter.

The interior of the pumps, and of all metal or copper fittings, *e.g.* valve chests used in connection with this service, are tinned.

SALT WATER SERVICE.

A 5-inch salt water main is fitted which supplies the sanitary tanks, fire service, washplaces, heads and urinals, and for washing decks, anchors, cables, etc. (p. 145).

The sanitary tanks are placed as high up as possible, *i.e.* on the boat deck, the tanks being also supplied by electrically driven plunger pumps, each capable of pumping 10 tons of salt water per hour.

The interior of sanitary tanks is coated with paraffin wax to prevent the disinfecting fluid attacking the plating. The tanks are fitted with an automatic arrangement, which starts the motor pumps when the level of the water has dropped to a certain point, and stops them when the tanks are full. When the tanks are arranged in pairs levelling pipes are fitted, and also overflows to take off any excess of water.

WASHPLACES.

Fresh water for the basins, and a heater for making water hot, also salt water sprays, supplied from the salt water service, and sponge baths, are fitted to the washplaces for the men. A bath is fitted in the officers' bathrooms.

The hand basins are fitted in frames inside galvanized troughs, the bottoms of which are slopéd to drain water to one end.

The drainage water runs on to the floor and into pockets, from which it is discharged by means of ejectors worked off the main

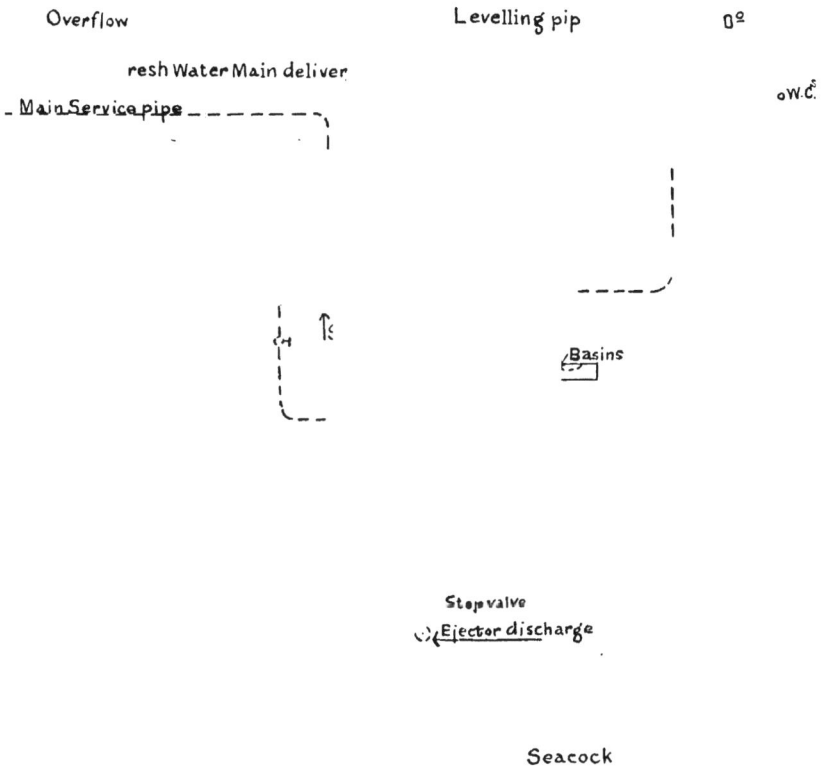

Overflow Levelling pip 0º

resh Water Main deliver

_ Main Service pipe _ _ _ _ _ _ _

oW.C.

Basins

Stop valve

Ejector discharge

Seacock

SALT WATER SERVICE

service pipe. The discharge from these ejectors is usually taken below the water line with a stop valve at the ship's side as shown above. A section through an ejector is shown on p. 146.

Galvanized racks are fitted in the stokers' washplaces, two for each stoker, for the storage of clean and dirty clothes, and cupboards are fitted in the C.P.O.'s washplaces, and also in the officers' bathrooms. The floors of the washplaces are tiled.

Ventilation supply is obtained by a trunk from motor fans, W.T. slide valves being fitted on bulkheads at orifice of trunks.

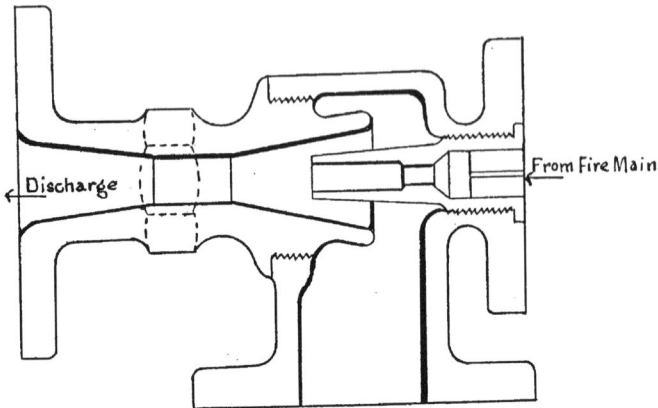

Ejector for draining Wash places.

W.C.'S AND URINALS.

There are two types of pedestal fitted, (1) seamen's (Figs. A and B), (2) officers' (Fig. C, p. 147), the former having a water trap with an air-pipe at the top of the S-shaped connection, while the latter are of the flooded-pan type, no air-pipe being necessary. The pedestals are mounted on a steel plate with a pad of indiarubber between the pedestal and the plate. Each pedestal is connected by a short branch pipe to a larger pipe called a "soil-pipe," the connection between pedestal and pipe being made by means of an indiarubber tube bound with copper wire or steel bands. To prevent the pedestals being broken by the concussion of the guns, the ends of pipe and pedestal are kept at a small distance from each other.

The water for flushing is obtained from the sanitary tanks by means of a pipe running along the top of the w.c.'s, from which branches are led to each pan, with a spring valve for flushing. The soil pipe can also be flushed direct through a branch from the main service pipe. The soil pipe passes through and down over the ship's side, and has a valve at its open end called a storm valve.

rated pipe

FIG. A.

Portable plate
for examination

Hose connection
for Flushing

Air
ply

₁₂

Flushing pi e

Rubber ti

This is a non-return valve to prevent water rushing up through and possibly flooding the w.c.'s. A plunger valve is also provided for the same purpose.

The floors of w.c.'s and urinals are usually covered with tiles laid in cement.

MESSING AND SLEEPING ACCOMMODATION.

Page 149 shows the usual fittings provided for the messing of the crew. The tables and stools are of pine, the former having a teak nosing, and made portable, or hinged, so that they can be triced up when washing decks, etc. The usual seating accommodation provided is about 20 inches per man. The support to the mess table at its inner end is provided with a hook for the mess kettles, and shelves are fitted on the side or bulkheads of the ship adjacent to the several messes for the stowage of the mess utensils, bread, etc. Racks are fitted overhead and on the sides of the ship, as shown in sketch, for the stowage of ditty boxes, cap and hat boxes, and boots.

The hammocks are slung from hooks attached to the beams, and where the height between decks is too great to permit of the hooks in the beams being easily reached, bars are fitted with indentations in them at intervals to which the hammock clew lines are secured. To save space the hammocks are arranged so as to dovetail into each other, the distance apart from centre to centre of hammocks being about 18 inches for the crew, and 24 inches for officers not provided with cabins.

Officers' Cabin. A typical cabin with the usual fittings is shown on page 150. Each cabin has a ventilation supply from fan with a throttle valve fitted to the pipe so as to regulate the supply of air, the exhaust being through perforated plates at upper part of cabin bulkhead as shown on sketch. In cases where the cabin bulkhead is attached to a deck girder, holes about 4 inches in diameter are cut in the girder and the opening fitted with portable wire netting to serve as the exhaust to the cabin.

The side of the ship is plated over from deck to bed berth, the "ceiling," as it is called, being attached to the frames by metal screws so as to be portable. In way of the side scuttle the ceiling is dished as shown.

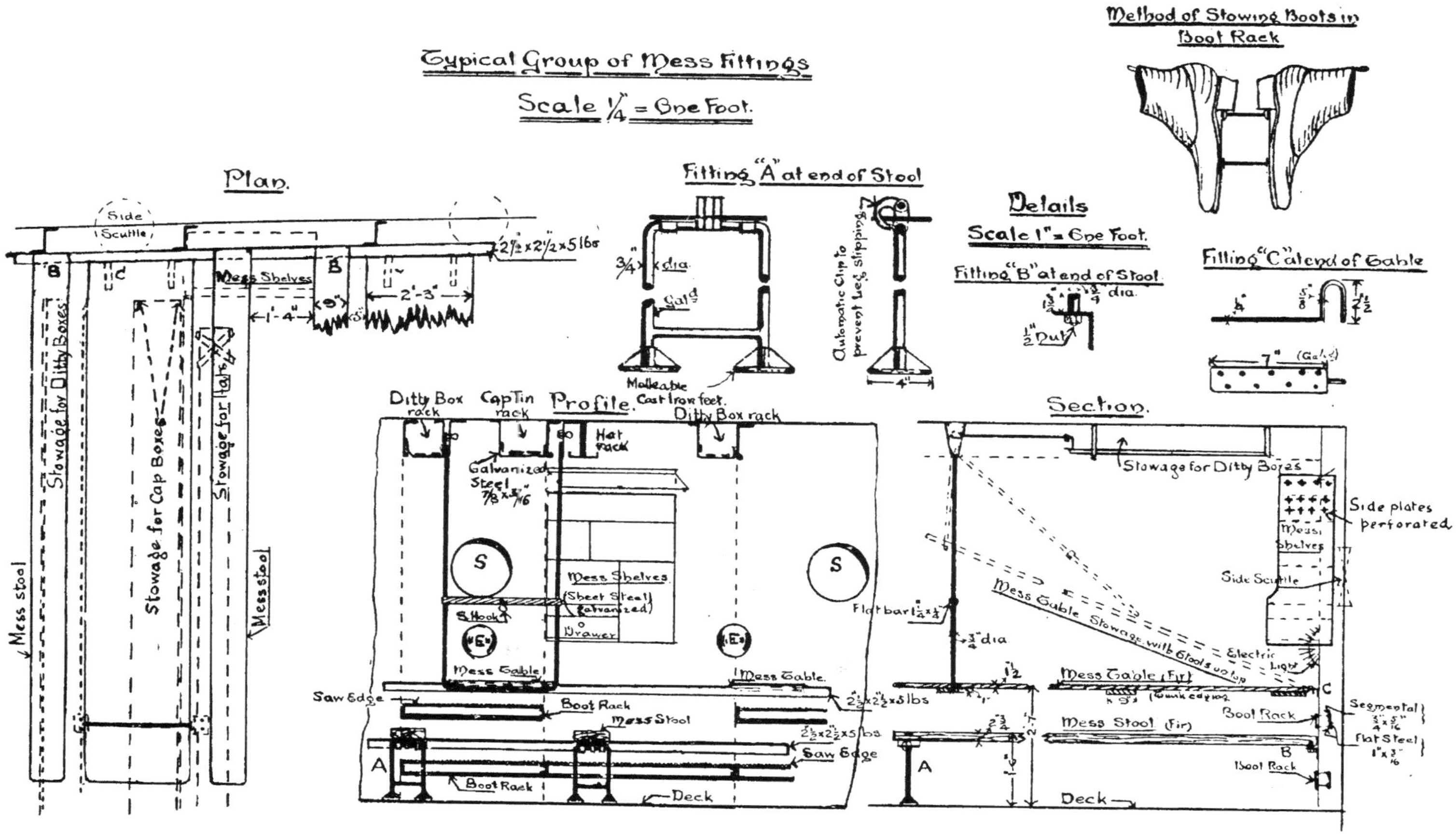

Typical Group of Mess Fittings
Scale 1/4" = One Foot.
Method of Stowing Boots in Boot Rack
Plan.
Side Scuttle
2½"x2½"x5 lbs
Mess Shelves
2'-3"
1'-4"
9"
Stowage for Ditty Boxes
Stowage for Cap Boxes
Stowage for Hats
Mess stool
Fitting "A" at end of Stool
3/4" dia.
Galvd
Malleable Cast Iron feet.
Automatic Clip to prevent legs slipping
4"
Details
Scale 1" = One Foot.
Fitting "B" at end of Stool
3/4" dia.
1/2" Nut
Fitting "C" at end of Gable
7" (Galvd)
Profile.
Ditty Box rack
Cap Tin rack
Hat rack
Galvanized Steel 7/8"x3/16"
Mess Shelves (Sheet Steel Galvanized)
S Hook
Drawer
Mess Table
Saw Edge
Boot Rack
Mess Stool
2½"x2½"x5 lbs
Deck
Section.
Stowage for Ditty Boxes
Side plates perforated
Mess Shelves
Side Scuttle
Flat bar 1¼"x¼"
3/4" dia
Mess Table Stowage with Stools up top
Electric Light
Mess Table (Fir)
Mess Stool (Fir)
Boot Rack
Segmental
Flat Steel
Deck

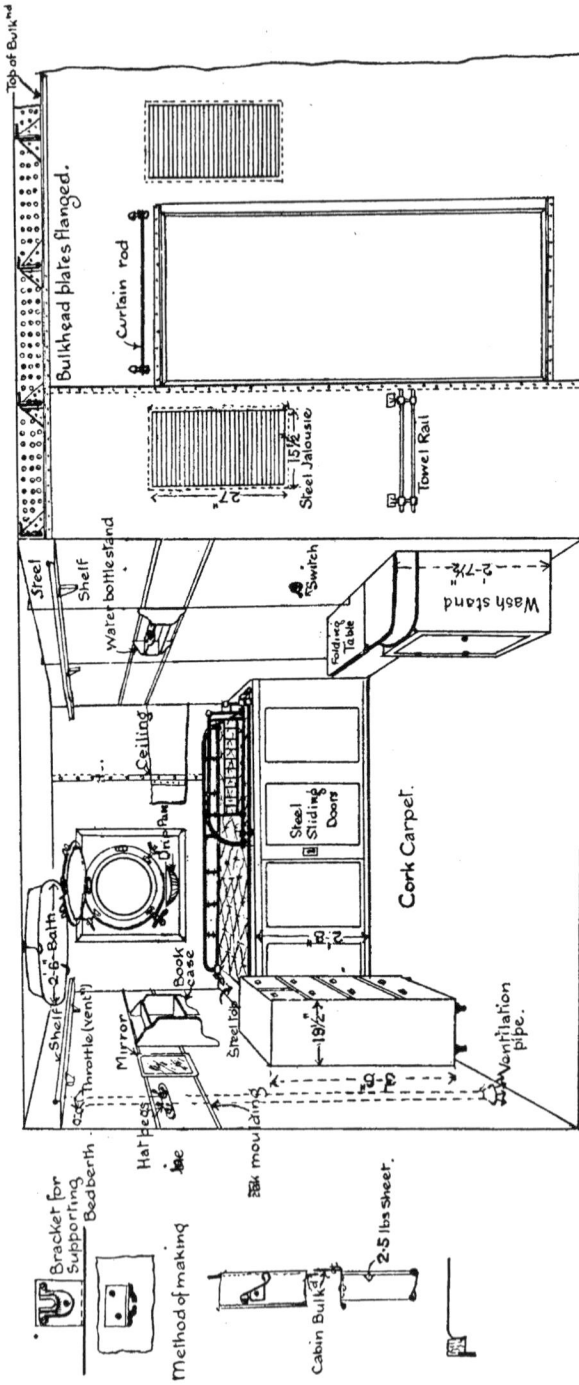

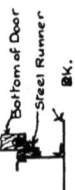

Officers Cabin.

Top of Bulkhd

Bulkhead Plates Flanged.

Curtain rod

Bottom of Door
Steel Runner
BK.

Steel Jalousie
15½"
27"
Towel Rail

Steel
Shelf
water bottle stand
Switch
Ceiling
Oil flame
Steel Sliding Doors
Cork Carpet.
Folding Table
Wash Stand
2'-7½"
Shelf 2'-6" Bath
Mirror
Throttle (vent.)
Bracket for Supporting Bed berth
Hat pegs
Book case
Steel top
Blk moulding
2'-0"
19½"
Ventilation pipe.
3'-9"

Method of making

Cabin Bulk⁴
2·5 lbs sheet.

Method of Stiffening
Bulk⁴ plating

HATCHWAYS AND SKYLIGHTS.

The hatch covers to the thick decks are of the same thickness as the deck, and lap at least the thickness of the cover round the edge of the opening. These covers are divided into two classes, (1) those which would always be closed in action, and (2) those which could be used as escapes in case of necessity. In the former the clips are arranged to be worked from above the deck only, and opened by means of a purchase, while in the latter the clips are arranged to be worked both from above and below the deck, and counterbalance weights fitted to ensure the cover lifting with

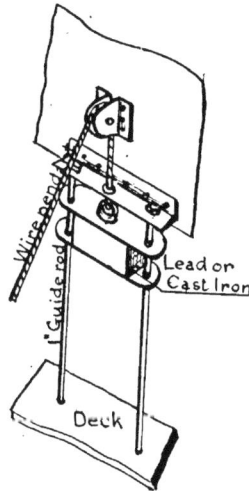

slight assistance from below, the counterbalance being about $\frac{2}{3}$ the weight of the cover, and fitted as shown above. Coamings are placed round covers resting directly on the deck except where the fittings in the vicinity render this unnecessary, as, for example, a cover near a storeroom, where the sill at the entrance to the latter fulfils the same purpose. In such a case the clip at the side from which a person would step off to the ladder leading below, is placed on the cover itself and not on the deck, as are all the other clips, to prevent the risk of tripping over it, this special form of clip and the ordinary clip being shown in Fig. on p. 152.

When access is provided by means of a series of vertical ladders to compartments, as on p. 152, footholds must be fitted on the underside of the cover, and the ladders stop just short of the top of cover. when open. To ensure the cover bearing equally all round its edge, to obtain watertightness, the hole in the hinge is

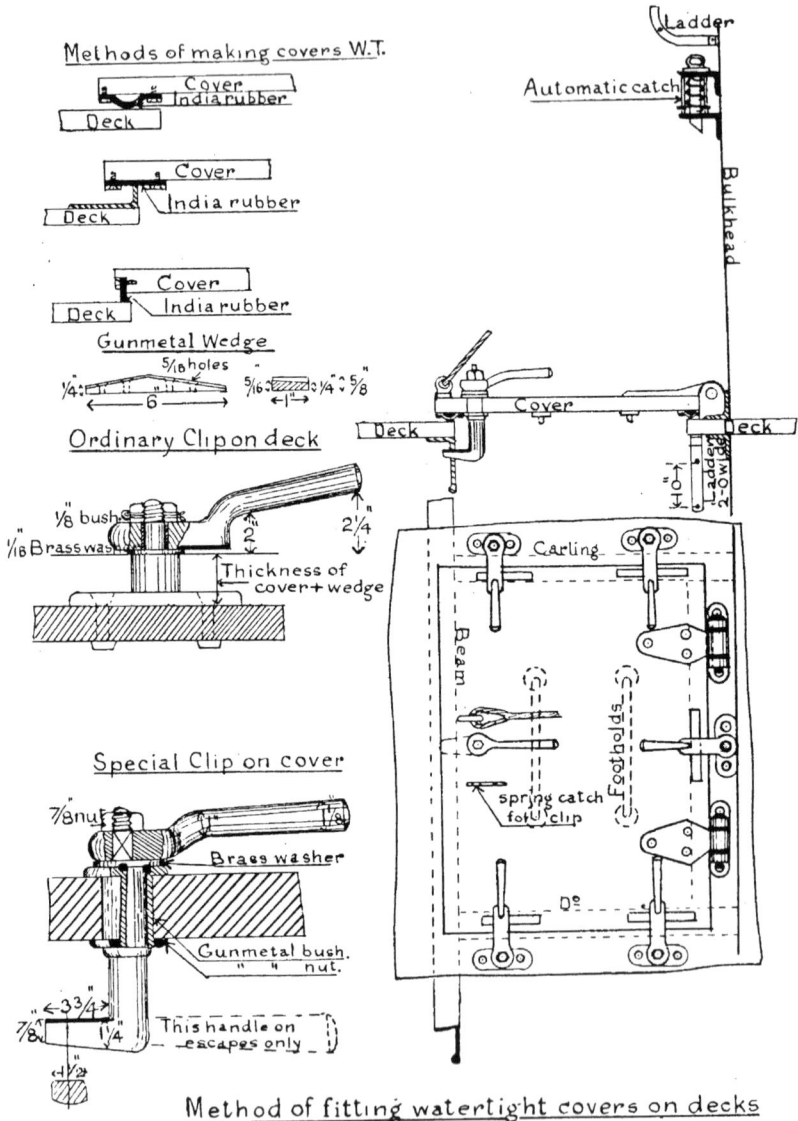

Methods of making covers W.T.

Cover
India rubber
Deck

Cover
India rubber
Deck

Cover
India rubber
Deck

Gunmetal Wedge
5/16 holes
1/4"   6"   5/16" : 1/4" : 5/8"
1"

Ordinary Clip on deck

1/8 bush
1/16 Brass washer
2 1/2"   2 1/4"
Thickness of cover + wedge

Special Clip on cover

7/8 nut
Brass washer
Gunmetal bush.
"   " nut.
3 3/4"
7/8   1/4
1/2"
This handle on
escapes only

Ladder
Automatic catch
Bulkhead
Deck   Cover   Deck
Ladder 2-No-2'6"
HOT
Carling
Beam
Footholds-5
spring catch
for U clip
Do

Method of fitting watertight covers on decks

made oval and metal bushed, to prevent it becoming bound on the steel pin. The cover is kept open by the automatic clip shown.

There are several methods of fitting the indiarubber to covers which rest directly on a steel deck, and these are shown above.

In some cases a cam lever has been fitted in the hinge to raise the cover from its seating before attempting to open it, this preventing possible damage to the rubber.

For other than thick decks the covers to hatches are fitted on top of raised coamings (see below). These coamings are usually

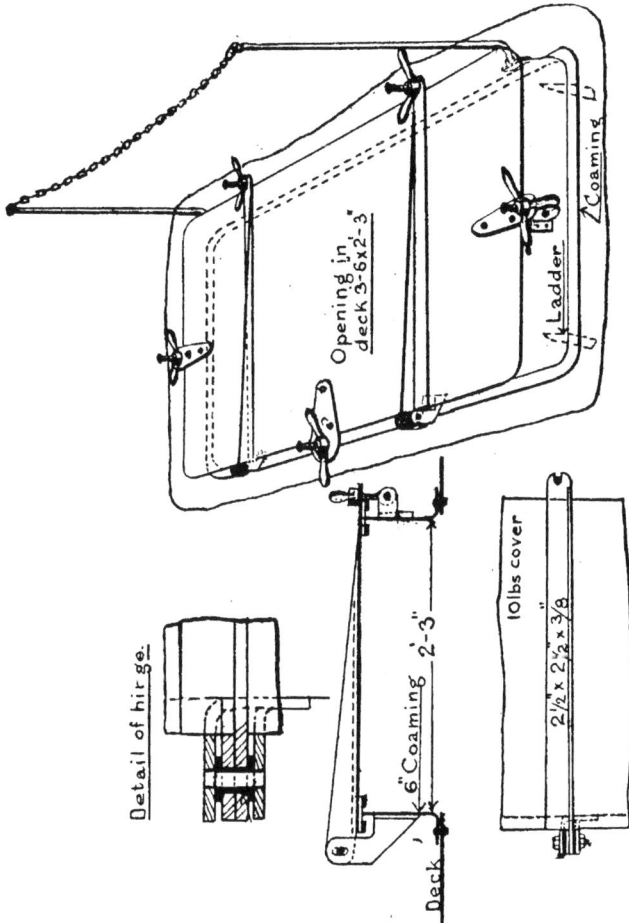

6 inches in depth at all except weather decks, where they are 12 inches deep.

Certain hatches are so large as to necessitate the covers being fitted in two pieces, and two methods of making such watertight are shown on pp. 154, 155. Hatches which might have to be open in action are fitted with coffer dams.

A special form of hatch, shown by Fig. B, p. 155, is fitted at the top of ammunition trunks. The covers are fitted with springs which automatically close when the charge has passed through.

Skylights. These are for the purpose of giving light and ventilation to spaces below the weather deck, and means must be pro-

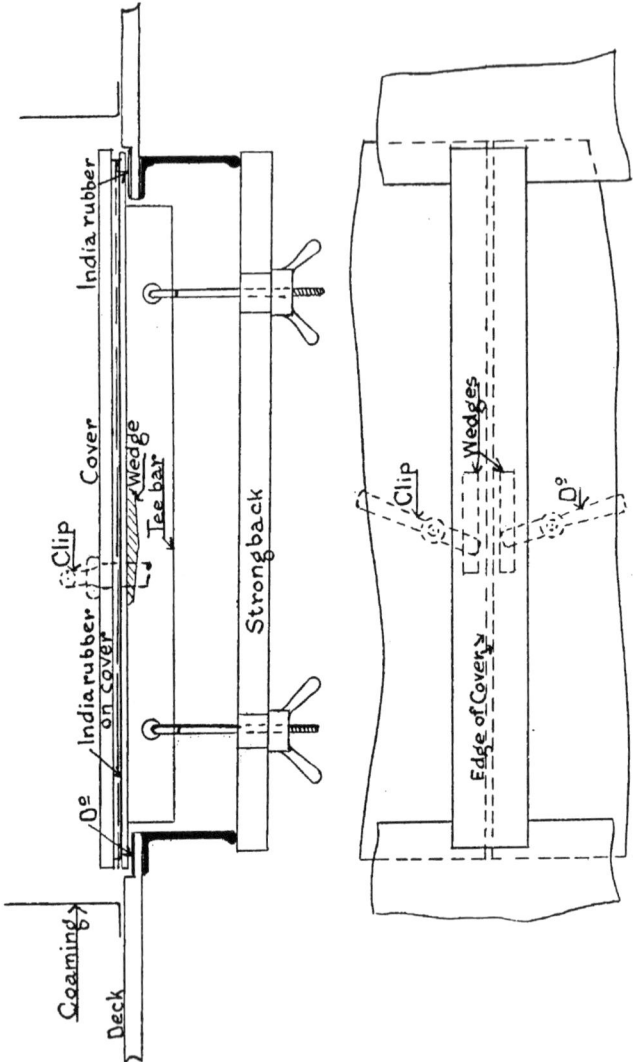

MAKING W.T. A COVER MADE IN TWO PIECES.

vided for making them watertight. A sketch of a skylight is given on p. 156, Fig. A, and it will be noticed that the same clip is used to secure both the watertight cover and the glass frame. Certain skylights have to be arranged with portable tops, as, for example, those vertically above a torpedo hatch.

Ladders. These are of steel for access to the mess spaces, and are constructed as shown on p. 156, Fig. B. They are arranged

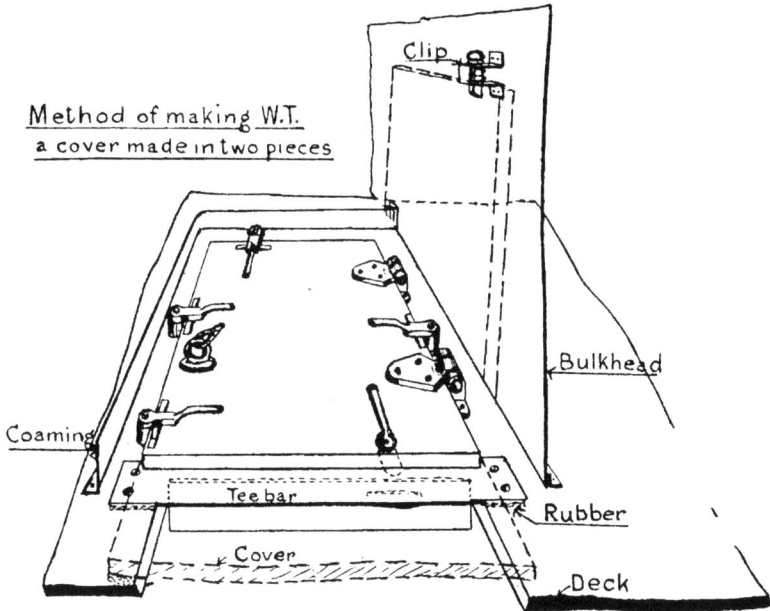

Method of making W.T.
a cover made in two pieces

Clip

Bulkhead

Coaming

Tee bar

Rubber

Cover

Deck

FIG. A.

to hinge up, and, to protect the corticine at the foot of the ladders, galvanized plates are fitted. The vertical ladders are constructed of flat bars for the strings, and round for the steps, the strings

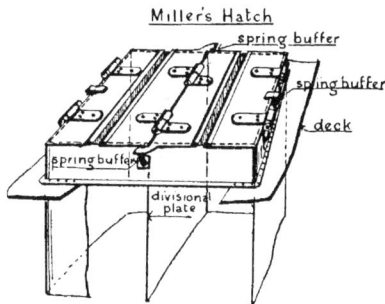

Miller's Hatch

spring buffer

spring buffer

deck

spring buffer

divisional plate

FIG. B.

being curved at top and bottom to stand off from the bulkhead. Vertical ladders are constructed of brass near a compass.

The ladders for access to officers' compartments are of wood with brass strips on the treads.

The accommodation ladders are constructed as shown on p. 157, and must be provided with fittings to allow of them being adjusted to the varying draught of the ship. Davits are fitted to assist in

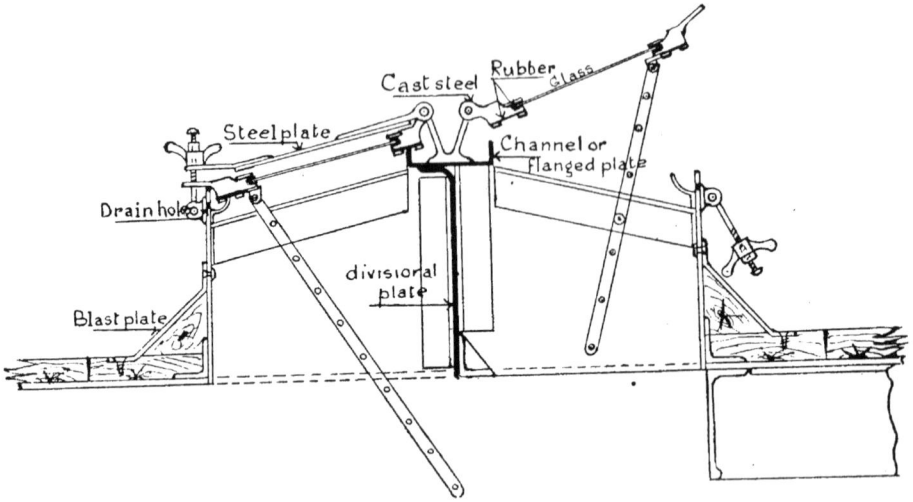

Skylight

FIG. A.

supporting these ladders and for tricing them up, and the guard stanchions arranged for the necessary gangway.

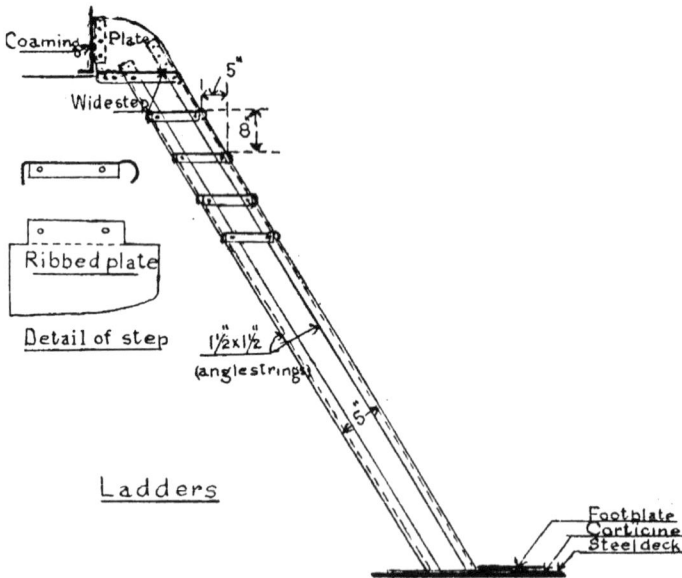

Ladders

FIG. B.

Escape Trunks. These are fitted to compartments below lower deck in which men would be working during an action. They are rectangular in shape and watertight, the lower end having a water-tight sliding scuttle (p. 158) which can be closed from the main deck, and also in the compartment in which it is situated. Armour gratings, provided with balance weights, are fitted at the lower

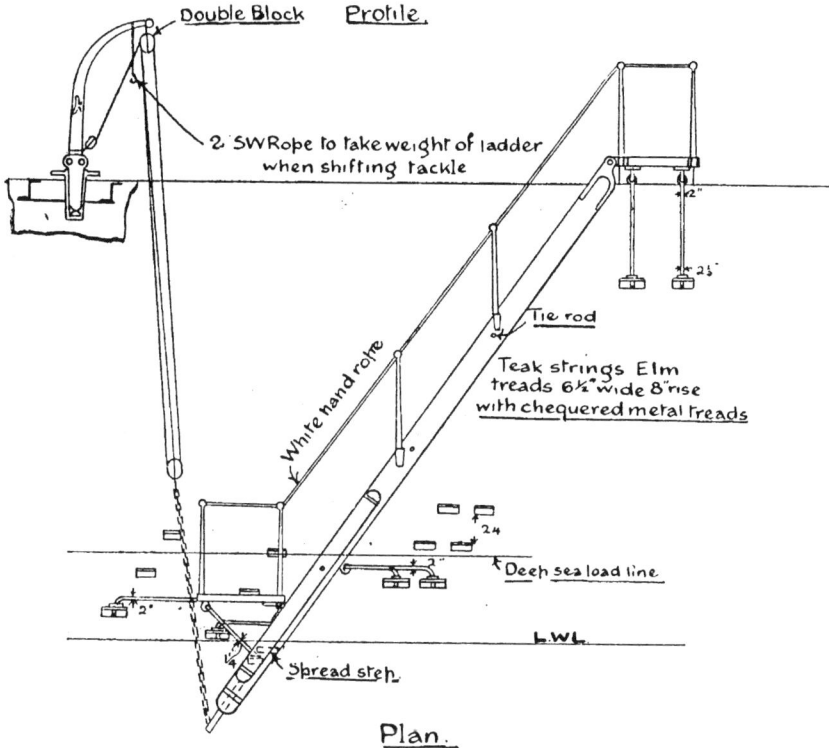

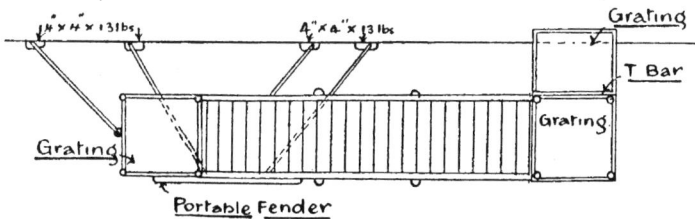

ACCOMMODATION LADDER.

deck to preserve the efficiency of the deck as regards protection, splinter nets being also fitted under the armour gratings to escape trunks communicating with auxiliary machinery compartments.

Sketch on p. 159 shows a typical case.

Stove Funnel. The details of this are shown on p. 160. When the funnel is in place a gunmetal covering which screws into a socket in the deck prevents water getting below, and being of larger diameter than funnel prevents risk of damage to wood deck by heat.   The funnel is of copper pipe, and when required to be carried up to clear an awning, etc., the joints are made as at B.

The cover C makes the opening in deck watertight when the coaming and funnel are unshipped.

The funnel is supported by metal stays attached to a band A.

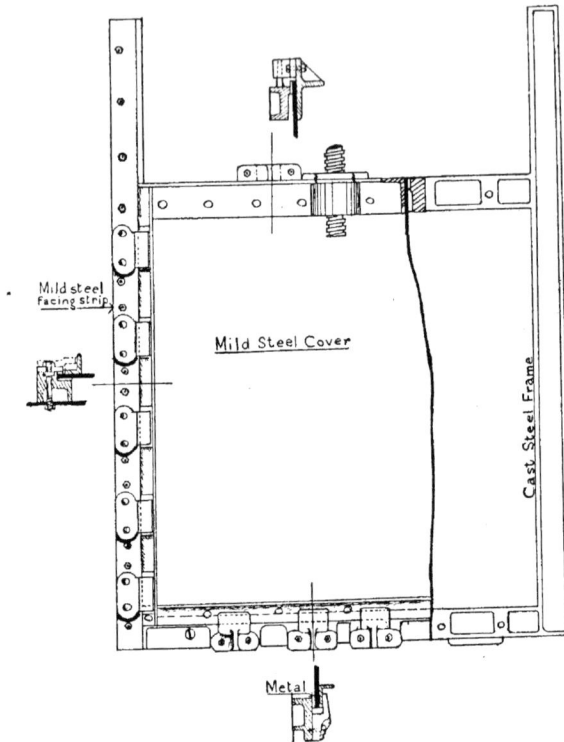

Mild steel facing strip

Mild Steel Cover

Cast Steel Frame

Metal

Torpedo Net Defence.

The general arrangement of booms, guys, etc., is shown on p. 162.   The nets are suspended from the jackstay, which is attached by shackles to the heads of the booms, each boom being supported by two topping lifts with slips and screws fitted at their upper ends.   The eye plates for the topping lifts stand vertically and are attached to the ship's side.

The heel of each boom has a hook welded into the tube, and engages with a swivel eye, working in a socket attached to the

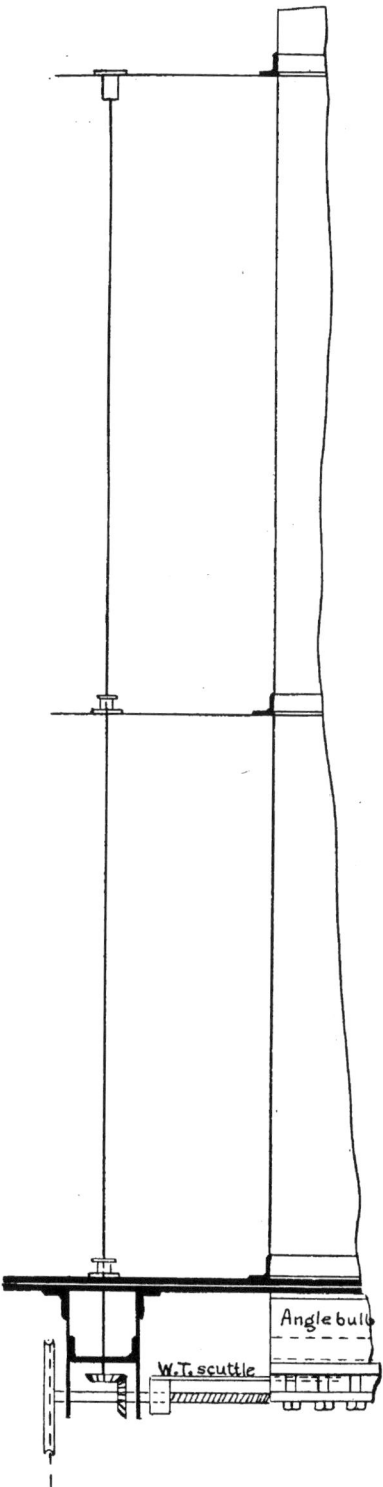

W. C

Escape trunk

Clip

Angle bull

W.T. scuttle

wheel

ship's side. The fore side of the swivel eye is faced off to permit of the boom being shipped, and to prevent the boom from becoming unshipped when it is in its stowing or working position, the opening of the hook is less than the thickness of the swivel eye. The swivel eye is capable of revolving, but its movement is limited by means of the two pins shown, half the diameter of the pin being in the socket and the other half in the swivel eye. The score taken out of the stud of swivel eye is made to allow of such movement as will permit of the booms being readily shipped and stowed.

Method of fitting Stove Funnels.

The fittings on all the booms are not the same the forward, middle, and after booms having additional lugs, etc., worked on them, to which the working and standing guys are attached.

The defence is hauled in and out by means of the working guys led to the capstans, and by additional working guys fitted to the middle boom, bearing-out spars being also provided to assist in getting defence out. When the defence is out it is kept in position by the standing guys, the after one being secured to an eyeplate on the ship's side, and the forward one led through a block and over a roller on ship's side, and secured by means of a slip and screw to an eyeplate on the deck.

DETAILS OF FITTINGS FOR TORPEDO N

Scale ½" =

Locking pins

Hook tulip.

Elevation of Ordinary Boom.

Plan.

Eye plug & Pad

Topping Lift.

Do.

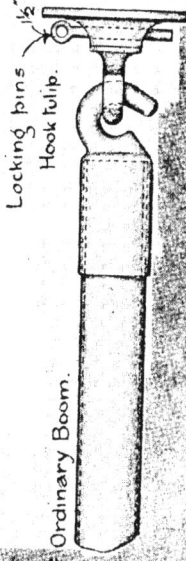

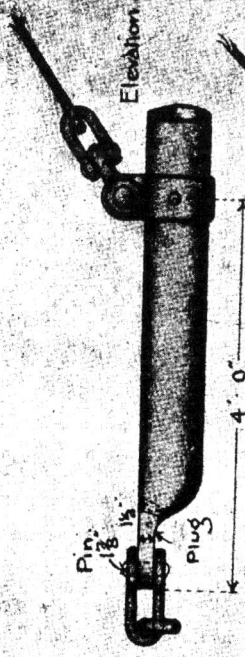

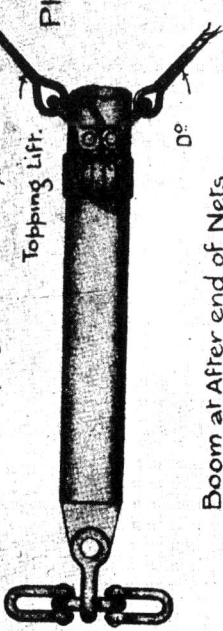

Boom at After end of Nets.

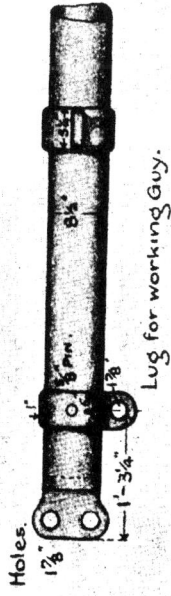

Pin

Plug

4' 0"

Holes.

Pin.

Lug for working Guy.

The method of working the brails which roll up the nets is shown on below.

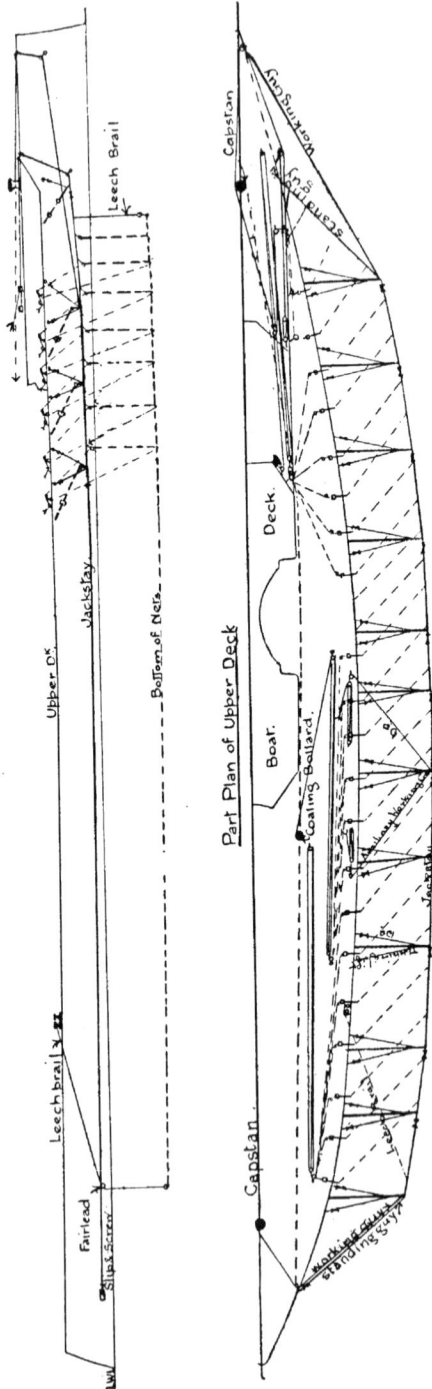

GENERAL ARRANGEMENT OF TORPEDO NET DEFENCE.

The nets are stowed on shelves built round the sides of the ship, the shelves being of perforated plate to allow water to drain off and galvanized to prevent rust.

A leech brail is provided which automatically rolls up the after-piece of net as the booms are hauled into their stowing position.

The booms when in their stowing position rest in chocks attached to the ship's side, placed about one-third of the length of the boom from its head, the booms being secured to the chocks by hinged semi-circular hoops. To further secure the booms, chains attached to eye plates on the ship's side and fitted with slips and screws are passed round the heads of the booms.

COAL SHIP ARRANGEMENTS.

This is the operation of transferring the coal from the collier to the deck of a battleship. A wire span is rigged, as shown on p. 164, right fore and aft of the ship on the main derrick and wood coaling derricks. At intervals along this span there are lifting plates, from which are suspended the blocks for coaling whips, the latter being worked by coaling bollards and winches.

Referring to the sketch, it will be seen that there are two positions for the coaling span (which can be rigged on either side of the ship), viz. when coaling from a collier and when coaling from a lighter. In the former case the outhaul will be worked by the collier's winches, and in the latter case from the lighter.

The lifting plate is a steel plate about 10 inches long and 1 inch in thickness.

Coaling Arrangements. These are the arrangements for getting the coal from the deck of the ship into the bunkers and trimming in the bunkers.

A typical sketch is shown on p. 165. The coal bunkers are at the sides of the ship above and below the protective deck, being divided up by the wing bulkheads and the transverse watertight bulkheads. Between the protective and main decks is a fixed non-watertight trunk which surrounds the opening in the protective deck, the latter having a hinged watertight scuttle of the same thickness as the deck. The sides of the trunk are provided with doors for trimming coal from upper and lower bunkers.

The coaling scuttles are fitted at main and upper decks

vertically over the fixed trunks with shoots between decks, these in some ships being portable, and in others fixed. Details of canvas coaling shoot and coaling rims are given on p. 166.

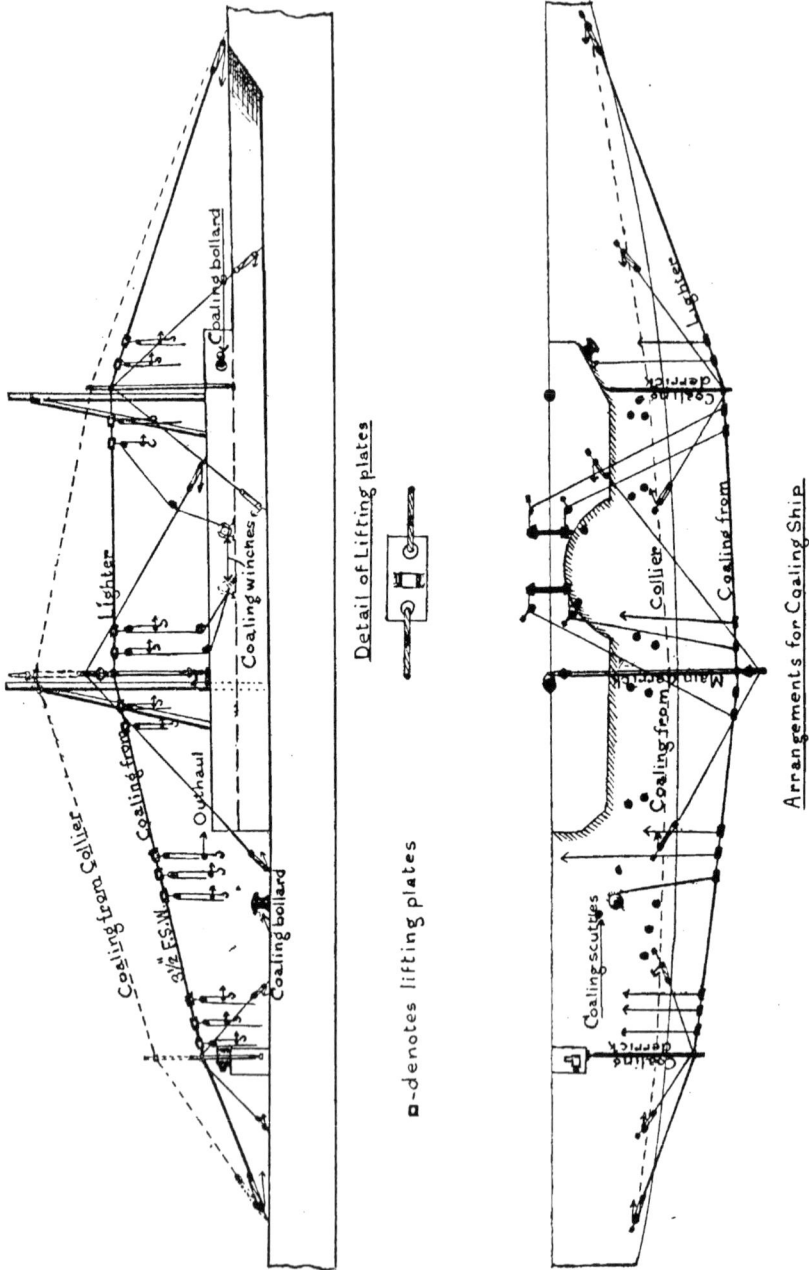

Arrangements for Coaling Ship

The following requirements have to be met :—

Coal upper and wing bunkers direct ; trim from upper to lower and wings, trim from wing to lower, and from one upper to another. These operations are indicated by the arrows. To trim along the upper bunkers, rails are fitted on the slope of the deck on which

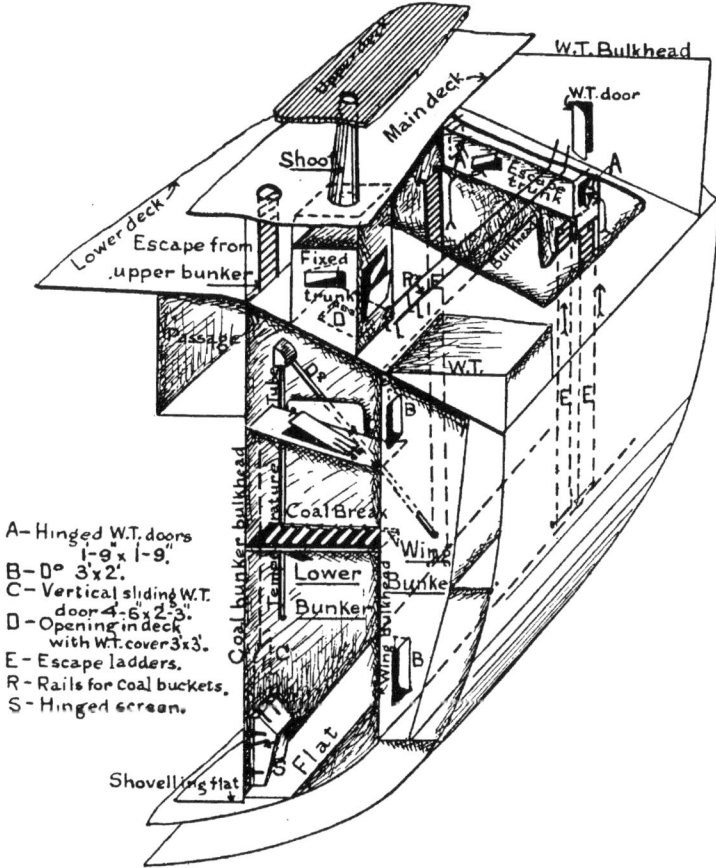

buckets can be run, but in ships in which no watertight doors are fitted on the bulkheads no rails are fitted.

Means of escape from the upper, lower, and wing bunkers are provided as shown in sketch. The escape from the upper is by a ladder on bulkhead through an escape scuttle in main deck, and from lower bunkers by way of ladders in through the doors marked inside of escape trunk, and from wing bunkers through a watertight door marked A into escape trunk.

The escape trunk is not watertight, and has footholds on flat and handholds on its side, and a watertight door in coal bunker

Portable Canvas Coaling Shoot.

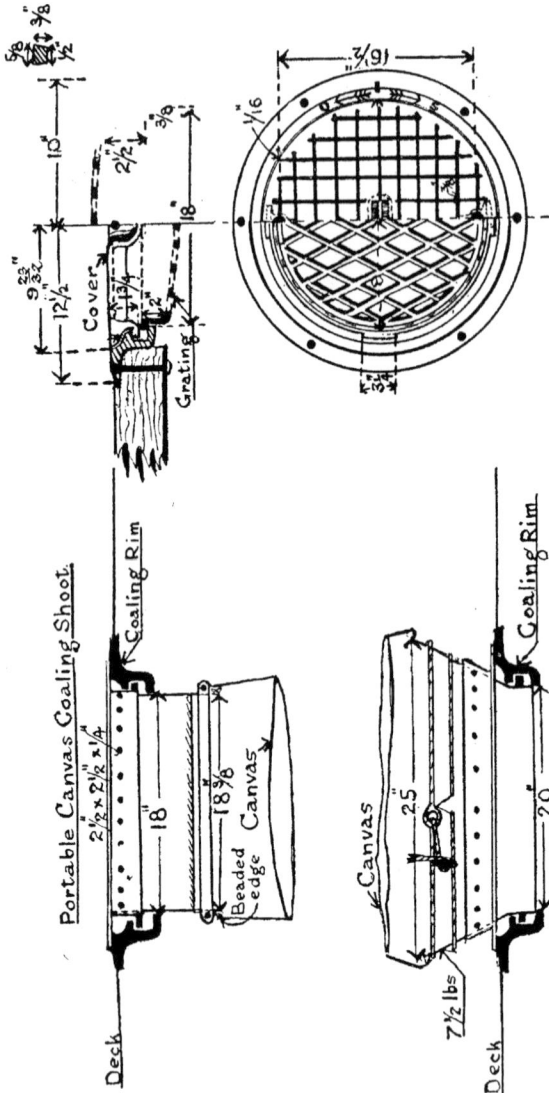

Special Coaling Rim

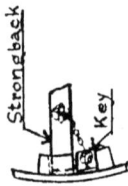

Coaling Rims.

bulkhead providing access to the ammunition passage. The coal is got into the stokehold through vertical sliding watertight doors on coal bunker bulkhead, these being provided with screens in the bunkers to prevent coal blocking up the opening and the door being closed. The sliding doors can be worked from deck-plates on main deck, and also by a chain and sprocket wheel from the stokehold flat.

The coal is trimmed to the underside of the beams, and proper ventilation is ensured as described under bunker ventilation.

REFRIGERATING CHAMBER.

This is an insulated chamber in which meat, vegetables, etc., can be stored. The compartment is lined with teak and well insulated from the plating by silicate cotton or compressed cork. The teak is attached to fir pieces secured to the plating on floor and sides, and at the crown to short pieces of fir connected to the beams, the planks being tongued together (p. 168).

The floor is covered with lead, which is carried up about 1 foot round the sides of the chamber.

There are separate compartments for meat and vegetables, with doors for access.

Round two sides of the floor teak trunks with sliding doors are fitted, and overhead round the other two sides there are suction trunks, as shown in sketch.

On the wall AB is a series of pipes through which the brine from CO_2 machine in the refrigerating machinery compartment circulates, a small motor fan being provided for circulating the air through the chambers. The supply orifice of this fan is at the bottom of the cooling pipes (section AD), and its discharge orifice to the trunks on the floor of the chamber. Thus the cold air is discharged from the lower trunk, circulates through the room and into the suction trunks, which take the air back to the top of the cooling pipes.

There are baffle plates between the coils of pipes to ensure thorough circulation of the air over them, and drip trays for the water produced when thawing off the coils.

Watertight slide valves are provided to the trunks through

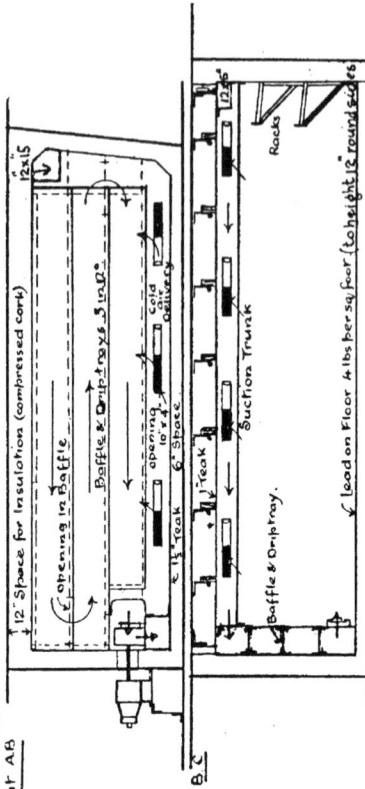

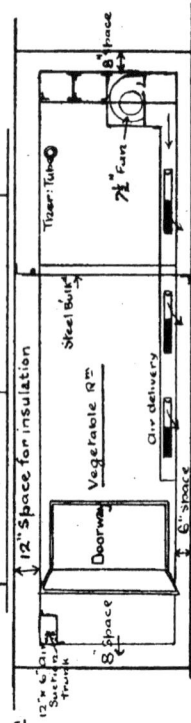

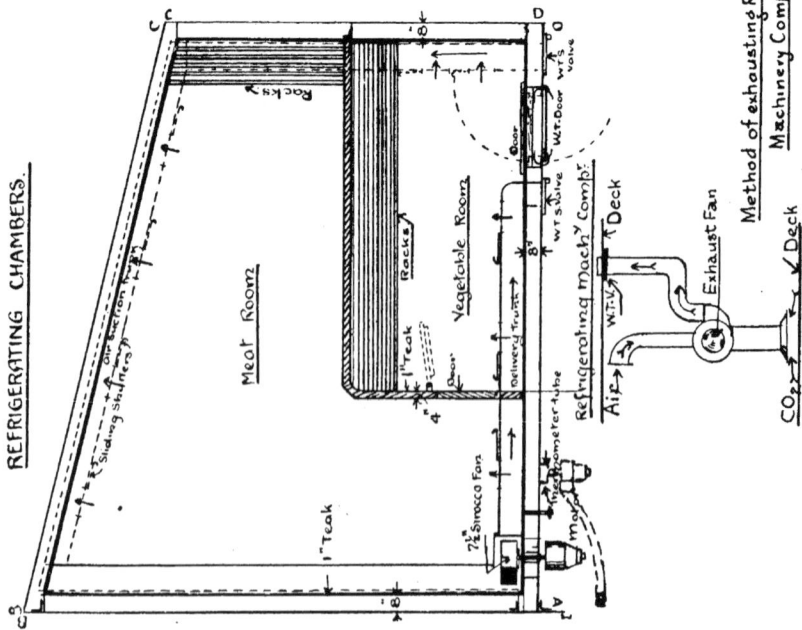

which air can be supplied if necessary.   These will be closed when cooling.

The refrigerating machinery compartment contains the CO_2 machine, the brine tank, and the ice tank.

STEEL MAST BUILDING.

Sketches of the mast are prepared, showing the disposition of the butts, riveting, extent of doubling, etc.   The butts of the plates must be arranged so that there shall not be more than one butt in the same cross section as shown.   The edges are

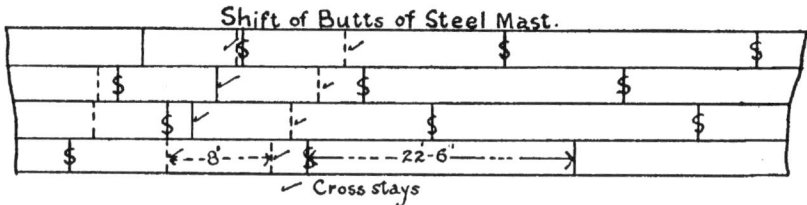

Shift of Butts of Steel Mast.

connected to T bars, the several lengths being strapped as shown, the thickness of the straps being $\frac{1}{16}$ inch more than that of the mast plates, the riveting being of the special description shown.   Cross stays, connected to the T bars, are worked at intervals of two feet, alternately, these also being of T bar (p. 170).

The plates are lined off to the correct dimensions, this being one-third or one-fourth of the circumference for the width, according as this is made up of three or four plates, and the holes in edges and butts marked off.   The holes are punched and countersunk, and the edges and butts planed, after which the plates are rolled to the correct form, as shown by section moulds made to the inner surface of the plate.   All the plates are of the same length, except those at the top and bottom of the mast.   In order to build up the mast, a series of bearers is prepared, their upper surfaces levelled, and on these the plates forming one strake of the mast are laid with their edges in line, as checked by a chalk line, and the butt straps placed in position, marked off, drilled and bolted. The T bars are next placed in position, having previously been bent, and held temporarily in place by means of boiler screws, and the holes drilled off from those already in the edges of the plates. Wooden moulds are used to support the other T bars while the remaining strakes of plating are being worked and the holes in the edges of the T bars drilled off. Before the holes in these latter T bars can be drilled, it is necessary to close up the

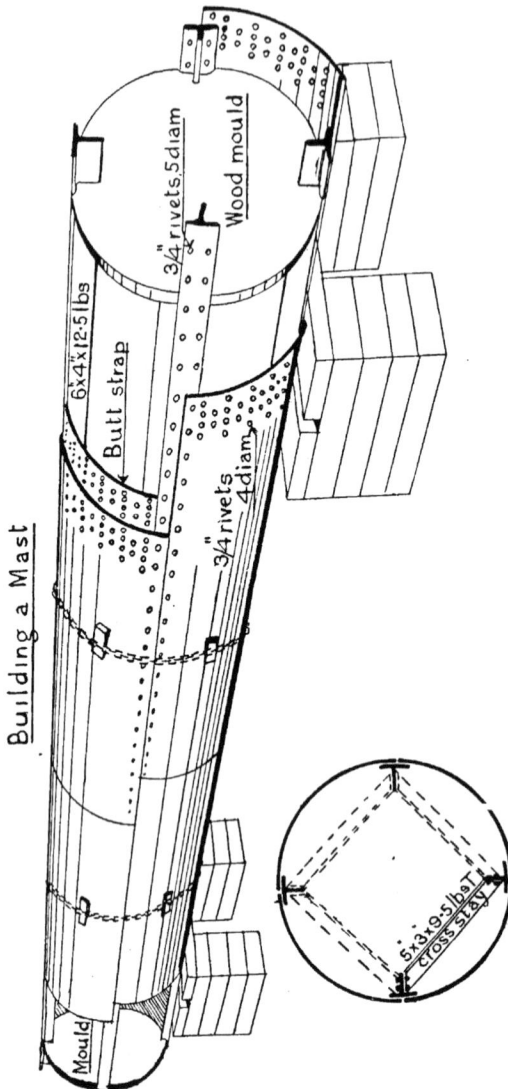

Building a Mast

work by means of chains round the plates, tightened by wedges, until the edges of the plates are in contact.

Doubling plates are worked on the mast where it passes through the decks, at the derrick pivot post and the link plates, the edges of the doubling plates being kept clear of the edges of the mast plates.

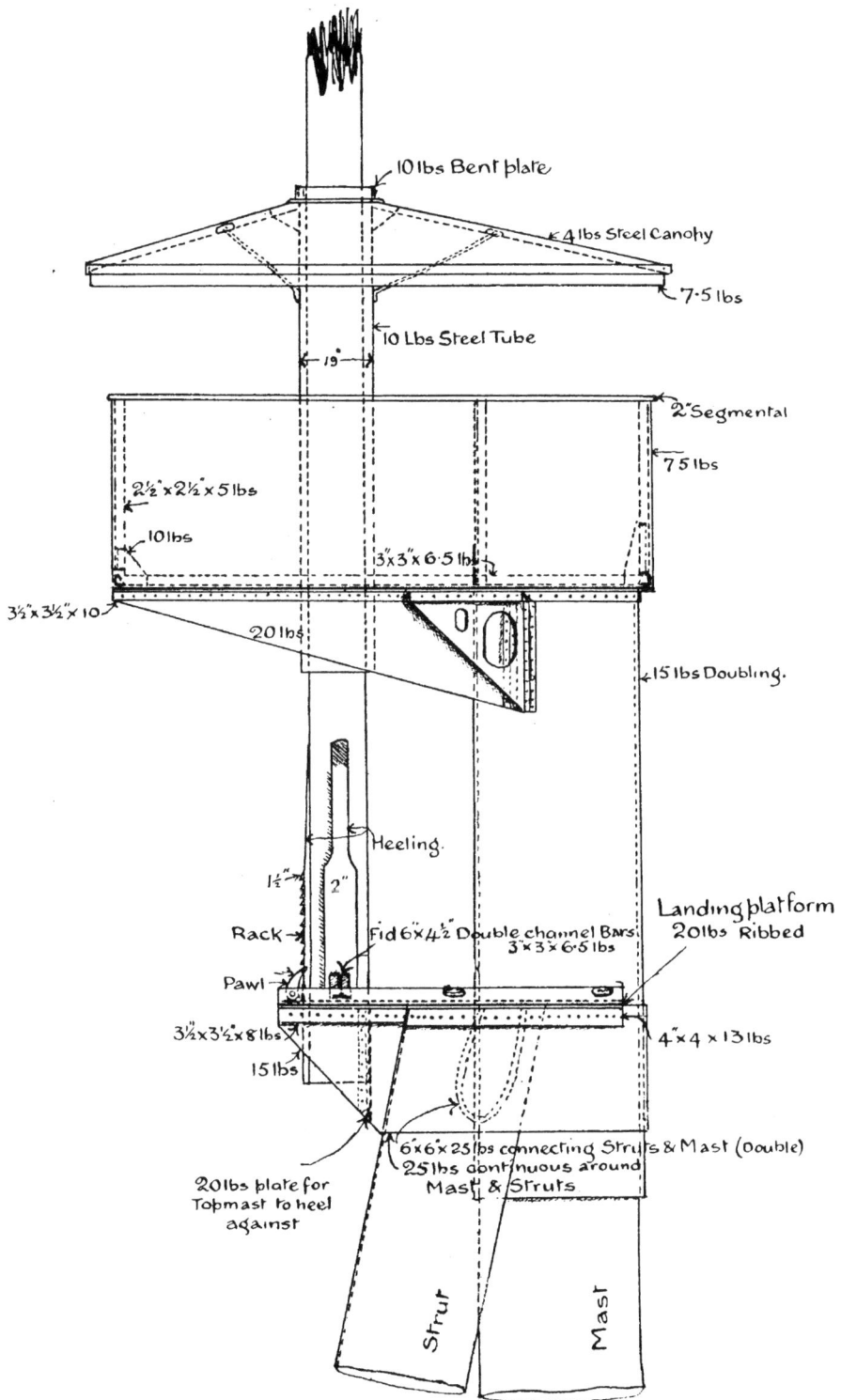

10 lbs Bent plate

4 lbs Steel Canopy

7·5 lbs

10 Lbs Steel Tube

19"

2" Segmental

7·5 lbs

2½" x 2½" x 5 lbs

10 lbs

3" x 3" x 6·5 lbs

3½" x 3½" x 10

20 lbs

15 lbs Doubling.

Heeling.

1¼" 2"

Rack →

Fid 6" x 4½" Double channel Bars
3" x 3" x 6·5 lbs

Landing platform
20 lbs Ribbed

Pawl →

3½" x 3½" x 8 lbs

15 lbs

4" x 4 x 13 lbs

20 lbs plate for
Topmast to heel
against

6" x 6" x 25 lbs connecting Struts & Mast (Double)
25 lbs continuous around
Mast & Struts

Strut

Mast

FIRE CONTROL PLATFORM.

In recent ships the masts have been fitted with struts, and the upper ends of the struts cut away to fit against the mast, the two

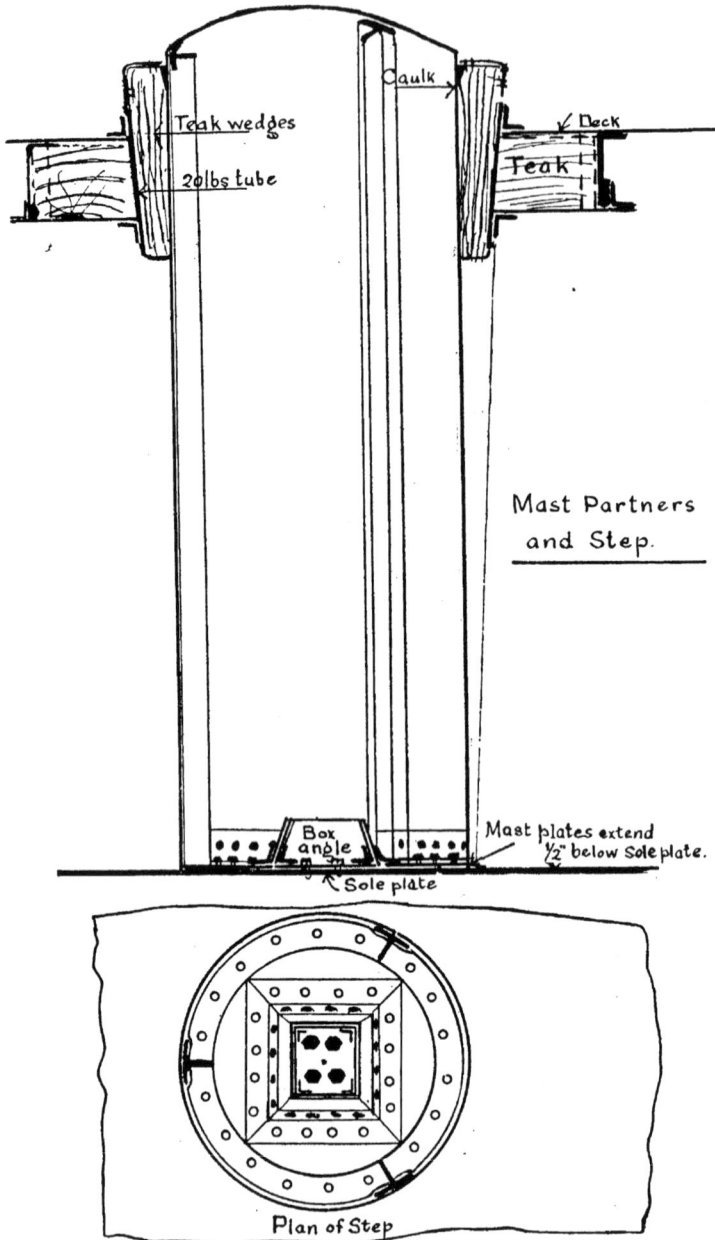

Teak wedges

20 lbs tube

Caulk

Deck

Teak

Mast Partners
and Step.

Box
angle

Mast plates extend
½" below Sole plate.

Sole plate

Plan of Step

being connected by angle bars, as shown on p. 171. The heel of the mast, and also that of the struts, is connected to the deck by angle bars. The method of obtaining the shape of the top and heel plates of the strut is shown on pp. 257, 258.

If the struts are connected to the mast before placing the mast on board, the deck plating will have to be left loose, or elongated holes cut to permit of shipping.

For a mast fitted with shrouds the heel is fitted as shown on p. 172, which also shows the mast partners.

The method of connecting the shrouds to the shroud plate is given below.

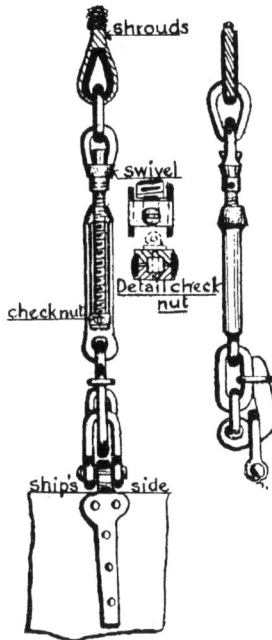

BOAT-HOISTING STEEL DERRICK.

The derrick is constructed of mild steel plating $\frac{3}{8}$ inch thick, the circumference being formed of two plates of equal width, joined together at their edges by T bars, 5 inches by 3 inches, of 9·75 lbs. per foot, and at their butts by straps $\frac{1}{16}$ inch thicker than the plates, riveted as indicated on the sketches. The diameter of the derrick is 21 inches in the middle, tapering to 17 inches at the ends, with a parallel portion of 5 feet at each end. To stiffen the ends, a web plate $\frac{3}{4}$ inch thick and 2 feet long is worked and connected to the T bars by five $\frac{3}{4}$-inch rivets along

Details of Main Derrick.

each edge, and the derrick is further stiffened along its length by cross stays 4 inches by $\frac{3}{4}$ inch, spaced 3 feet 6 inches apart ; the cross stays being worked on alternate sides of the T bars (p. 170).

The head and heel castings are of standard pattern, so as to be interchangeable with other derricks, and are made of quality " A "

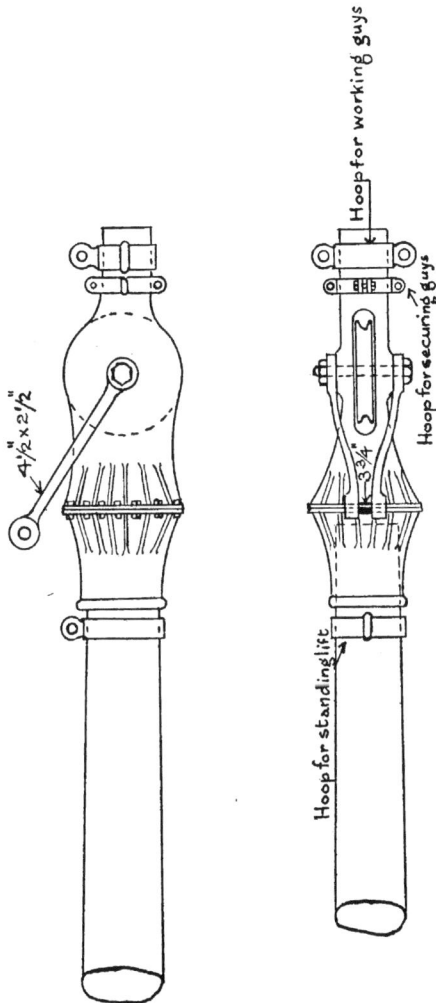

steel. They are secured to the tube by $\frac{3}{4}$-inch tap rivets, which pass through the flanges of the T bars, the heel casting having four rivets top and bottom, and the head casting having five such rivets top and bottom. The sheave at the head of the derrick is of cast steel with a $\frac{3}{4}$-inch phosphor bronze bush, and is tested to 80 tons proof stress ; the sheave pin is $4\frac{1}{2}$ inches diameter. The hoops for the securing and working guys are shrunk on and secured by $\frac{3}{4}$-inch countersunk head tap rivets, and the link for topping lift is of forged steel tested to 80 tons proof stress.   The trunnion is of

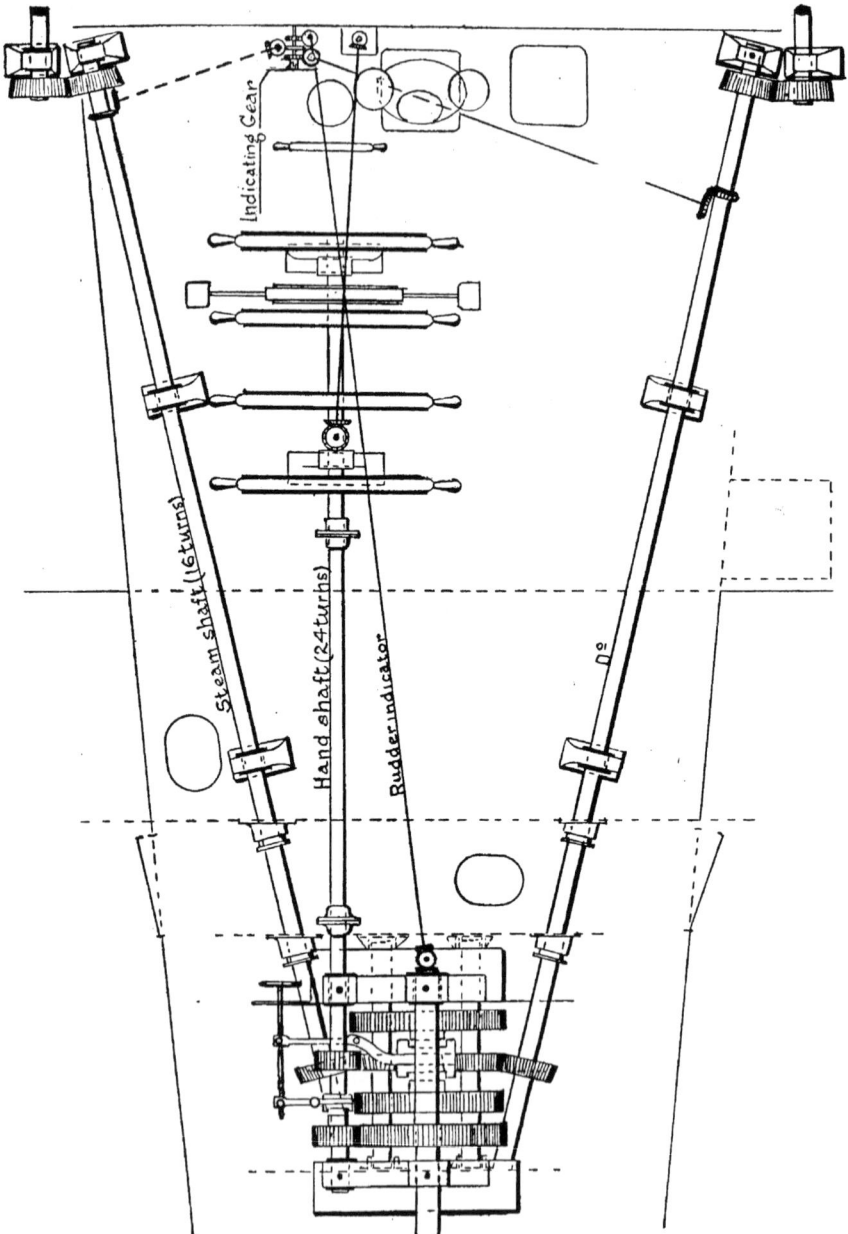

forged steel, $4\frac{1}{2}$ inches in diameter at the top, and turned to this diameter over a length of 4 inches, the diameter at the bottom being 6 inches, turned to this diameter over a length of $7\frac{3}{4}$ inches.

It will be noticed that there are 14 bolts for connecting the head and heel castings to those on the derrick tube, and between each pair of bolts there is a web. Hence for interchangeability the fastenings must be similarly situated in the castings, *i.e.* if there is a bolt at the top of the heel casting a bolt must also occur on the

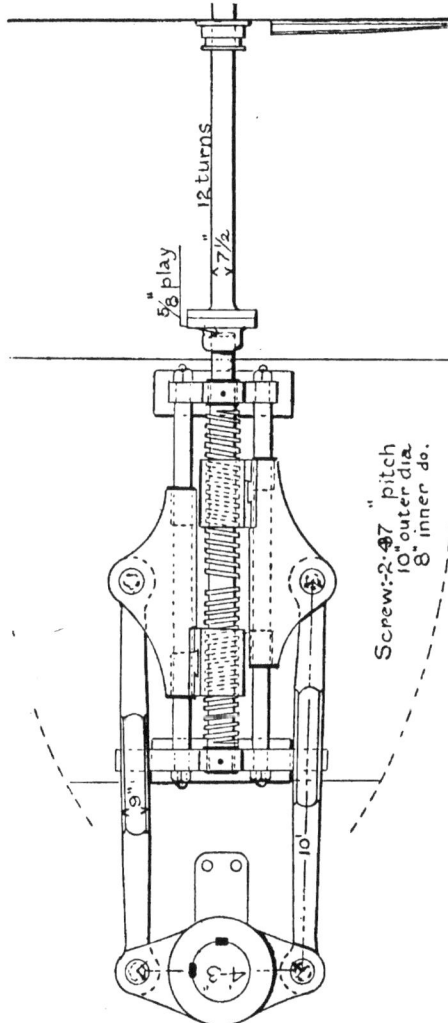

head casting, not a bolt in one case and a web in the other, as the castings would in that case not stand correctly on the derrick.

Another method of fitting head of derrick is shown on p. 175.

STEERING GEAR.

The various types of gear will be described in order, commencing with the **Screw Gear.**

N

Pages 176-77 show this in detail, arrangements being made
to steer either by hand or steam.   In all cases the steam steering
engine is duplicated, and in the present case a shaft runs along
the port and starboard sides of the ship respectively, to the steering
engines placed in the engine rooms.   The gearing has to be such
that either of the engines or the hand gear can be used, and further
that there shall be no possibility of more than one being in gear
at any time.   The connecting rods from the crosshead to the
sleeves on the guide rods are not parallel, and it is necessary to
allow for motion of the screw-shaft endwise.   The amount of this

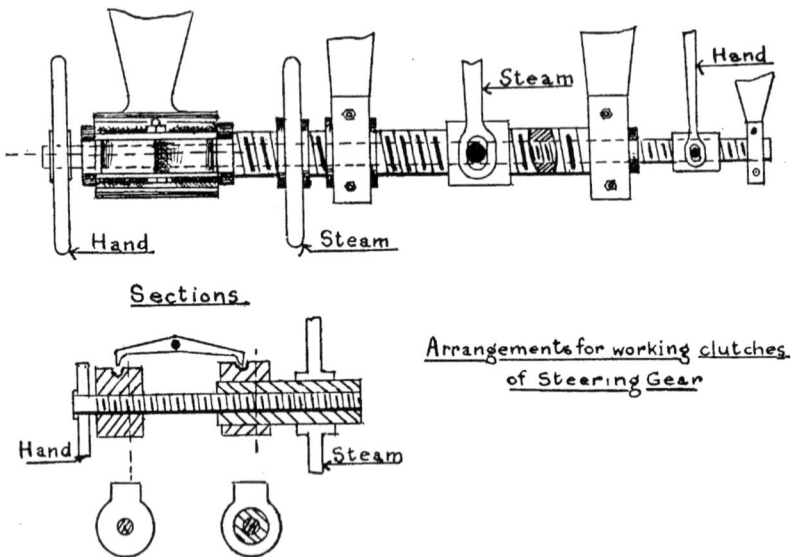

Sections.

Arrangements for working clutches
of Steering Gear

allowance will depend on the dimensions of the gear, and in the
present case it is about 0·204 inch.   This motion is allowed for at
the coupling on the end of the screw shaft, and the method of
calculating it from known dimensions given on p. 179.   The gear
is not "compensating," that is, as the rudder goes over it is
necessary for the engine to exert a greater power, but it has the
great advantage of being non-reversible, that is to say, it will not
run back should the rudder be struck by a sea when the turning
force is removed, as the screw thread would have to be stripped
before this could occur.   A further advantage is that the gear can
be fitted to ships which are very fine at the after end, as is the case
with cruisers.

To ensure that only one gear shall be in at a time a safety device is fitted to the gearing which is shown on p. 178. This consists of two screw shafts, one working through the other actuating the clutch to the hand gear, and the other, that to the steam gear. Another device is shown below.

On p. 180 is shown a modification of the previous arrangement by which only one shaft is required from the steering engines to the gear itself. It consists of a worm wheel actuated by a worm which can be driven by either steering engine. The arrangements for shifting the clutches to put engines in gear are shown, and it is clear only one engine can be in at a time.

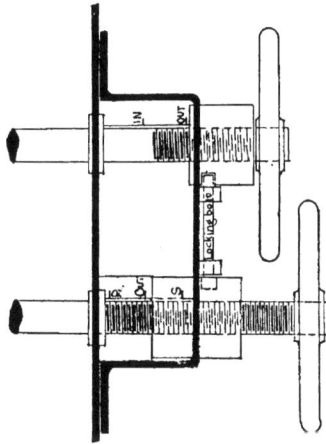

An arrangement of screw gear for double rudders is given on p. 181. The length of shaft B is not continuous with the lengths A and C, but there is a small clearance at its ends. The clutch at the foremost bulkhead is a double one, and arranged so that the hand and steam gear cannot be in at the same time. There is another clutch to connect shafts A and B.

Screw Steering Gear. In this gear it is necessary to allow for motion of the screw shaft endwise, as indicated on the drawing of this gear (p. 177).

The amount of this motion can be calculated when the dimensions of the gear are known. Referring to p. 182, *ab* denotes the positions of crosshead on the rudder, and *ac*, *bd* the connecting rods when the helm is amidships, and *a'b'*, *a'c'*, *b'a'* their respective positions with the helm at an angle θ. The points *a* and *b* move

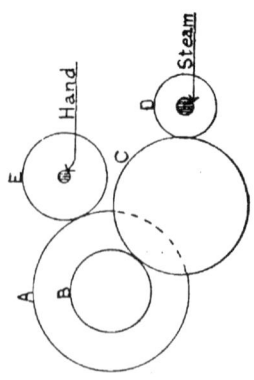

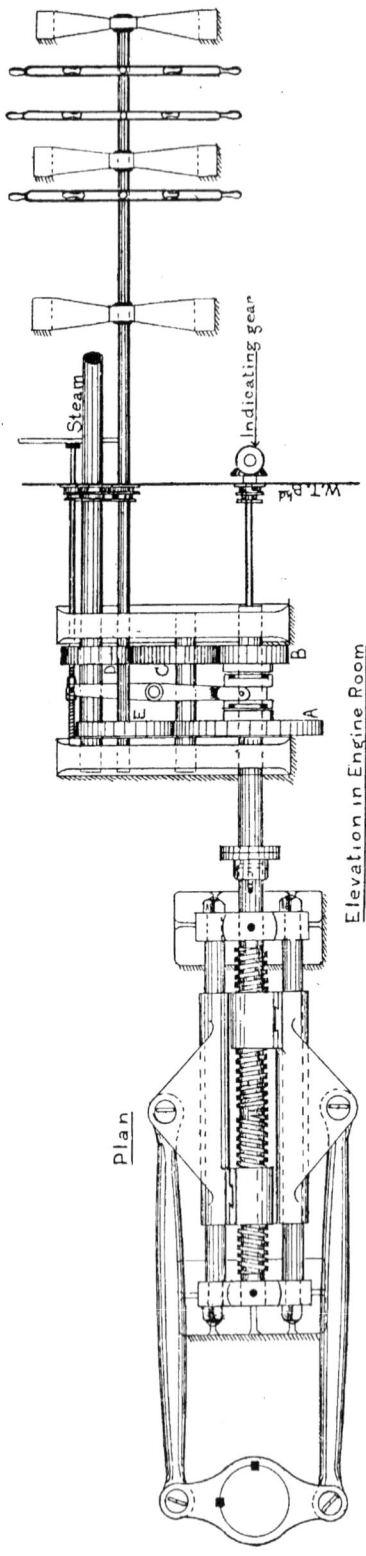

Steam

Indicating gear

W.T. B'hd

Elevation in Engine Room

Plan

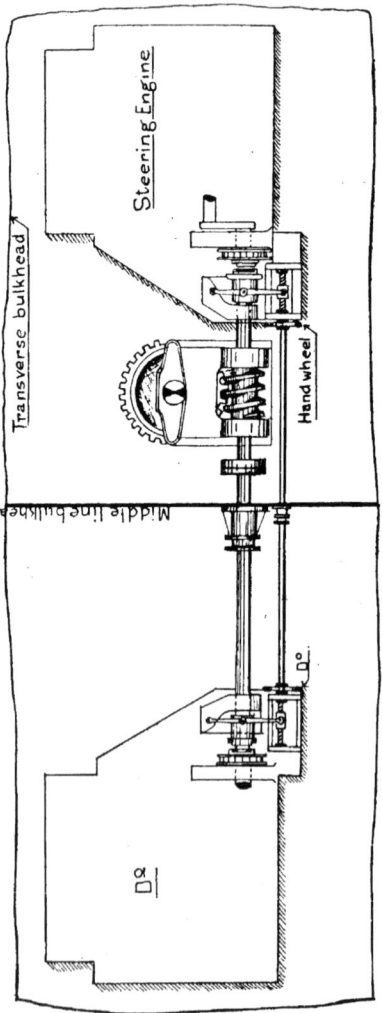

Transverse bulkhead

Steering Engine

Hand wheel

Middle line bulkhead

Dº

A B C D E
A B C D E

Hand

Steam

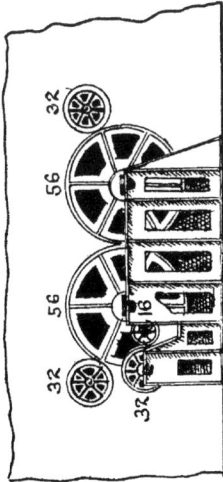

W.T.Bulkhead

Hand Wheels

Telemotor Gear

8 6"

7" F

40lbs doubling

I.B.kd

Elevation on Bulkhead.

32   56   56   32
16
32

The Screw Gear on this side will be to reverse hand as the shafts move in opposite directions.

in a circle, while c and d move along lines cc', dd' parallel to the centre line of screwed shaft ef. The line joining the points c', d' will not pass through f, but through a point f', ff' then being the endwise motion of the screwed shaft.

To take an example, let $ae = eb = 18''$, $cf = fd = 24''$, $ac = bd = 120''$, and $\theta_1 = 35°$.

$$\text{Then } \sin\beta = \frac{cf - ae}{ac} = \frac{24 - 18}{120} = 0\cdot05$$

$$\text{and } \cos\beta = \sqrt{1 - \sin^2\beta} = 0\cdot9987 \ ;$$

$$\text{also } \sin\phi = \frac{cf - a'e\,.\,\cos 35}{a'c'} = \frac{24 - 18\,.\,\cos 35}{120} = 0\cdot077$$

$$\text{and } \cos\phi = \sqrt{1 - \sin^2\phi} = 0\cdot9907$$

$$cc' = ac\,.\,\cos\beta + a'e\,.\,\sin 35 - a'c'\,.\,\cos\phi$$

$$= 120\,.\,\cos\beta + 18\sin 35 - 120\,.\,\cos\phi$$

$$dd' = b'd'\,.\,\cos\phi + b'e\,.\,\sin 35 - bd\,.\,\cos\beta$$

$$= 120\,.\,\cos\phi + 18\sin 35 - 120\,.\,\cos\beta$$

$$ff' = \frac{cc' - dd'}{2} = \frac{2\,.\,120\,.\,\cos\beta - 2\,.\,120\,.\,\cos\phi}{2}$$

$$= 120\,(\cos\beta - \cos\phi) = 120\,(0\cdot9987 - 0\cdot9970)$$

$$= 120 \times 0\cdot0017$$

$$= 0\cdot204 \text{ inch.}$$

It will be noticed that values of cosines must be calculated to four places of decimals for a result to be obtained.

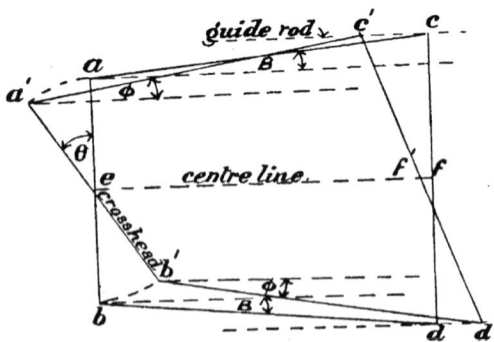

Harfield's Gear. This differs in design from either of the preceding. To make the gear compensating, a specially designed rack is fitted to the dummy rudder-head which gears with a pinion placed eccentrically on a vertical spindle. As the pinion

volves, the distance from its centre of motion to its circum-
rence decreases, and, consequently, that from the centre of the
mmy head to the rack increases, *i.e.* to say, the force on the

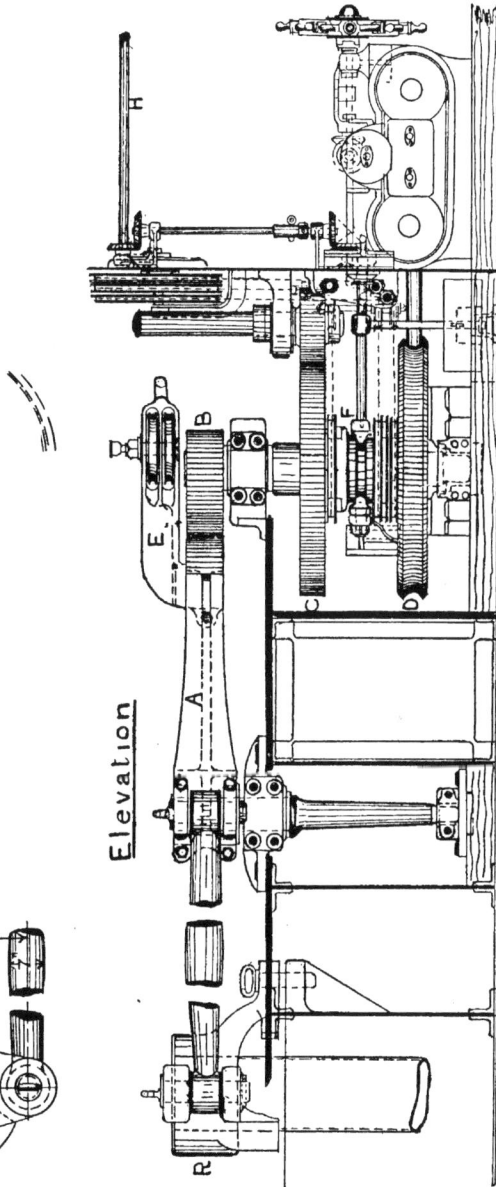

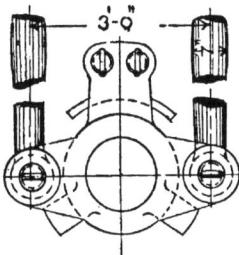

mmy cross-head is working at a greater leverage, giving a
irly constant movement. The method of changing from hand
steam will be clear from the sketch. This gear being reversible,
brake has to be fitted. A friction brake is shown on p. 184.

Livingstone's Patent Clutchless Gear. In this gear, clutches are done away with altogether. The gear is of the screw type, and consists of two shafts, one with a right and left-hand screw cut on it, and a second parallel to it with a key-way along its whole length (p. 185).

To each sleeve on the screw shaft a pinion is connected which is free to revolve. When steering by steam, the screw shaft is revolved by the engine, and the sleeves move and so turn the rudder, the pinions on the screw shaft simply sliding along, while still gearing with the pinions on the screw shaft.

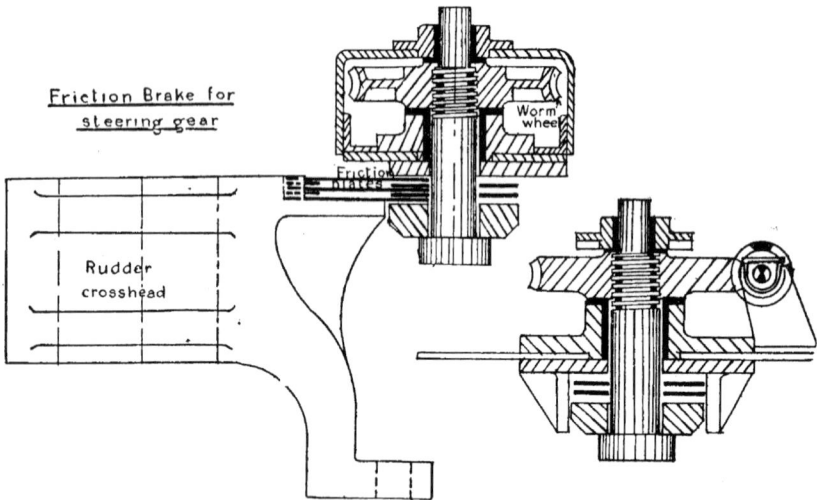

When steering by hand, the screw shaft does not revolve, but only the hand shaft and the pinions with it, and these pinions cause those on the screw shaft to revolve also, and the latter thus move along the screw and carry the sleeves with them. This gear is fitted in some destroyers.

Rapson's Slide. This is a simple form of gear, but it can only be fitted in ships very full aft owing to the large space required for the sweep of the tiller. The general arrangement of the gear is shown on p. 186, and consists of a tiller, the after end of which is formed to serve as a dummy crosshead to which the parallel connecting rods from the rudder crosshead are connected.

To do away with the necessity of shipping the tiller over the dummy head, which requires a good deal of room overhead for the lifting appliances, the upper end of the latter has a square projection on it, and the crosshead is formed to slide on to this, and the tiller is kept in place by the key shown. The tiller passes

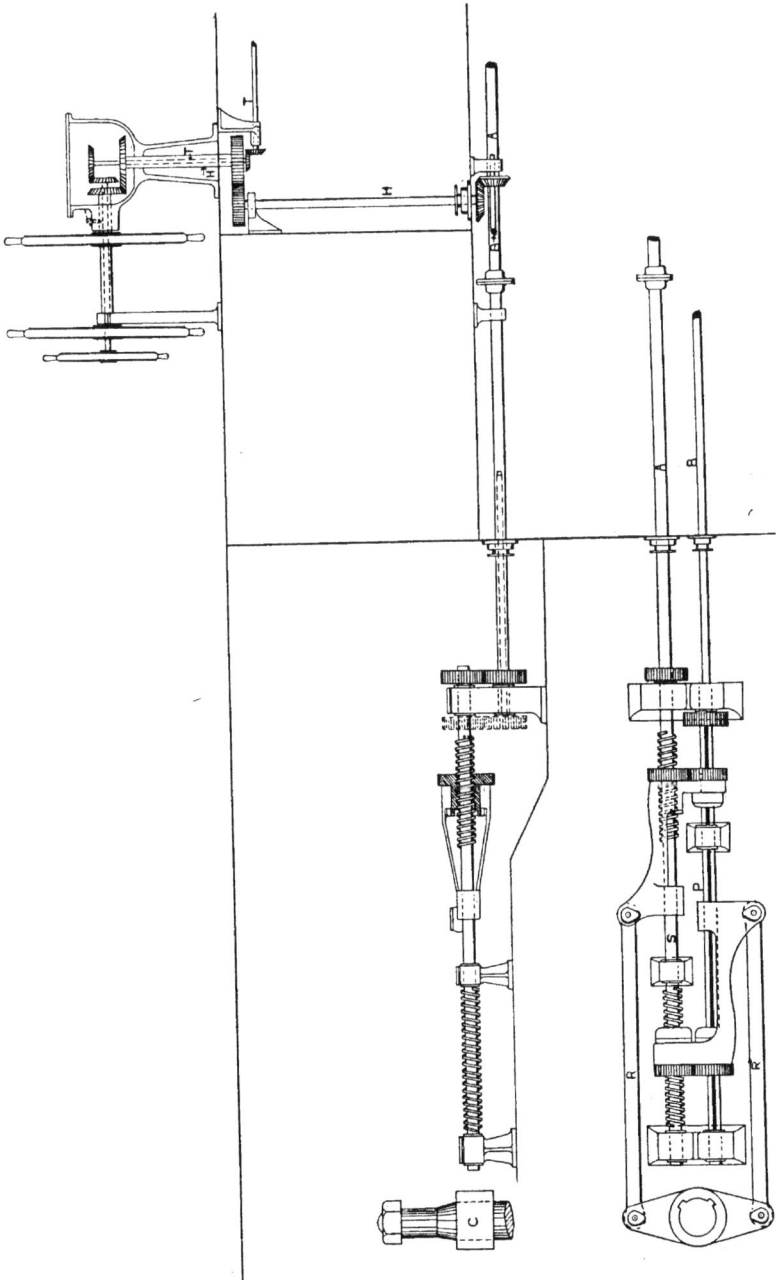

through a swivel piece which slides on metal guides, and the sprocket chains are connected to this, and gear with the sprocket

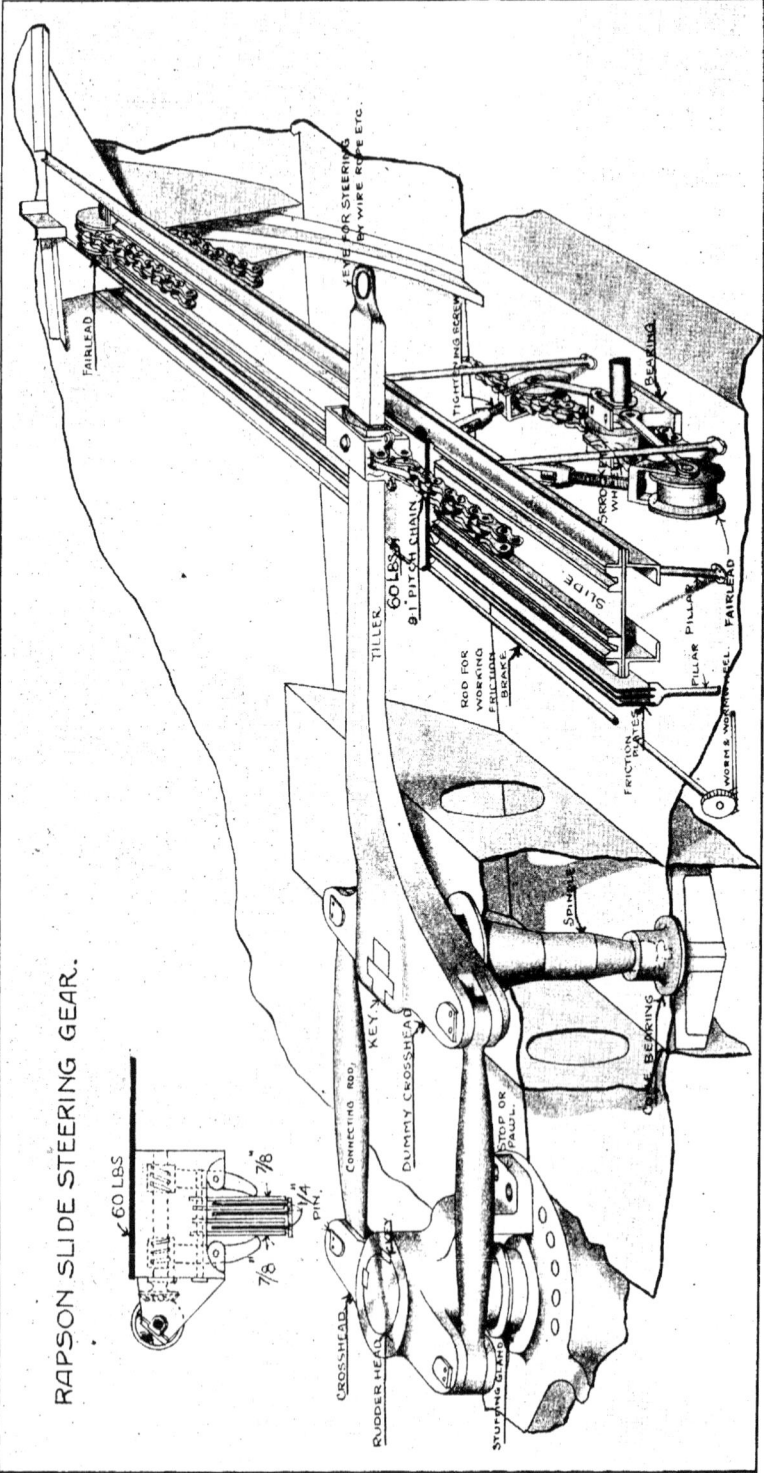

RAPSON SLIDE STEERING GEAR.

wheel ; to allow for adjusting the sprocket chain, jockey pulleys are fitted. As this gear is reversible, a brake has to be provided, which consists of a series of friction plates connected to a thick steel plate moving to and fro with the sliding block, and engaging with another series of fixed plates running transversely across the ship. The brake is put on by moving the wheel, which causes the jaws to press the two sets of plates together, and can be applied in any position of the tiller.

This gear is compensating (see p. 186). The gear wheels are not shown, since they are similar to those already described for the screw gear. An important detail to be noted is the design of the sprocket chain. It was found that with the ordinary design of the sprocket chain it gave way under the strain by the shearing of the pins, and this was remedied by fitting the extra pieces marked.

In all types of steering-gear means must be provided for indicating the angle of helm with both the hand and steam gear. This necessitates three indicators, viz. one for each of the hand and steam gear, and one for the rudder. It is clearly possible to disconnect both the hand and steam steering shafts from the gear itself, and connect up again with either steam or hand showing no helm, say, and the rudder any degree of helm. Thus a rudder indicator is always indicating whether hand or steam be in. These indicators are shown in sketch of screw gear.

All the shafting must be easily accessible, and the gear so arranged that the ship's head always moves in the same direction as the top of the steering wheels, both when using hand or steam. The pointers on the pedestals show the angle of helm, and will therefore move in the opposite direction to the ship's head.

There are several positions in the ship from which she can be steered by steam, and the shafting has to be arranged to permit of either position being used.

Theory of Rapson's Slide. Let ac denote tiller at an angle θ to the middle line ab, and the forces on the sliding block P, Q, R as in the figure, friction being neglected.

The turning moment $= Q \cdot ac = Q \cdot ab \cdot \sec \theta$, also $P = Q \cdot \cos \theta$ (resolving along bc).

Hence turning moment $= Q \cdot ab \cdot \sec \theta = P \cdot ab \cdot \sec^2 \theta$, and if P, the pull of the chain, is constant, the turning moment varies as $\sec^2 \theta$, which increases as θ increases, so that the gear is compensating

The above result can also be arrived at readily thus—

Energy exerted = P . dx, if x denotes movement along bc,

But $x = ab$. tan θ

and $dx = ab$. sec$^2\,\theta$. $d\theta$,

∴ energy exerted = P . ab . sec$^2\,\theta$. $d\theta$,

and turning moment = P . ab . sec$^2\,\theta$ as before.

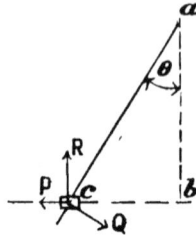

Controlling Shafting. This is the shafting actuating the valve on steering engine. It is of steel tube, except the vertical portions near a compass, which are of metal. Owing to the long lengths of shafting necessary and its indirect lead, expansion and universal joints are necessary.

The controlling shafting junctions must be so arranged that the two lengths of shafting at a junction can only be connected up at the position where the indicators on steam steering wheel and engine agree. This is effected by means of the disconnecting coupling shown on p. 189, which only permits of the pin being inserted when the two portions occupy same relative positions.

Telemotor for working Helm Signals. The gear comprises a lower cylinder (the piston of which is moved either directly by the rudder, as on p. 191, or by means of gearing, see p. 181) and an upper cylinder, the piston of which responds to any motion of the lower one. The motion is communicated from one to the other by means of pipes, and filling the whole system with a fluid. This fluid is one that will not freeze at ordinary temperatures, and is generally composed of one part glycerine and two parts water.

The piston rod of the upper cylinder is connected by cross pieces to two powerful springs arranged so as to be in compression in whichever direction the piston moves, and which tend to bring the piston back to its central position (the figure shown ticked shows how this is brought about). At the end of the upper piston

Upper Conning Tower

Steering Pedestal

Controlling Shaft.

Lower Deck

Disconnecting Coupling

Lower Conning Tower

Dis.ᵍ Coupling

Steering Pedestal.

Platform Deck.

Expansion Joint.

Disconnecting Coupling.

Pin

Keyway.

Metal Nut.

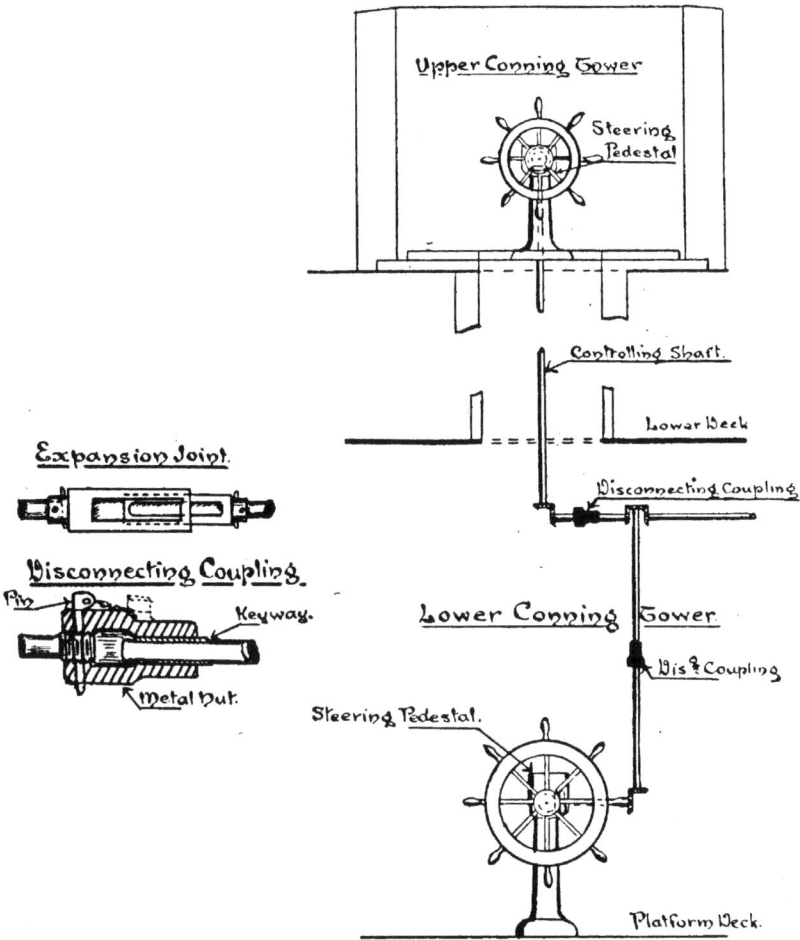

<u>Plan shewing arrange.ᵗ of Controlling Shaft in Engine Room</u>
<u>Scale ¼" = One Foot.</u>

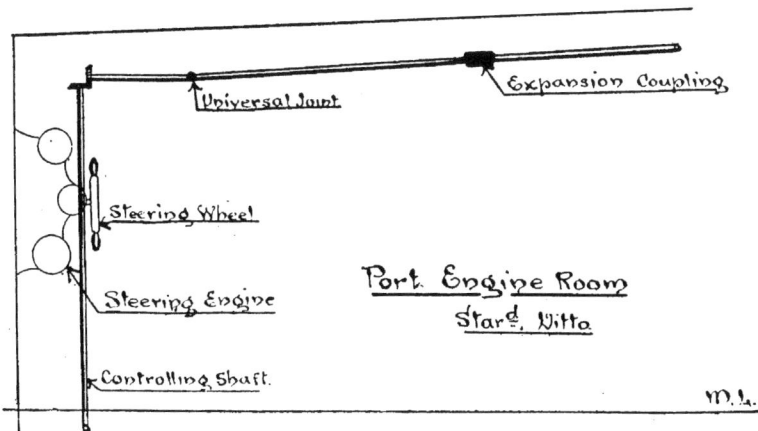

Universal Joint

Expansion Coupling

Steering Wheel

Steering Engine

Port Engine Room
Starᵈ. Ditto

Controlling Shaft.

M.L.

rod a rack works a small pinion on the same shaft as a large wheel, round the rim of which is the steel wire working the ball and flag. To take up any stretch in the steel wire there is a "taking-up screw," and to prevent damage to the wire, etc., by sudden strains being brought on it, a spring box is fitted. At the upper part, the steel wire passes through two blocks attached to an outrigger from the mast. In some cases the spring box has been omitted from the steel wire, and placed at one of the upper blocks, as shown on p. 191, Fig. B. This method has the advantage that there is less likelihood of damage occurring, as the extension of the spring will be much less than that of the steel wire.

It is necessary to consider the details of the lower cylinder to understand the working of the gear. Referring to the figure, it will be seen that in the central portion there is a portable piece made in half, connected by means of ribs to the short cylinder surrounding the piston, this cylinder being also necessarily in half. With this arrangement, and with the piston in its central position, as in Fig. A, the fluid can pass to all parts of the lower cylinder and through the pipes to upper cylinder also. Hence, to charge the system it is necessary to always bring the piston of the lower cylinder to its central position. The charging is done by means of the force pump, which gets its supply from the charging tank shown, and it is vital to the working of the gear that all air be expelled. This is ensured by opening all the valves, and pumping until the fluid issues at them. To allow for expansion there is a small tank provided into which the fluid passes through the relief valves shown.

When fitting up the gear in a ship it is of the utmost import-ance to arrange it so that it shall be readily accessible at every point, in order that leaky joints can be discovered and put right at once. If the fluid does not fill the whole system, the upper and lower pistons will not work in unison, with the result that the amount of helm will not be correctly indicated by the ball and flag.

Mechanical Method of working Helm Signals. This is shown on p. 191, Fig. C, and consists of a drum with a spiral groove for the steel wire; and a clutch for putting the drum in and out of gear with the shaft. As the steel wire always remains in the same plane, the drum, and consequently the shaft, must move endwise, which is ensured by the feed gear shown. For the same reason a large keyway must be cut in the shaft for the mitre wheel.

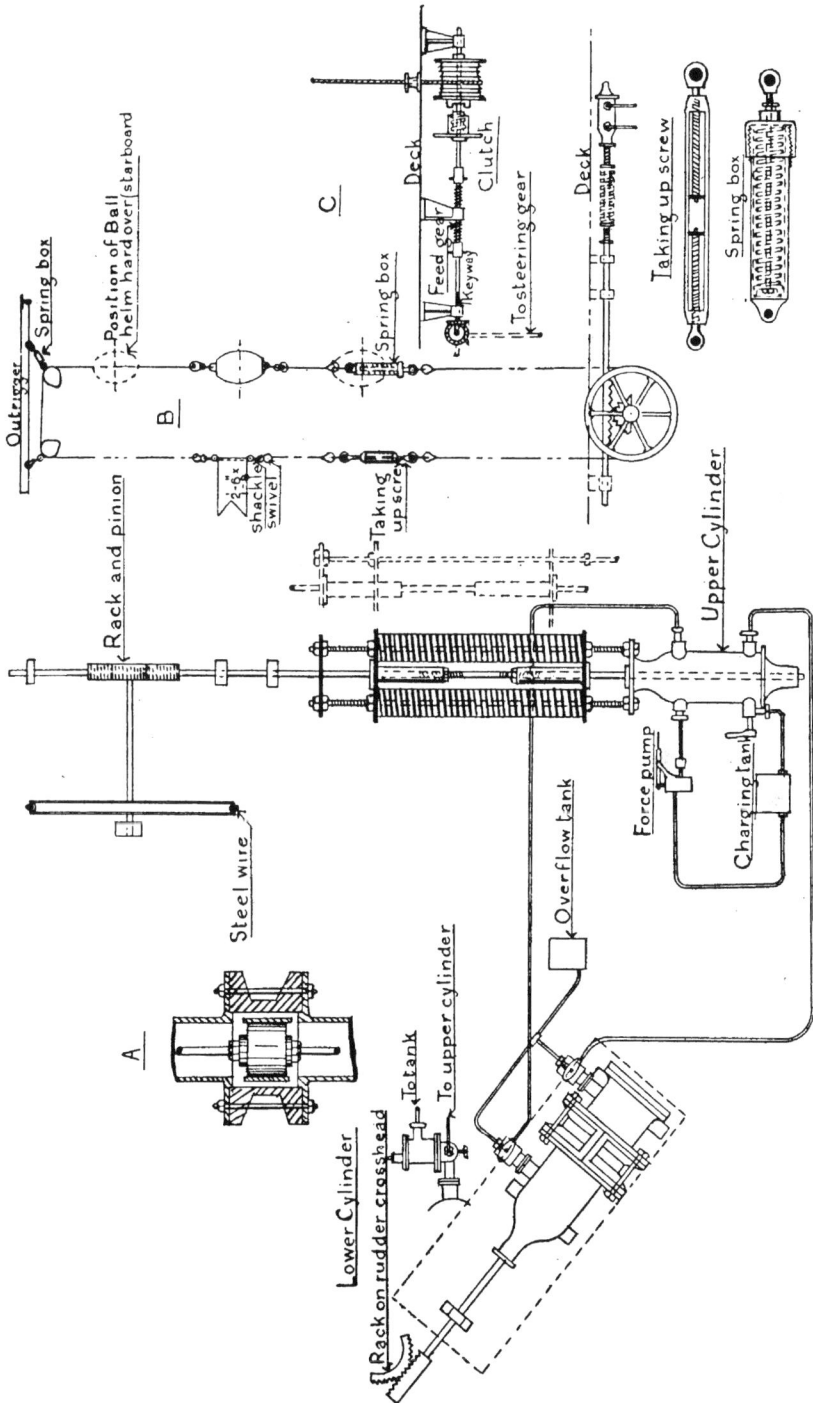

The details of taking-up screw and spring box are given on p. 191.

The travel of ball and flag is about 21 feet, and they must be placed high enough up to be seen at all positions.

CAPSTAN GEAR.

The information required is a sketch showing general arrangement of gear, and after ship has been built far enough, the actual distances between decks, to enable lengths of spindles to be obtained.

Harfield's Cable Holder. Two types of this fitting are shown on p. 193. In the one (shown in elevation only) when hauling in the cable the drum D is made to revolve with the cable holder spindle by turning the top casting A, which lowers B, jamming down the friction plates F. There are two sets of friction plates, made in the form of rings, one half being connected to the holder by projections on drum D, and the remainder to the steel casting C, which is keyed to the spindle. When letting go the anchor B is raised, freeing the friction plates and allowing the drum to revolve freely around the spindle. To ensure the friction plates being free when B is raised, small springs are fitted to the plates as shown.

In the other type (shown in plan and elevation) there are no friction plates, the drum being made to revolve with the spindle by lowering B, which slides on a nut, N, secured to the spindle, and engages with stops in the drum D (see plan). A separate brake is fitted to this type, and consists of a band round the base of the drum, the gearing for working this being shown in plan.

Napier's Cable Holder. This differs from the above, both in the method of freeing and connecting up the drum, and also in the brake. Referring to p. 194, it will be seen to consist of three parts— the drum A, free on the spindle, the casting B, carrying two " bolts," which are moved by the scroll plate C.

The scroll plate has two grooves formed in it, eccentric with respect to the spindle, and in which work the pins in the bolts. Thus when the scroll plate is turned the bolts are caused to move outwards into the tapered notches in the drum.

The brake is shown in plan, and is of the differential type, actuated by means of a wheel. The brake is so arranged that

Plan.

Sectional Elevation.

web

deck

14"

F

web

key

C

D

li"

deck

casing

spindle

Spring

F

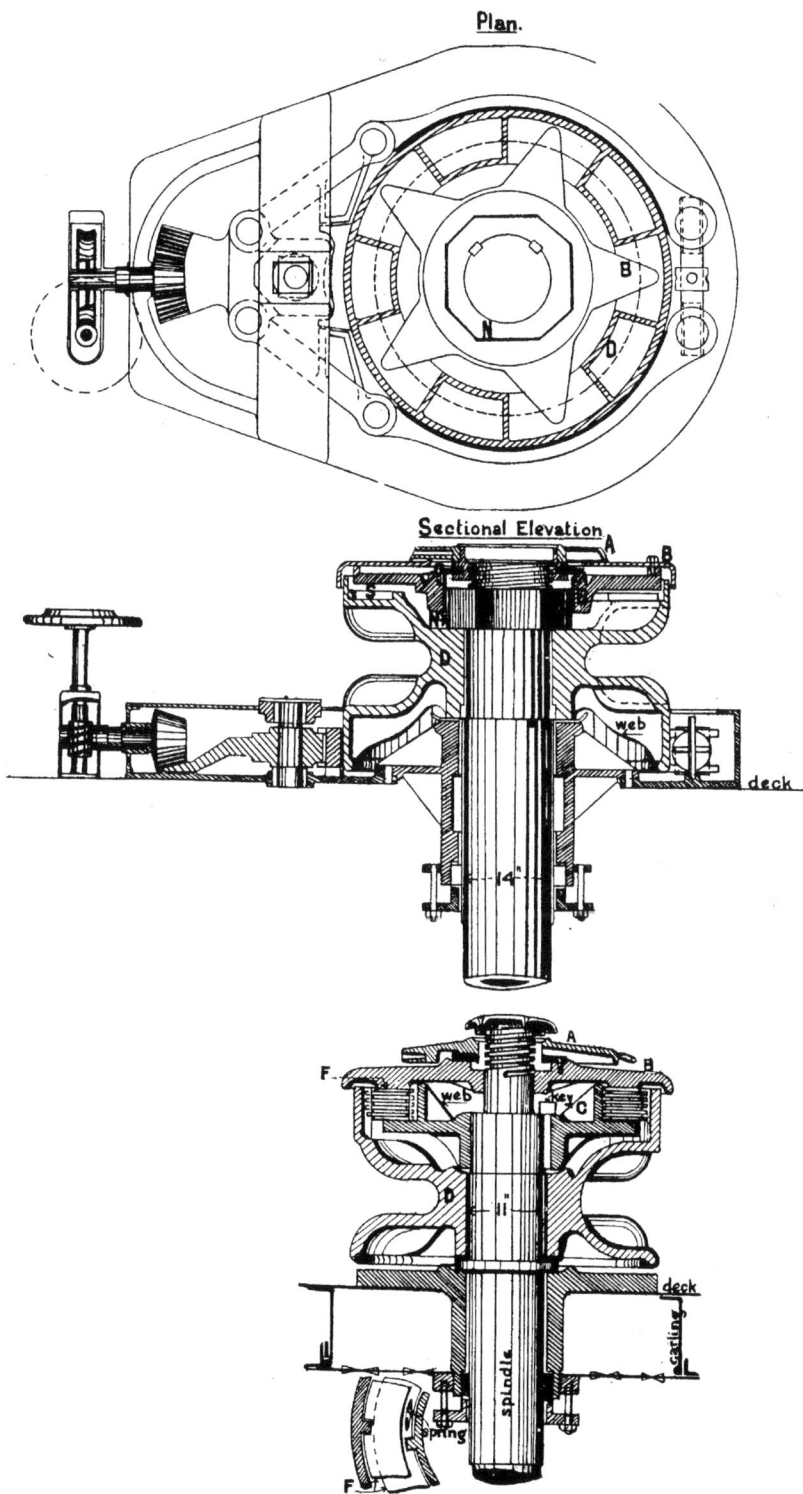

HARFIELD'S CABLE HOLDER.

when applied any tendency of the drum to revolve automatically increases its power. This is done by allowing a small motion of the cams, worm, bevel gear and spindle to hand wheel.

Middle Line Capstan. This is shown on p. 196, and is provided

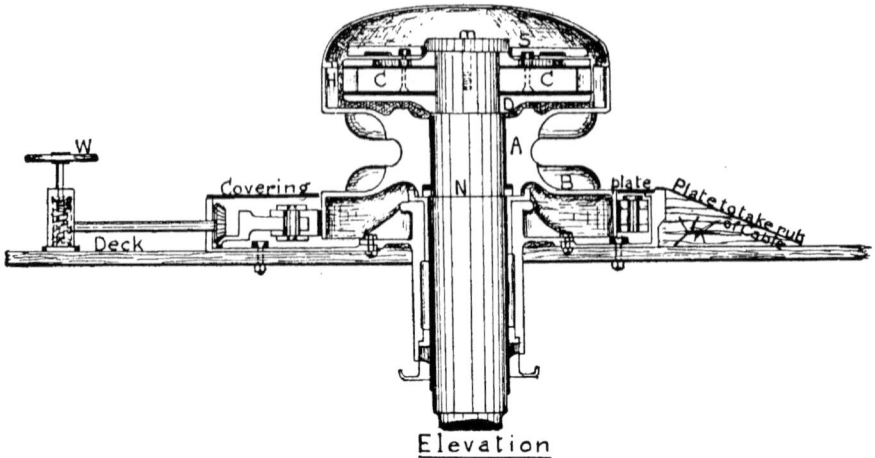

Elevation

Napier's Cable holder

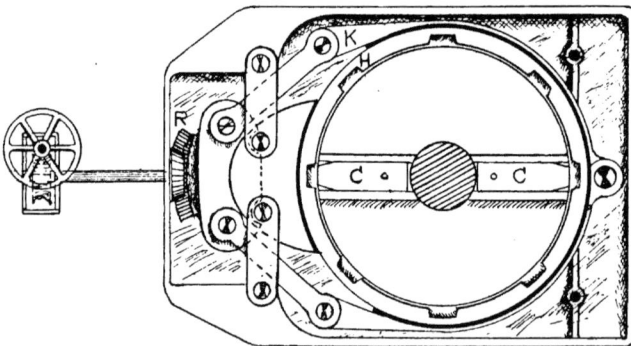

Plan (covering plate removed)

with portable whelps, which are shipped when the capstan is to be used for a steel wire rope. Capstan bars are also provided for working this capstan by hand, these being of ash. The general arrangement of gear in capstan engine room is shown on p. 195.

ANCHOR AND CABLE ARRANGEMENTS

On pp. 196, 197 is shown, in plan and elevation, the fittings for working the anchors. The cable is stowed in the cable lockers

Starb⁴ Cable holder spindle.   Port -do-

Worm-wheel 40 teeth 4" pitch 12½ broad.
Phosphor bronze rim
Cast steel centre.

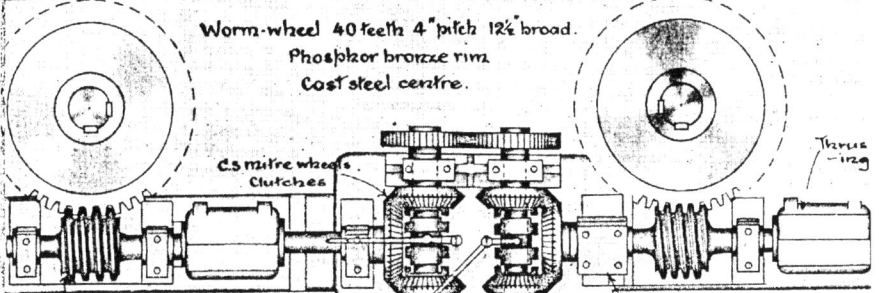

Thrust -ring

C.S mitre wheels
Clutches

Forged steel worm.

Cap bearing.

Clutch lever.

← Engine.

Worm-wheel 50
teeth 4" pitch
10" broad.

P.B. Rim
C.S Centre.

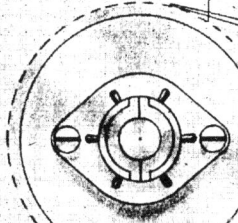

Clutch.

Bearing (cap removed.)

Thrust bearing
(cover removed).

Capstan spindle.

Section thro' Capstan
Spindle.

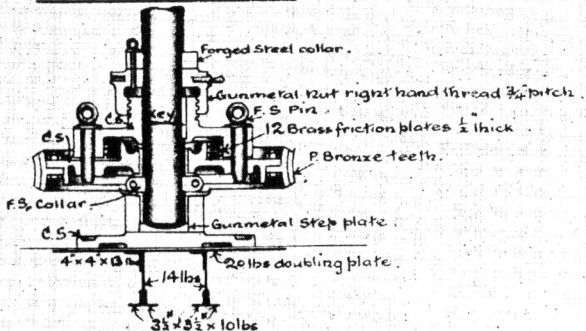

Forged Steel collar.

Gunmetal nut right hand thread ¾ pitch.
F.S Pin.
12 Brass friction plates ½" thick.
P. Bronze teeth.

C.S key

C.S.

F.S. Collar.

C.S.

Gunmetal Step plate.

4"×4"×13ℓ

20 lbs doubling plate.

14 lbs

3½"×3½" × 10 lbs

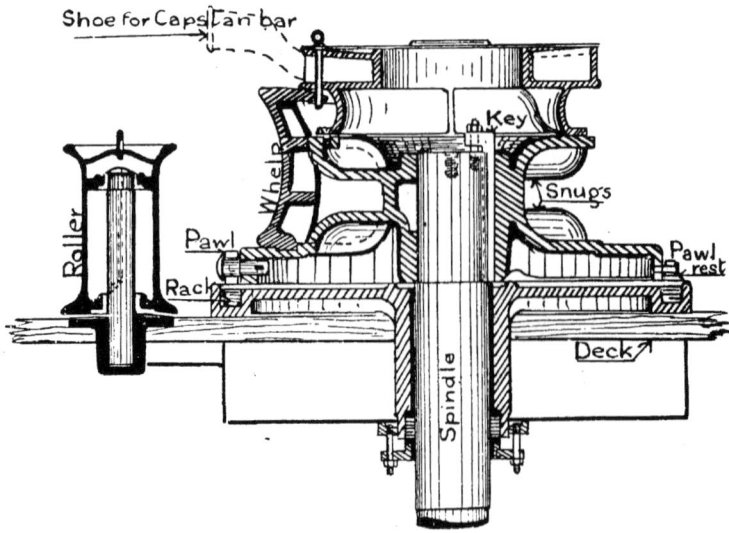

Shoe for Capstan bar

Key

Roller

Wheel?

Snugs

Pawl

Rack

Pawl
rest

Deck

Spindle

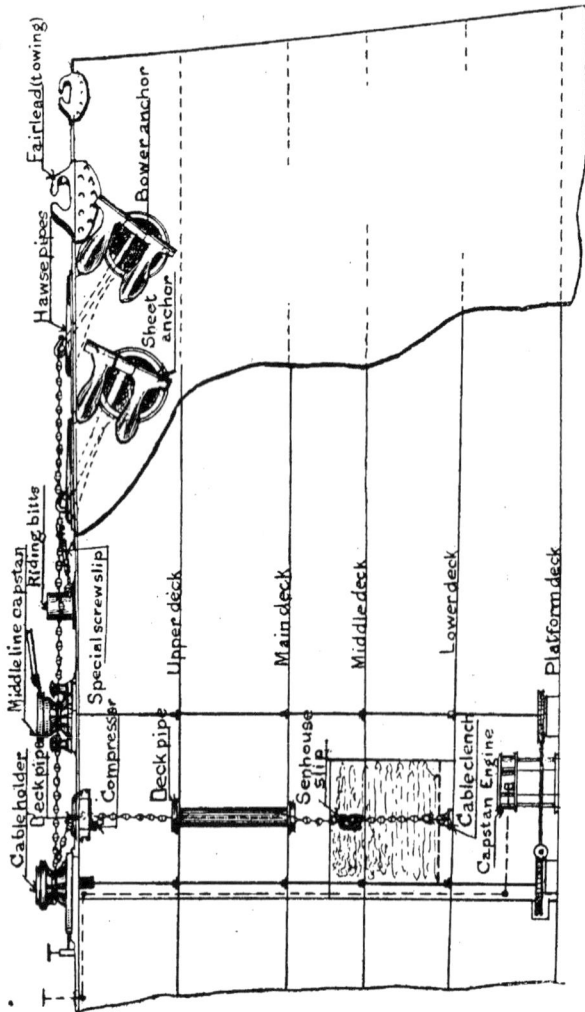

Fairlead (towing)

Hawse pipes

Bower anchor

Sheet anchor

Middle line capstan

Riding bitts

Cable holder

Deck pipe

Compressor

Special screw slip

Deck pipe

Upper deck

Main deck

Middle deck

Lower deck

Platform deck

Senhouse slip

Cable clench

Capstan Engine

GENERAL ARRANGEMENT OF ANCHORS AND CABLES (*Elevation*).

on the lower deck, and led up through deck or navel pipes to the weather deck, and then round the cable holders to the hawse pipes, the upper end of the cable being shackled to the anchors. The lower end of the cable is connected to the Senhouse slip, which is attached to a cable clench secured on the lower deck. The length of chain between the clench and the slip must be sufficient to permit of the lower end of the cable coming up clear of the locker, so that the cable can be slipped if necessary ; it must not, however, be so high that the slip will strike the deck pipe when slipping the cable..

C.—Cable holders.
D.—Deck pipes.
P.—Compressors.

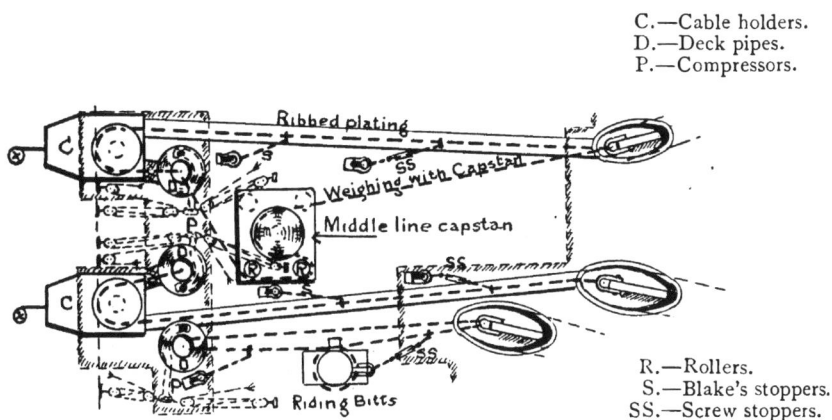

R.—Rollers.
S.—Blake's stoppers.
SS.—Screw stoppers.

Beneath the deck pipes on the weather deck, cable compressors are fitted, which are worked by tackles. The remaining fittings are the Blake and special screw slips.

In ships fitted with stockless anchors, these are stowed in the hawse pipes, being hoisted up into the hawse pipes by means of the cable holder, and then hove up tight and secured by the special screw stopper. In other ships the anchors are stowed either on the anchor beds or on the side of the ship, as shown on pp. 198, 199. With both of the latter methods a cathead is necessary for placing the anchors in their stowing position. The method of hoisting the anchors on to the anchor bed is indicated on the drawing, a chain, called a cat or ground chain, connected to the gravity band, being used. The anchor is secured by means of chains passing round its shank, the free ends being passed over the tumbler. This tumbler

is a simple arrangement for slipping the anchor, and consists of a
rod with projections on it, over which the elongated links on the ends

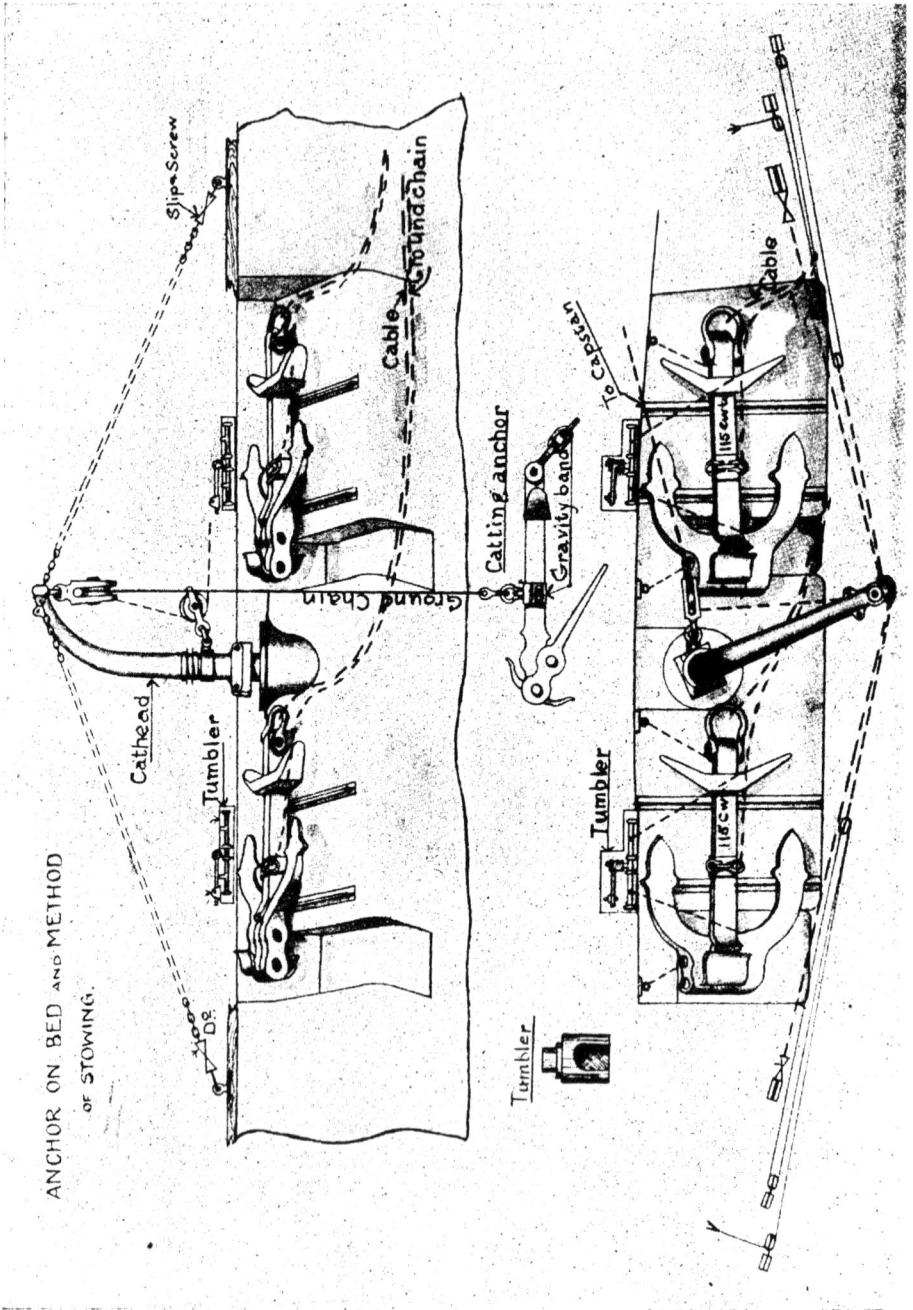

ANCHOR ON BED and METHOD of STOWING.

of the securing chains pass. The releasing arrangement consists of a hollow cylinder, one side of which is cut away. Thus between the holding and releasing position there will be about half a turn of the tumbler.

The anchor beds are built at a slope, the·anchors resting on short girders secured to the beds.

The anchors are hauled in or let go by the cable holders, there being usually two of these for working the three anchors, viz. two bowers and one sheet. The sheet cable does not connect

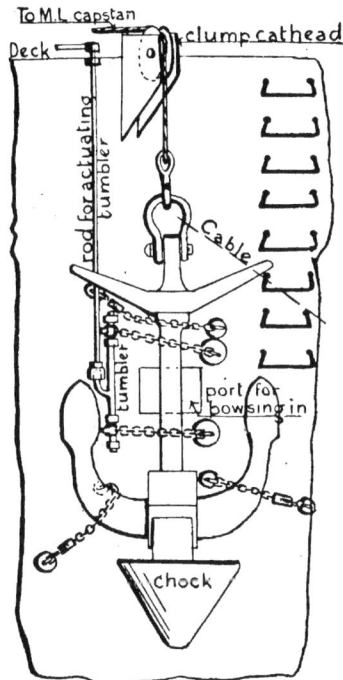

Anchor stowed vertically

to a cable holder, but to a riding bitt, though in some ships a third cable holder has been provided which is used for veering only. It is a dummy cable holder, fitted with a brake only, and not connected to the capstan engine. In addition to the cable holders, there is a middle-line capstan placed forward of them, which can also be used for hauling cable, and for this purpose portable rollers are fitted so as to guide the cable when working middle-line capstan, the lead of the cables in such cases being shown on the drawing. This capstan is provided with portable whelps, which are shipped when it is desired to work a hawser.

A clump cathead is provided in ships fitted with stockless anchors, to which the anchor is hung when ship is moored to a buoy, the cable being either passed out through the hawse pipes or over the large fairleads fitted at the bow.

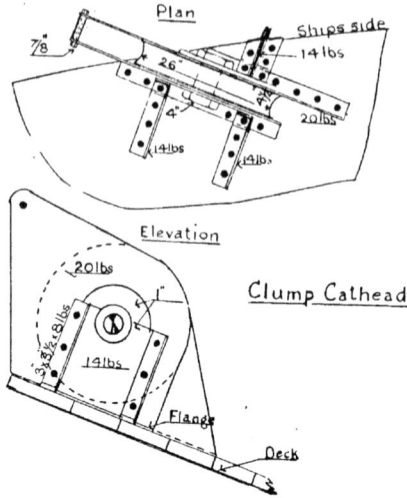

Clump Cathead

The deck pipes are shown under, and also the method of connecting them to the deck. The pipes project above the deck to prevent water getting below. Bonnets are fitted on the weather deck to the deck pipes, the vertical plate at the after end being

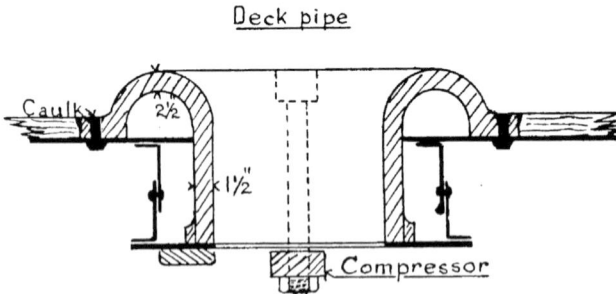

Deck pipe

slotted, so as to fit close down to the link of the cable and prevent a large quantity of water finding its way to the deck below.

On the side on which the cable is compressed a cast steel piece is riveted to take the rub of the cable.

The internal diameter of the deck pipe is equal to 8 times diameter of cable iron.

The cables are made up of 12½-fathom lengths, being 15 in number for each of the bower cables and 10 lengths for the sheet cable.

FIG. A.

A.S.—Anchor shackle.   S.S.—Senhouse slip.
J.S.—Joining shackles.   1—Long link.
S.P.—Swivel pieces.   2—Enlarged stud link.
C.C.—Cable clench.   3—Common link.

The method of assembling the different portions of a complete cable is illustrated by Fig. A, above.

In the case of stockless anchors a lengthening piece is required

Special Screw Slip.

FIG. B.

as the slip of the special screw stopper will not fit over the links in the swivel piece, the opening of the Blake slip being 1·1 times the size of cable iron, hence the slip must be placed at a common link.

To equalize the wear on the cable, the swivel pieces, inboard and outboard lengths of cable, are interchanged from time to time.

The details of the various slips used in connection with anchors and cables are given on p. 201, Fig. B, A being Blake stopper bolt, to which is connected the Blake slip B ; C is a Senhouse slip, and D a clear hawse slip. The proportions shown are in terms of cable·

The form of the hawse pipes will depend on the type of the anchor carried, that is, whether stockless or not. In the former case the whole pipe is in one casting, which has to be formed to ensure the anchor entering the pipe smoothly, and also being stowed in the pipe. In the latter case the pipes will be cylindrical, and generally consist of two castings, one on the deck, the other on the ship's side, with a steel tube connecting them. The pattern for this casting has to be mocked up.

GUARD RAILS AND STANCHIONS.

These are fitted round the edges of decks, bridges, etc. They are spaced about 8 feet apart, and are 3 feet 6 inches in height,

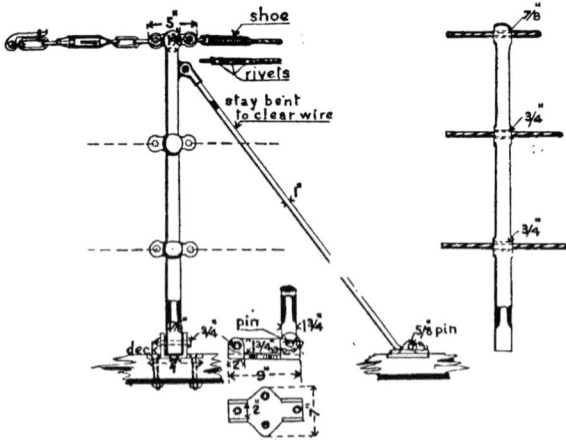

GUARD STANCHION.

and formed so as to be readily turned down when clearing for action. The heel fitting which permits of this being done is shown above, and is of stamped steel. There are two fastenings for the stanchion, one a bolt and the other a pin, and by removing the latter the stanchion can be turned down, and where possible in a fore and aft direction, with toes up.

The fastenings for the shoe (p. 202) are metal screws, but two fastenings should go right through the steel deck, with nuts on the point.

All guard stanchions and chains are galvanized.

At the gangways, and in way of the fairleads and bollards at the side of the ship, stanchions with stays are fitted to admit of the chains being set up, the chains having slips and screws on them for this purpose.

The guard rails and stanchions round compass platforms are of metal, so as not to affect the compass, the heel fittings being different to the ordinary ones, as they are not required to turn down (see p. 204).

All guard stanchions are placed with the heel fittings about 6 inches clear of the spurwater to provide a clear waterway.

BRIDGES.

These have to be constructed to stand the strains caused by the rolling and pitching of the ship. The construction is shown on p. 204, and consists of a platform of wood secured to a framework of angle bulb, the whole being supported by pillars. Diagonal stays are fitted to the end pillars and elsewhere, as shown, to take the racking strains.

The guard rails and stanchions to bridges are carried up higher than ordinary ones, to take the canvas weather cloths.

The chart house and shelter are fitted on the bridge, and the roof of the former serves as the standard compass platform.

No steel is used in the construction of the platform for the standard compass, but all plating, beams, etc., are of metal. This also applies to hand rails, stanchions, ladders, etc.

BOAT'S DAVITS.

On p. 205 is shown a pair of davits for a life cutter, with all the fittings and their various uses. The distance between the davits must be sufficient to permit of the boat being turned inboard between them. The griping bands are arranged diagonally, and the griping spars so as to be portable.

All davits in way of gun fire are hinged, so as to be capable of being turned down and stowed against the ship's side clear of the

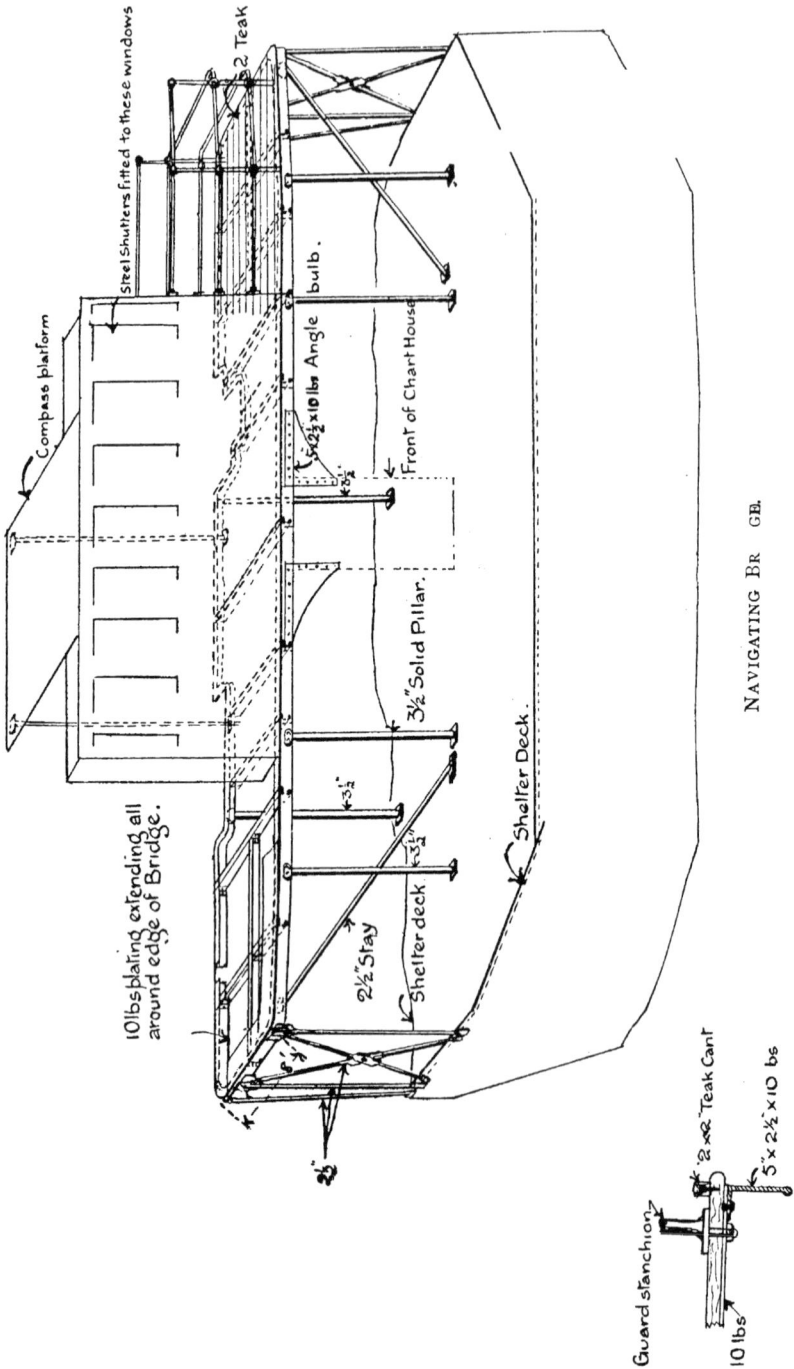

Compass platform

Steel Shutters fitted to these windows

2 Teak

4 2½×10 lbs Angle bulb.

Front of Chart House

3½" Solid Pillar.

3½"

5"

10 lbs plating extending all around edge of Bridge.

Shelter Deck.

Shelter deck.

2½" Stay

NAVIGATING BRIDGE.

2 ×2 Teak Cant

5" × 2½" ×10 lbs

Guard stanchion

10 lbs

deck, locking arrangements consisting of a short piece of angle bar

and a locking pin being fitted to the lower part, as in Fig. B, to
prevent the davit slewing while being turned down.

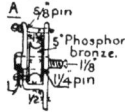

Fixed b

Fairlead

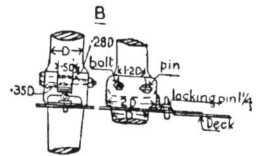

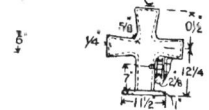

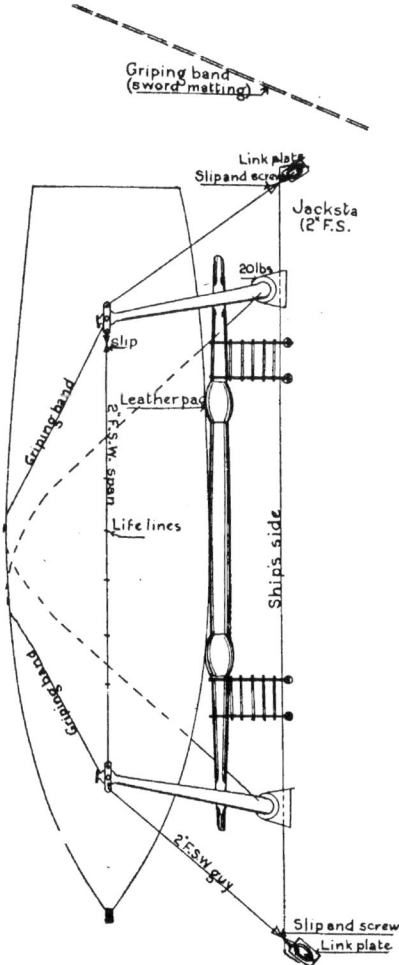

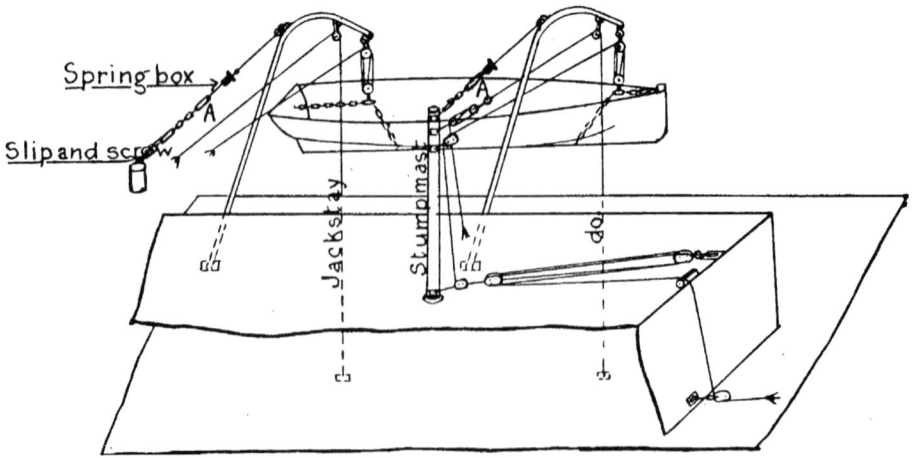

Spring box

Slip and screw

A

Jackstay

Stumpmast

do

do

do

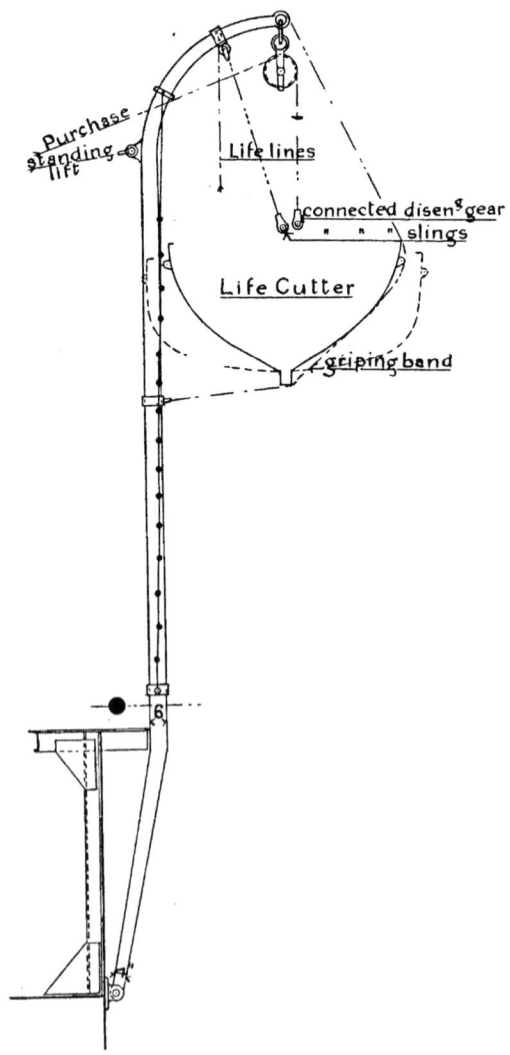

Purchase
standing
lift

Life lines

connected diseng gear

" " slings

Life Cutter

griping band

6

The suspending chain (Fig. D, p. 205) is for taking the weight of the boat when turning in and out.

Another arrangement of davits for a life cutter is one in which the davits have to be arranged not to obstruct the gun fire (p. 206). In this method the davits must be long enough to lower the boat clear of the ship's side, and permit of the boat being stowed in the sea, harbour, and stowing positions respectively. To do this a preventer stay is fitted to each davit, with long links in it corresponding to the sea and harbour positions, which connect to a slip and screw on the stump mast. To prevent undue strain coming on the stays when the ship rolls heavily a spring is fitted in each stay.

BOAT'S CRUTCHES.

These are the supports in which the boats rest, and are constructed as shown on p. 208. The lining or padding is of teak, and has to be carefully fitted to the shape of the boat, and is of sufficient thickness to permit of adjustments being made should a boat of slightly different section be carried subsequently.

The process of pricking down the boat is carried out as follows: the boat is plumbed on the boat slip or floor of boat-house, and moulds made at the positions of the crutches, the bevellings of these moulds against the bottom being taken at several points and marked on the moulds at the positions at which they were taken, the bevelling taken being that between the plane of the mould and the bottom of the boat in a fore and aft direction. While all the moulds are in position a line is sighted in on all of them, which enables them afterwards to be correctly pitched on the crutches. This line is marked "sighting line" on the drawing.

As the bevelling is considerable for the end crutches, it is necessary to leave sufficient wood on the lining when roughly sawing them out.

The moulds are made to the smaller section of the boat, as this will clearly correspond to the greater thickness of lining. Hence in pitching the moulds they must be placed on the side of the crutch which corresponds to the smaller section of the boat, that is, in the fore body on the fore side, and in the after body on the after side.

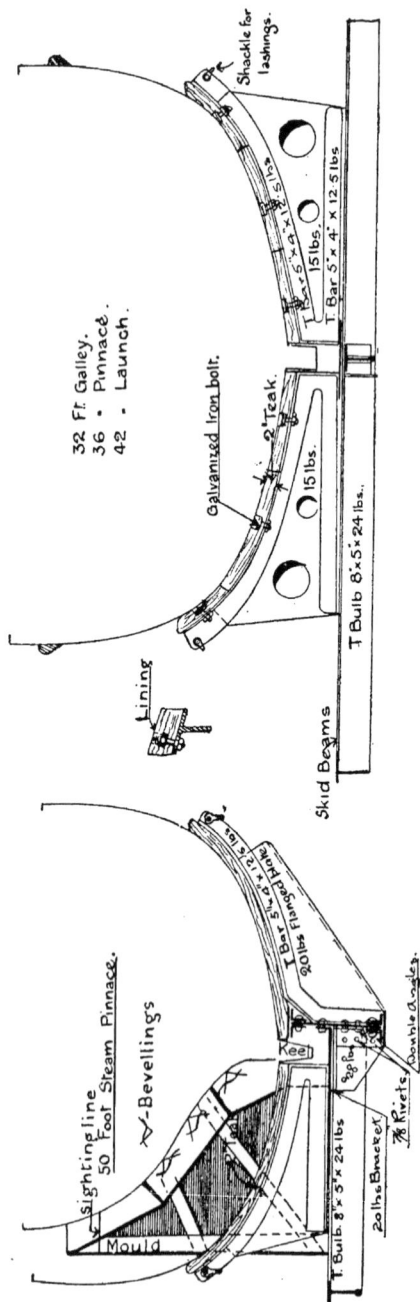

Sections shewing details of Crutches for Pinnaces and Launches

32 Ft. Galley.
36 " Pinnace.
42 " Launch.

Shackle for lashings.

15 lbs.

T. Bar 5 × 4 × 12·5 lbs

Galvanized Iron bolt.

2" Teak.

15 lbs.

T. Bulb 8 × 5 × 24 lbs.

Lining

Skid Beams

Sighting line
50 Foot Stream Pinnace.

Bevellings

T. Bar 5 × 6 × 12·5 lbs.
20 lbs Flanged plate.

Keel

Double Angles.

7/8 Rivets

Mould

T. Bulb 8 × 5 × 24 lbs

20 lbs Bracket.

The final adjustment of the lining is made by lowering the boat into the crutches and pricking off the boat with a pair of compasses, taking care that the points of the compasses are kept vertical, when the lining can be trimmed as necessary. To avoid

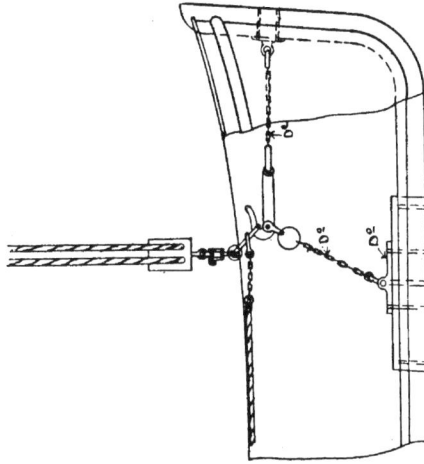

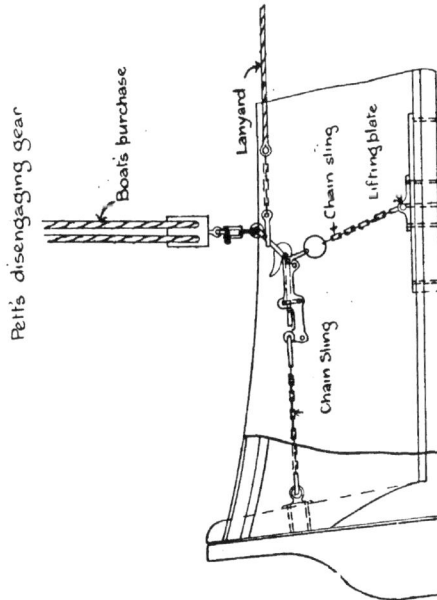

subjecting the boat to undue strains the lining requires to be carefully fitted.

The supports on which the boat's crutches are carried are called the skid beams, and are constructed of angle or tee bulbs, beams

P

and carlings, the latter being intercostal. The whole is stiffened up by fitting pillars to the beams from the deck below, and by strips of plating on the carlings. as shown in Fig. p. 208.

Fore and aft stays are fitted to the crutches to take the strains caused by the surging of the boat when the ship pitches, and shackles for lashing boat and boat's covers.

It should be noticed that the crutches are so constructed that the whole of the weight of the boat shall come on its bottom and not on the keel.

Boats which are stowed in crutches and lifted out and in by means of their own davits, and not the main derrick, have the crutches arranged to hinge down on one side, so that they shall swing clear when the weight is taken. In the case of boats lifted in and out by means of the main derrick fixed crutches are fitted, except in cases where the drift which can be obtained between the main derrick blocks and the boat is small, in which case hinged crutches will be necessary.

Disengaging apparatus, types of which are shown on pp. 209, 211, is fitted to lifeboats only.

SHEATHING.

The wood sheathing fitted on sheathed ships is often in one thickness, with a minimum thickness of 4 inches. A sketch is prepared of a development of the bottom on which the edges and butts are lined off, the former following the run of the plate edges as far as possible, the length of plank worked being not less than 20 feet. Using this sketch as a guide, the edges and butts are lined off on the bottom plating, and then the bolt holes are marked off. The butts are placed between the frames, and there are two fastenings in every frame space for each plank. No bolts should come on single riveted laps, but they can be placed on double riveted laps between four rivets. The holes are drilled through the plating, using the special square to ensure their being normal to the bottom, after which the planks are trimmed to the width and length as measured from the lines on bottom, hollowed out to fit the bottom if necessary, and put in place, allowance being made when trimming the edges and butts for the caulking seam. The holes are now marked on the plank, using the special tool, shown, after which the planks are taken down and the holes bored through with a screw auger, the recess for the head of the bolt being cut out with a ring engine. The plank is now ready for final erection,

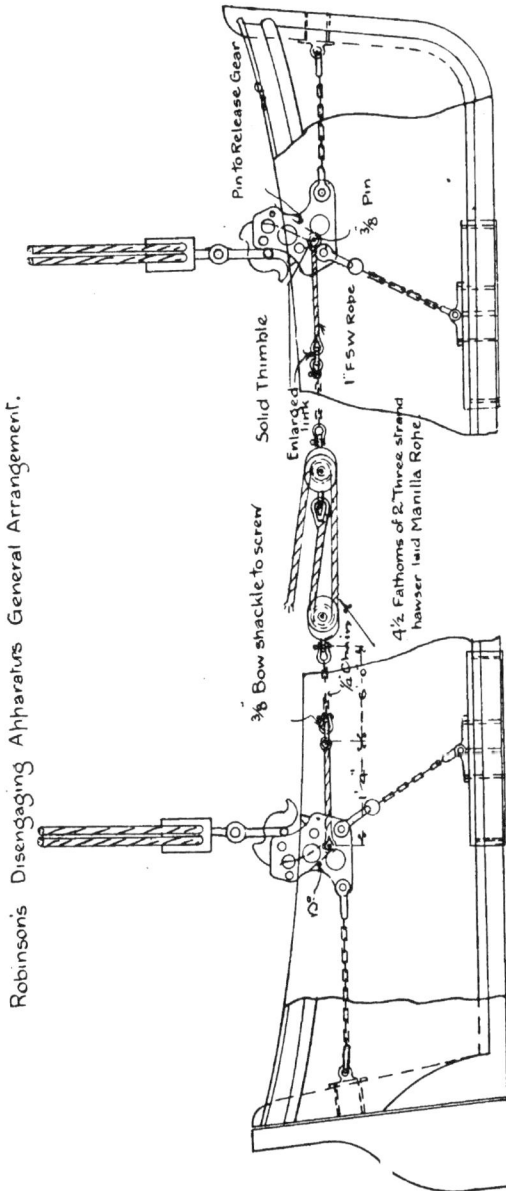

and when it is in position the holes in the plating are tapped, using a special tap which is passed through the hole in the plank. The sheathing bolt, which is of naval brass, is next hove in with a ratchet

Plating

Teak 4" over raised Strake

Copper sheathing with tarred paper under

⅛ Steel washer & Grommet

Method of using special centre bunch for marking planks.

Grommet

¾ N.B. Bolt

1⅝"

Plating

Point of Bolt upset with centre bunch marks

Portland cement

½ ⅝

Holes in plating tapped after plank is in place through the plank

Red lead injected between wood & steel

1⅝" Muntz Metal nails

Sheathing Composite Ship

Metal Washer

⅛ Copper

Frame

Grommet & Washer

Method of fitting Main & False Keels

Flat Keels

Red lead

1" Naval Brass

Copper

Copper clench

Main Keel Teak

False Keel, Elm

Metal Dumps

Copper

Taking account of a Closer

Length Batten.

Steater.

Plating

Taking bevel of Butt.

Batten for Taking length.

Trimming Closer.

Width

Faying Surface

Allowance for seam for caulk'g

Measuring Width

Taking bevel of edge.

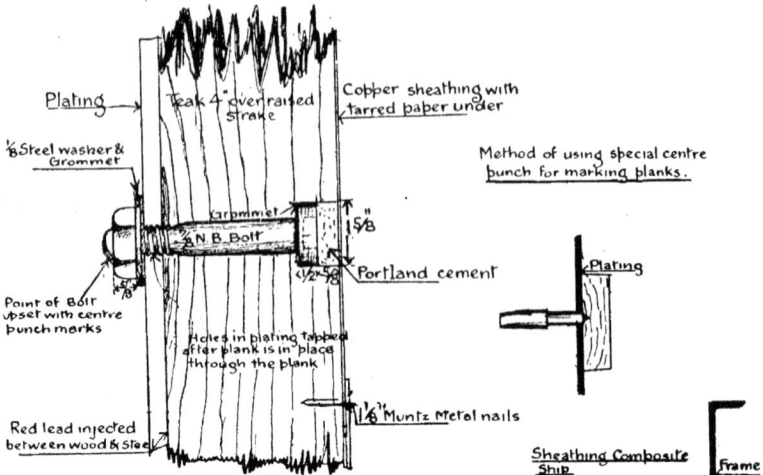

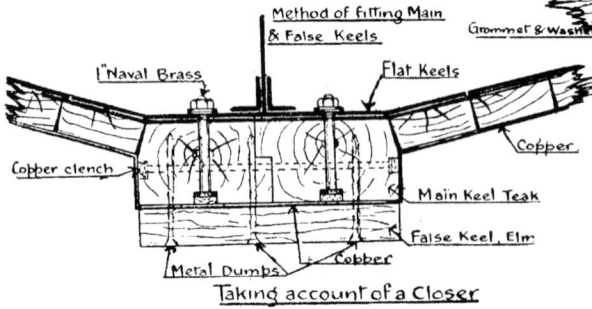

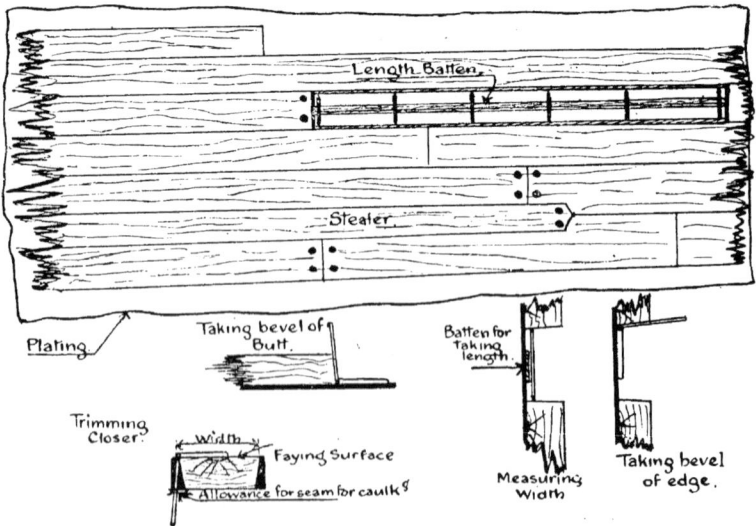

METHODS OF WORKING SHEATHING.

Hammer.

Ash Handle.

← − − 12" −−−−−−→

← ⅞ →

Punch.

½"
1⅜
← 2" →
5½"

Mallet.

Lignum Vitae

←−−−−−− 16¼" −−−−−−

Special centre punch

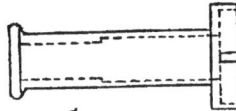

Gauge for testing bolts.

Lead Beater

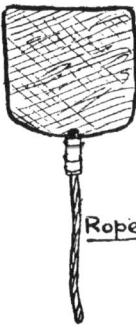

Square

Rope handle.

Gauge

Holes bored in wood
and afterwards plugged

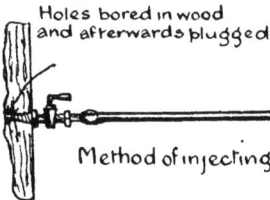

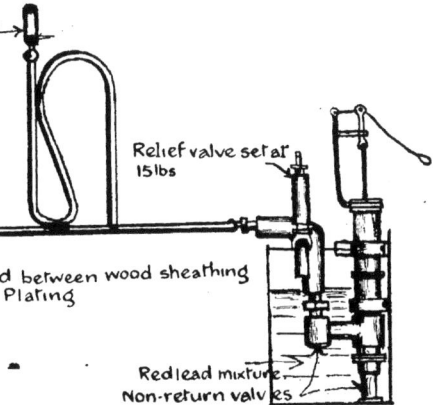

Relief valve set at
15 lbs

Method of injecting red lead between wood sheathing
and Plating

Red lead mixture
Non-return valves

brace sufficiently far to ensure the wood being compressed about $\frac{1}{8}$ inch under the head. A grommet and plate washer is placed on the point, with a nut to complete the fastening, and to prevent this coming off the point of the bolt is upset with centre punch marks. The space over the head of the bolt is filled in with cement, and the whole is now ready for caulking (p. 212).

Certain planks in every strake which complete the strake are termed closers, and the method of taking account of these is illustrated on p. 212.

A number of special tools are provided for working sheathing, some of which have been already referred to, and it only remains to draw attention to those for testing the size of the bolts and at the same time to ensure the shank and head being concentric, and the mallet for beating out puckers in the copper sheets when nailing them to the wood. A very useful tool for performing this operation is the lead "beater" shown, which consists of a piece of sheet lead about 1 foot square attached to a piece of rope.

Wood Caulking.

This must be very carefully done to ensure watertightness, and before driving in the oakum, wedges are placed in the seams to prevent the planks being set edgewise, and so straining the fastenings. The butts are first caulked, and then the seams, commencing with the narrower seams, as they might become too narrow for the caulking iron to enter if the planks happened to be set edgewise. The reason for caulking the butts first is to obviate the seam threads being injured by the caulking irons if the seams were first done. The ends of each thread of oakum is left exposed so that it can be readily seen whether the correct number of threads have been inserted. It should have been stated that any shakes should be caulked before commencing on the butts and edges, as by so doing the oakum is compressed in these, and so remedies as far as possible the defect. It might not be possible to caulk a shake at all if the edges were done first; and although they would close up they would not be watertight. The tools in general use are illustrated on p. 215, and consist of (1) caulking iron for driving the threads of oakum into the seams. (2) Crease or Meaking iron for dressing down the upper layers of oakum on

Spike Iron.

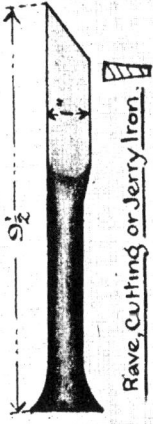

Rave, Cutting or Jerry Iron.

$9\frac{1}{2}$

1″

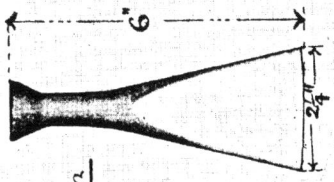

Crease or Meaking Iron.

6″

$2\frac{1}{4}″$

$\frac{1}{8}″$

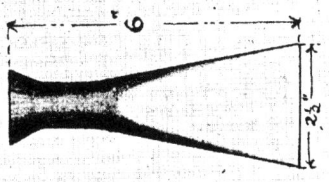

Caulking Iron.

6″

$2\frac{1}{2}″$

$\frac{1}{16}″$

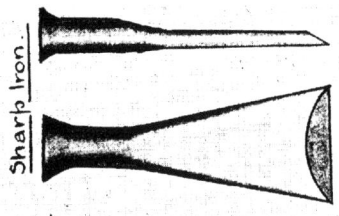

Sharp Iron.

Reeming Iron.

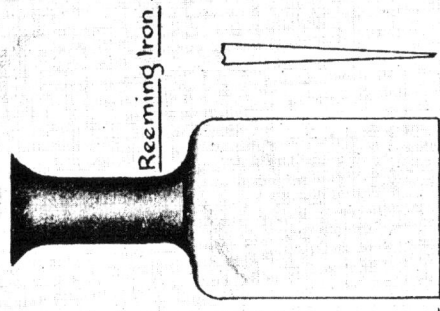

Double Crease
Single
Blind

Horsing

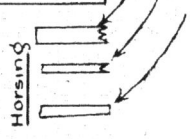

Horsing Iron.

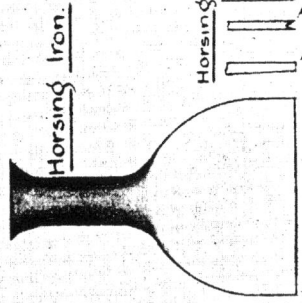

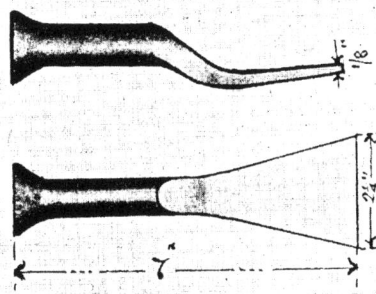

Crooked Iron.

$\frac{1}{8}″$

$2\frac{1}{4}″$

7″

thin decks. In the case of thick decks the Horsing iron would be used. (3) Crooked iron ; this is used in place of the ordinary caulking iron where there are any obstructions. (4) Spike iron is used around sharp corners, *e.g.* hatchways, shoes of guard stanchions, etc. (5) Sharp iron is used for cutting out defective threads. (6) Jerry iron is for removing threads of oakum ; the iron has a tapered section so as to clear itself when being driven along the seam. For caulking a shake and "piecing" planks a similar iron to the caulking iron, but with a finer driving edge, is used.

The number of threads in the seams depends on the thickness of plank, it being usual to allow one thread for each inch of thickness of the plank. The first thread is laid straight and the remainder coiled. The oakum is finished off by horsing down with a crease iron and beetle, leaving a depth of about $\frac{1}{2}$ inch for the pitch. After the whole of the wood sheathing has been finished holes are bored at intervals through it, and red lead mixture, composed of 2 cwt. of white lead, 1 cwt. of red lead, mixed with 3 galls. of linseed oil, is injected between the wood and the plating, the holes being afterwards plugged. To prevent the planks being forced away from the plating when injecting, the valve on the force pump is set at 15 lbs. per sq. inch, and when this pressure is reached the mixture simply flows out through the cock shown into the containing vessel.

The whole surface is now tarred, and on this tarred paper is placed, the sheathing being completed by the attachment of the copper sheets. These are secured by Muntz metal nails, the edges and butts of the sheets being lapped, the foremost lapping over the next abaft it, and the lower lapping over that next above it. The holes in the sheets are made with the special punch shown, the impressions for the nail holes being already marked on the sheets. The nails are driven at the centre portion of the sheet first, then working outwards, thus making sure of removing any puckering. The copper sheets are 4 feet long by 14 inches wide, and are 28 ozs. per sq. foot. Thicker copper, 60 ozs. per sq. foot, is used where the rub of the cable occurs, round the edge of the bilge keel, and at the wash of the propeller, these being secured with screws.

PITCH.

This must be of good quality, and the tests it is subjected to are given later ; but a rough test is carried out by burning a sample and seeing if there is any ash left, and also noting if it is brittle when cold. The brittleness is tested by means of a pricker driven into the pitch in the seams, which operation should not cause any of it to be chipped out. Pitch may be softened by the addition of tar, or, if this cannot be got, by oil or tallow.

PICKLING PLATES.

Steel plates as received from the manufacturers have mill scale on them, which must be removed before they can be coated with paint or other material, otherwise when the scale falls off bare places will be left, causing corrosion.

This scale is removed by " pickling," the plate being stood on end in a hydrochloric acid bath (19 water, 1 acid) for about 12 hours, and then removed and well washed with water. Unless plates are thoroughly washed after pickling the acid continues its action, and prevents the paint or composition getting thoroughly into the plate and adhering to it. In such cases if the paint be removed the plating will be found to have a damp feeling.

The pickling bath is placed conveniently near the plate racks, and should be deep enough to take the widest plate.

GALVANIZING.

All fittings exposed to weather or damp are galvanized, that is, covered with a thin coating of zinc, to prevent corrosion. Before galvanizing is done all paint, etc., is burnt off, and the articles placed in a bath composed of hydrochloric acid one part and water forty parts to remove rust, grease, etc. They are finally immersed in a bath composed of acid one part and water three parts, and put in a warm place to dry.

After immersion in the zinc bath they are hoisted up clear of the zinc and allowed to drain, superfluous zinc being got rid of by sprinkling with sal-ammoniac.

The zinc bath is covered with a sheet of iron at night, to prevent loss of heat, and the zinc is never allowed to set fast in the bath.

The following fittings are galvanized : guard rails and stanchions, coaling rims, covers and gratings, ladders and footsteps, lockers for wash deck gear, basins and troughs and racks for storage of clothing in washplaces, mess racks, fittings on mess stools and tables.

The additional weight due to galvanizing by the hot process is from $2\frac{1}{4}$ to $2\frac{1}{2}$ ounces, and by the electrical process about 1 ounce, per square foot of exposed surface.

LAYING OFF AND MOULD PRACTICE.

PROCEDURE ON RECEIPT OF BUILDING DRAWINGS.

The building drawings comprise the following : Sheer draught, plans of decks, sections, sections showing construction, and a table of "offsets." The sheer draught gives the geometrical form of the ship by means of three plans, namely, sheer plan, halfbreadth plan, and body plan. The sheer shows the longitudinal elevation of the ship, position of bulkheads, frames, decks, etc. ; the half-breadth plan, of water-lines and beam end lines ; and the body plan, the shape of transverse sections of the ship. The length of the ship is given " between perpendiculars," the fore perpendicular being the vertical line through the intersection of the load water-line and fore edge of stem, and the after perpendicular the axis of the rudder. This length is divided up into an even number of equal spaces, usually twenty, giving twenty-one stations called *displacement stations*, and the body plan shows the form of the ship at each of these stations. For convenience, half the number of stations is shown on each side of the middle line of body plan.

The first work to be done is to fair the " body." This may appear unnecessary at first sight, as the sheer draught has already gone through this process ; but in going from a $\frac{1}{4}$-inch scale to full size errors inappreciable on the small scale will become apparent when increased to full size. For this reason the table of offsets is given, as the displacement has been obtained from them and errors would creep in if they were scaled from the body plan.

First set off middle line of body and then the water-lines at the distance apart given on the table of offsets, and set off along the water lines in turn the " offsets," numbering each spot to agree with the displacement station to which it belongs. On attempting to get a flexible batten to pass through all the spots on a curve it will become apparent why fairing has to be done, as it will be found that the batten will not pass through all the spots and at the same time give a fair curve, and it will be necessary to give and take, passing inside some and outside others, as shown by the tendency of the batten. The batten is kept in position by floor nails, and in order to see if there is any edge set the nails are removed at intervals, when it will be found that the batten will spring slightly and must then be secured in its new position. When adjusting the batten to give a fair curve, no large departures from the spots should be permitted or the displacement will be altered.

Having copied the body plan *full size* on the floor, the half-breadth plan is next drawn. In a long ship a floor of great length would be necessary, and to get over this the displacement stations in the halfbreadth are usually spaced one-fourth their actual spacing ; but the offsets from the middle line, to give the water-lines, will be *full size* (see p. 220). This scheme will have the effect of increasing the curvature of the water-lines, but this is an advantage, as a flexible batten gives a better curve when bent than when nearly straight (see p. 220). It is found possible to fair practically the whole length of the ship by this "contracted method" as it is termed, leaving only the endings of the lines on stem and stern castings to be carried out by the ordinary method. It is necessary to do these parts by the ordinary method, as we require the exact shape full size of the sections of the stem and stern castings for ending the water-lines on and making the patterns for these castings.

In order to have no discontinuity where the two methods of fairing overlap, each must be continued for a few stations on either side, to ensure that at the station where the junction occurs the same spots are maintained.

When arranging the scale of the contraction it is advisable to choose such as will give a whole number of feet and inches for the frame spacing, as the introduction of fractions of an inch may lead to confusion.

Halfbreadth Plan. By means of a staff of section thus $\triangle$ measure along each water line on the body to where a displacement station crosses it and mark this on staff thus : $\dfrac{4 \text{ W.L.}}{1}$, $\dfrac{4 \text{ W.L.}}{2}$, etc.

Transfer these to halfbreadth and mark off along corresponding stations. As in case of body plan, draw fair curves, keeping as close to the spots as possible. Rub out body plan and make a new one, copying from the halfbreadth and continue the process until the two plans agree, giving fair curves in both. A further

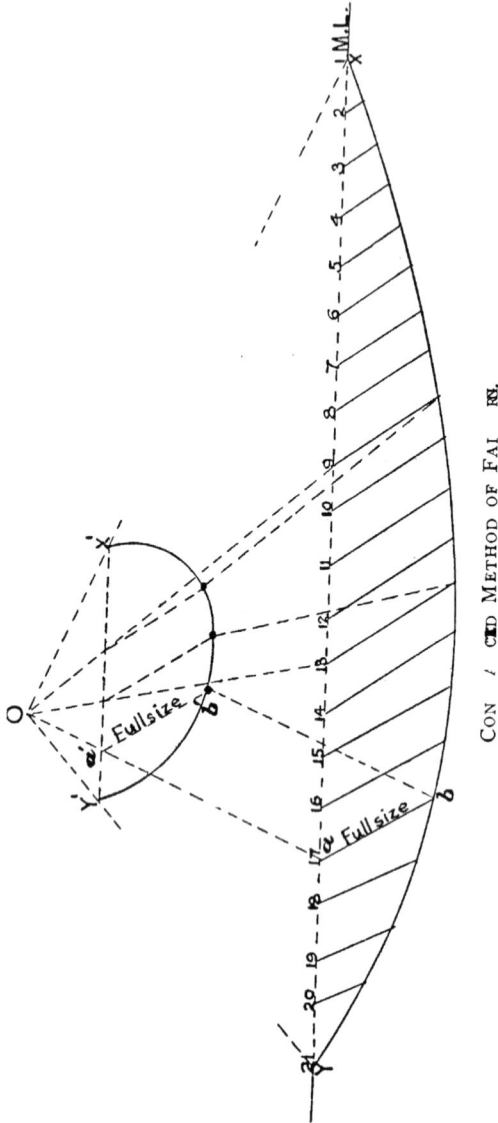

check is now put on the fairness by means of *diagonal lines*, which are lines drawn on the body plan to represent the traces of planes inclined to the sheer plan and at right angles to the body plan. These lines are drawn as nearly normal to the curves in body as possible, as by so doing we get the best results, the intersection of two lines being most definite when they cut at right angles.

To lay off a Diagonal. Measure on the body from the intersection of trace of diagonal with middle line to each station *along the trace*, and set off these along the corresponding stations in the halfbreadth, and draw fair curves as before. Here again it may be necessary to depart from the spots to obtain fair curves, and this will, of course, involve altering the corresponding points in the body plan.

There only remains now to put in a few *bow and buttock lines* at the ends of the ship. These are the intersections with ship's surface of planes *parallel to the sheer plan*, and being more nearly normal to the surface at these parts than the other planes, give a

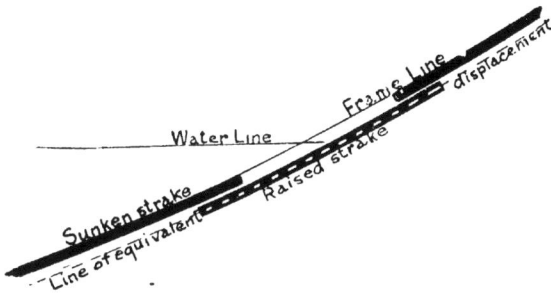

better check. When all the plans agree the ship is said to be " faired."

Equivalent Displacement Line. It has been stated that the spots must be adhered to as closely as is consistent with fairness, as shown by flexible batten; to keep near the displacement shown on the sheet, consequently all the lines dealt with so far must take account of the *total displacement ;* that is to say, they must include the displacement of the skin plating. Now with plating worked on the raised and sunken system, to obtain the frame line, which is the shape of the side of frame in contact with the inner strake of bottom plating, $1\frac{1}{2}$ times the thickness of the bottom plating must be deducted from the lines on the floor normal to the various lines. This line, shown ticked in the above figure, will clearly pass through the middle of the raised strake. For this reason the lines on the floor used so far are called *equivalent displacement lines*, and to obtain the frame lines measure in at each water-line $1\frac{1}{2}$ times the thickness of bottom plating at that particular place normal to the curve to cut the water line and then transfer these frame lines to the halfbreadth. A correct account of the surface of frames has now been obtained, and as the shape of every frame is required, these are measured from the sheer draught and set off in the halfbreadth on the contracted scale. It is only necessary to measure along each frame in the halfbreadth to where it intersects the several water-lines and diagonals, and set off these distances in body plan and draw in the curves with flexible battens. The frame lines are then scrieved in on the body only, all the lines dealing with the displacement stations and equivalent displacement lines being rubbed out.

When the body is fair the scrieve board can be got on with, the copying being done by taking off on battens the measurements along water-lines and diagonals, using bow and buttock lines in addition for the portion below the lowest water-line.

HALF-BLOCK MODEL.

The half-block model is made to a scale of $\frac{1}{4}$ inch or $\frac{1}{2}$ inch to a foot, depending on the size of the vessel, and represents one-half of the ship, and is for the purpose of obtaining a complete account of the shell plating to enable demands for this to be prepared at a very early stage of the work. As the model is on such a small scale, it can be prepared from the building drawings without waiting for the fairing of the body, since the differences which may be found would not be appreciable on the small scale on which it is prepared. The model is sometimes cut out from the solid, and at others the ends only are made solid, the intervening portions being planked as in a boat, the planks being attached to section moulds made at each displacement station. The following lines are drawn on the model:—Frame stations, displacement stations (for lining-off only), longitudinal sight edges, beam end lines, edges and butts of bottom plating, bilge keel, shaft swell and recess, all doubling plates, hawse pipes, sidelights, bow and stern protection (if any), additional framing in engine room, outline of stem and stern castings, shaft brackets, torpedo tubes, and L.W.L. (see p. 223). The model is secured to a base board,

and is so placed on it that the load water-line is parallel to the edge of the base. The frame stations are first drawn in by the aid of a sliding frame as in Fig. p. 223, which shows the fore part of the model only for clearness. The frame is arranged so that its plane is perpendicular to the edge of the base, and therefore to the load water-line, and a batten with a drawing pen attached to its end is moved round this frame as shown. The longitudinal sight or outer edges are next drawn in as follows :— Using the body plan on the building drawing girth round from the middle line at each displacement station to where the sight edge cuts, and transfer this to the model, obtaining in this way a series of spots on the model for each sight edge. A thin batten is then taken and placed with its edge passing through the points, the batten having no edge set ; but it will generally be found that it will not be possible to adhere to the spots, and it will be necessary to depart from some of them in order to obtain a fair curve. This curve is then transferred to the full-size body, when again it will generally be found that it will not be possible to adhere to the spots and at the same time obtain a fair curve, and the process must be repeated until the curves on the model and the floor agree. The longitudinals are run into watertight flats and platform where possible, so that a continuous girder, extending from end to end of the ship, is obtained.

The next lines to be got in are the plate edges, and these are shown on the midship section, and are arranged as follows : Commencing at the middle part of the ship, divide up the girth from edge of flat keel plate to armour shelf, so that the plates are as nearly as possible of equal width, at the same time keeping well clear of the longitudinals. Referring to the figure, it will be seen that two plate edges occur between any two consecutive longitudinals, and the width of plate will be about one-half the distance between these. It must be borne in mind that there is a limit to the size of plates which can be obtained, and this may have a bearing on the width of plate when lining off those plates coming above the longitudinals, that is to say, the strakes immediately below the armour shelf.

Since the girth decreases as the ends of the ship are approached, if there were the same number of strakes at the ends as at the middle portion of the ship their width would be very small, and some must consequently be stopped short of the end. In deciding where this shall. occur for the various strakes, one consideration is the width and the other is that they shall not all end in the same

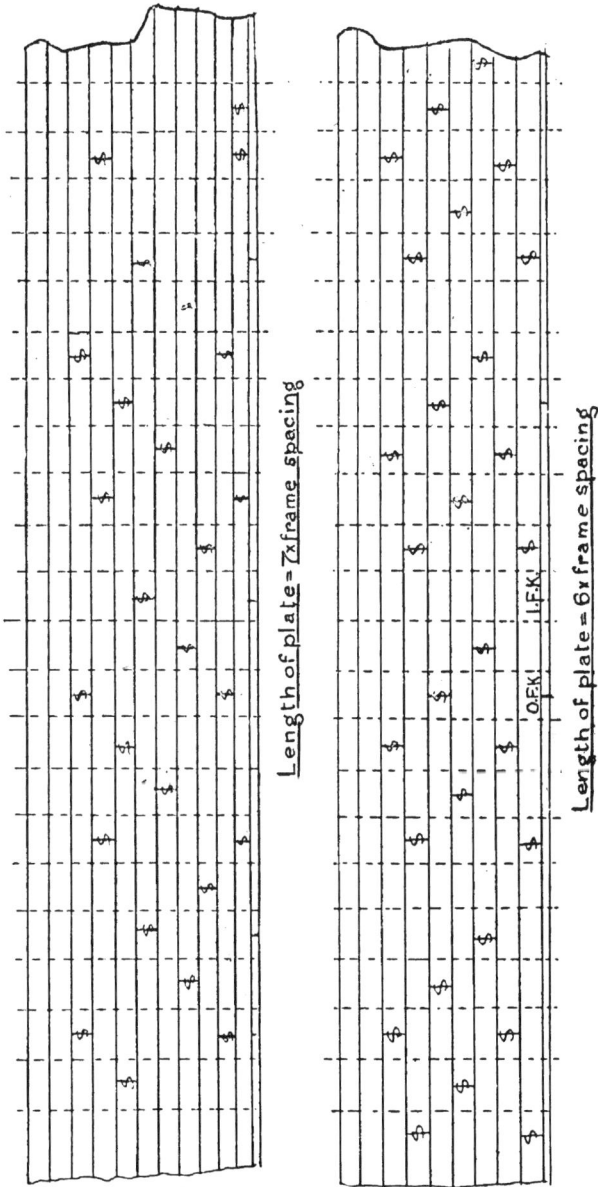

frame space, as this would give a weak section across the ship, hence it is necessary to stop the strakes at different frame spaces. This process is known as working "stealers," and several methods of working them are given on pp. 37 and 39.

It will be found that the plate edges and longitudinals cross one another as the ends of the ship are approached, and the former are diverted slightly, and run across the latter at right angles to ensure watertightness at these places, which would be difficult to arrange if the two cut obliquely. This is known as "joggling" the plate edge, and is arranged as shown on p. 223.

Shifting the Butts. This is the process of marking off the positions of the butts of plates, and the object aimed at is to get the greatest longitudinal strength. To ensure this, the butts of the plating occurring in any frame space must be as far apart as possible, that is to say, as many intervening strakes as possible must be obtained. This will depend on the length of plate which can be obtained, the length of the bending rolls, and also the spacing of the frames. The butts come in the middle of the frame spaces, consequently the length of plate must be some multiple of the frame space; for example, with frames 4 feet apart and bending rolls, say, 25 feet long, the length cannot exceed 24 feet. This case permits of 4 passing strakes between butts occurring in the same frame space, and the disposition is shown on p. 225. The case where the length of plate is seven times the frame spacing is also shown. A further condition for the arrangement of the butts is that butts in adjacent frame spaces must be at least two frame spaces apart, thus:

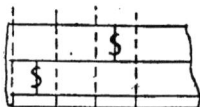

Stepped butts, as the following, are not permissible ·

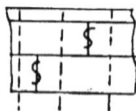

The shifting of butts is commenced by copying from the sketch of middle line work the positions of the butts of flat keel plates on to the model. It is very convenient to have as many of the bottom plates at the flat portion of the ship of the same dimensions, as this saves work in preparing the demands and subsequent lining-off of the plates. In preparing the middle line sketch, the butts of longitudinals must be properly shifted with respect to the butts of bottom plating, to avoid weakening the ship, the same also holding for the relatively large holes in the non-watertight longitudinals for lightening these and for providing access to double bottom spaces. After all the butts have been disposed, the strakes are lettered, commencing from the garboard strake, with the letters A, B, C, etc. Each plate in a strake is numbered, and the dimensions marked on each, *e.g.* a plate in the garboard strake would be shown : 24' 1" × 4' 6" × 25 lbs. O.B. A2; that is to say, length 24 feet 1 inch, width 4 feet 6 inches, weighing 25 lbs. per square foot, and plate number 2 in strake A. The lengths of all plates are taken from the model, and the widths from the full size body on the floor. It will be clear that the bare measurements would not be sufficient to allow of any small adjustments being made when working it, and so it is necessary to allow a certain amount on the length and width of each plate. This allowance varies according to the position of the plate, it being about an inch on the length and width for plates at the middle portions of the ship, increasing to two inches on the width and four to five inches on the length for plates at the other portions of the ship. The amount allowed in each case is settled as the plate is taken account of. Plates which are required to be furnaced are ordered of increased thickness, generally of about 2·5 lbs. to 5 lbs. per square foot.

Demand Forms. These forms give particulars of every plate, with the marking that is to be put on it by the contractor before despatch from the works. As these particulars are the same on the model, it is easy to pick out from the racks any desired plate.

Certain plates are not of rectangular shape, and further particulars of these must be given ; for example, the width of some is not the same at both ends, and such would be entered thus : $24' 6'' \times \dfrac{4' 6''}{4' 3''} \times 20$ lbs. O.B. D5 "taper one edge," or "taper equally," as the case may be. Other plates require a small sketch to be given on the demand form to completely describe them, such being called "sketch" plates. The reason for giving detailed information of such plates as the preceding is to save waste of material, which would result if all were demanded of rectangular form.

The plates on receipt are all stowed in racks, each plate as it is stacked being checked by aid of the demand form, and particulars inserted on the form as to the number of the rack in which it is stowed, so that at any time it can readily be picked out, and when removed struck off.

A proper system of racking material is important, as much time and labour may be lost if plates cannot be readily obtained from the racks

Laying off a Longitudinal.

The longitudinal will be shown on the body plan by its sections with the frame stations and inner and outer bottoms, that is, a series of lines inclined to one another and two curves as Fig. A, p. 229.

Before describing the process it will be advisable to refer to Fig. B, p. 229, which is intended to show in perspective the method adopted. We take the surface which the longitudinal represents, and place between three planes, two of which are parallel and at right angles to the remaining one, and imagine the surface continued sideways (by producing the straight lines or rays) to cut the parallel planes, which it will do in curves.

Now take these curves and straighten them out, that is to say, transform these curves into straight lines, and along them mark off the positions at which the ends of the rays in the longitudinal surface come, and so get the "development" of the surface between the parallel planes, and that of the actual longitudinal itself is formed by measuring from the parallel planes to the two edges of the longitudinal.

Thus the process is :— Draw a line AB parallel to the middle ray—this is for convenience only, and is not essential—and set up lines AC, BD at right angles to AB. These three lines will represent three planes mentioned above, and must be imagined as standing at right angles to the paper. Now produce the rays as shown to cut AC and BD in points a_1c . . . b_1d . . . respectively, and mark off on a batten the distances Aa_1, Ac_1 . . . and Bb_1, Bd_1

Draw two parallel lines along the floor the distance AB apart, and set off along them the frame stations 60, 62, etc. Measure the distances Aa_1, Ac_1, etc., and set these out on the frame stations 60, 62, etc., pass a curve through the points a_2, c_2, etc. Pen a

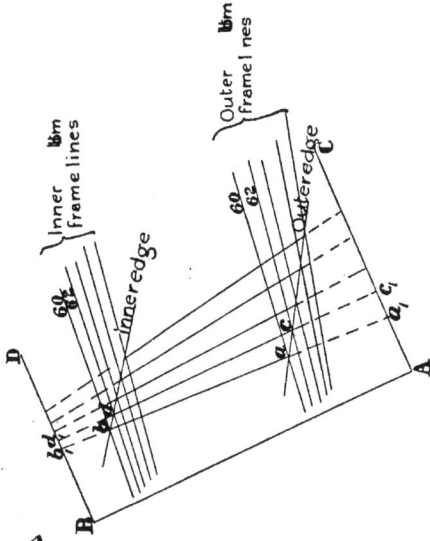

FIG. A.

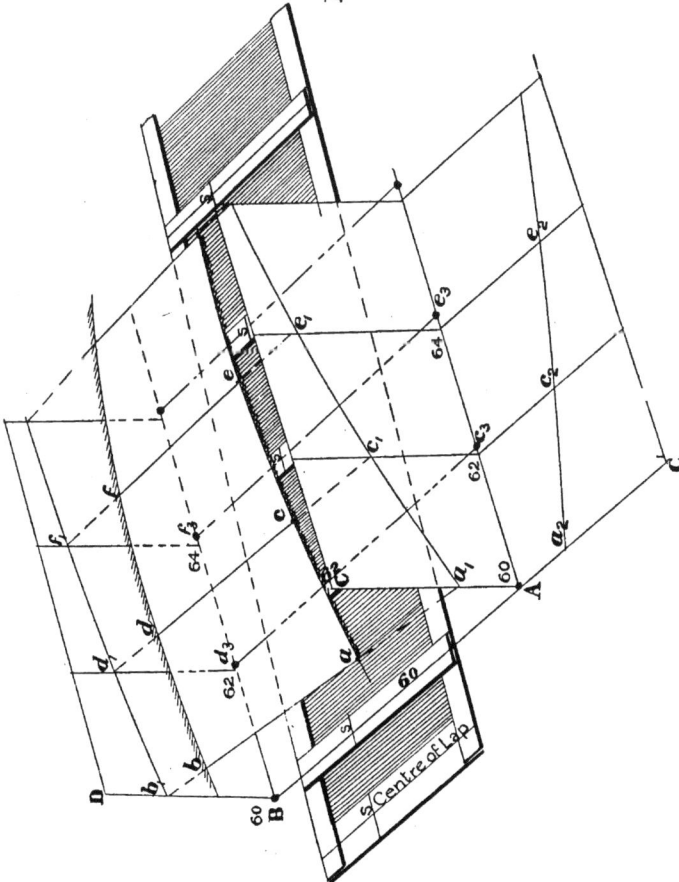

FIG. B.

batten to the curve $a_2c_2e_2$ (the plane AC' representing the plane AC rabetted on to the floor), and mark on it these points, then place a_2 to coincide with A, and let the batten spring straight along the line Ae_3 and transfer the points marked on it, thus obtaining the points c_3, e_3 . . .

Proceed similarly with the curve $b_1d_1f_1$, obtaining the points d_3, f_3 . . . These points will represent the ends of the position of the rays when developed or, as it is termed, the developed frame station. It only remains to get the developed shape of the two edges of the longitudinal, which is done by measuring the distances aa_1, cc_1 . . ., also bb_1, dd_1, and set these along their respective developed frame stations.

It is important to notice that in doing the work frame lines have been used, so that if the longitudinal comes on a raised strake the thickness of the bottom plating at this place must be set off from the lower edge of longitudinal to get the correct depth of the longitudinal.

The mould is made to the points obtained, cross battens being placed at the developed frame stations and at the butts of the longitudinal. The developed frame stations are marked on the mould, also position of the centre of lap, and before removing the mould a straight line SS . . . is struck along it to provide a check on the form of the mould.

The following method is sometimes adopted where there is not much twist in the longitudinal :—

Let *abc*, *ade* represent trace of longitudinal on outer surface of ship, and *fgh, fkl* the trace of inner edge, on p. 231. Draw base line AB and set off the frame stations 1, 2, 3, 4, etc., at their proper spacing and erect perpendiculars at these points. In the body starting from *f* draw *fm* perpendicular to *af* and *mn* perpendicular to *bg* to meet *ch*, produced if necessary in *n ;* repeat same for after body starting from *f.* From base line AB on No. 2 station, set down distance *fm* on No. 1 a distance equal to *fm* + *mn*, and repeat same process for stations 4 and 5 ; pen a batten to these spots giving curve *aa*. While batten is thus penned transfer the spots on 1, 2, 4, and 5 to the batten, and let batten fling straight along AB, keeping it fixed at 3, and transfer the spots to line AB and erect new ordinates through these spots, which will give the expanded stations as indicated by ticked lines. Starting from 3 set off on 3 f_1a_1 equal to fa, and on 2 (expanded) set off m_1b_1 equal to mb, and on No. 1 (expanded) set up n_1c_1 equal to nc in the body, and similarly for stations 4 and 5, then a curve through $c_1b_1a_1d_1e_1$ will give true shape of outer edge of longitudinal, and the inner edge can be obtained by setting off on expanded stations the

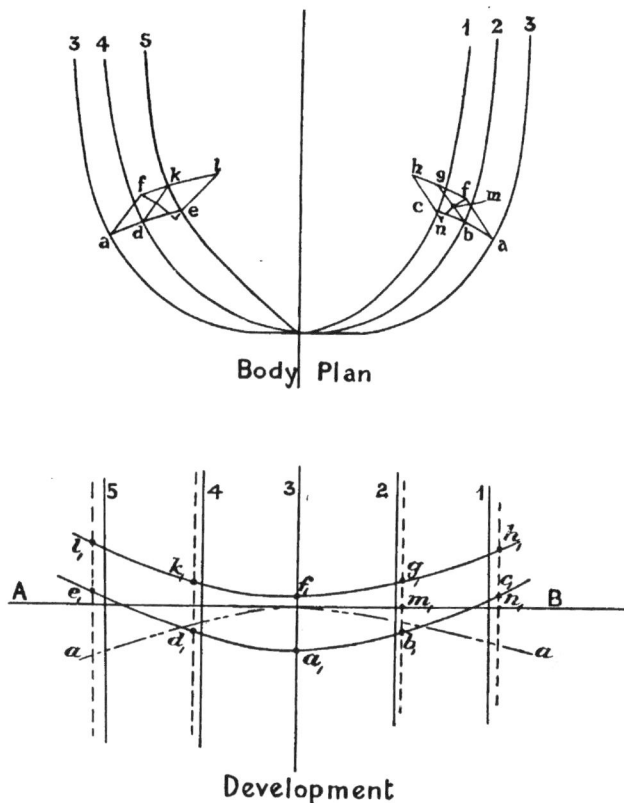

Body Plan

Development

depths of longitudinal at the several stations, giving the curve $h_1f_1l_1$ and the space between used for making moulds.

The laying-off method is only reliable when there is not much twist in the longitudinal, that is, when the lines ab_1cd . . . are nearly parallel. This will be clear on reference to the perspective diagram, as in obtaining the mould the projected length c_3d_3 has been assumed equal to the true length c_1d_1. When this is not the case, either the mocking-up method described below or that on p. 234 can be adopted.

Hurdle moulds are made at each square station from any convenient base line, the movable batten is then set to the section of

the longitudinal as in figure, and the inner and outer edges of the
longitudinal marked on it. These moulds are then erected on
the floor at their respective square stations, being correctly pitched
by means of a straight line square to the base line which is marked
on the base of hurdle moulds. To avoid making the hurdle moulds

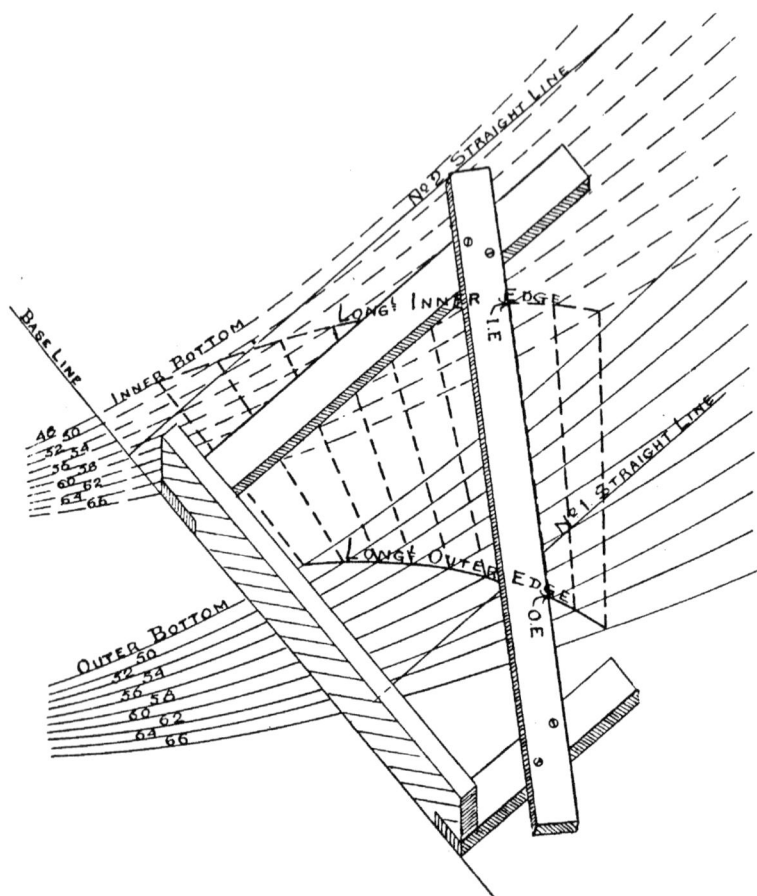

excessively large, two straight lines are used as pitching lines.
Battens are then attached to the top of hurdle moulds so as to pass
through the points giving inner and outer edges of longitudinal, the
battens being so placed that the mould for the longitudinal can be
made between them. In the case where the outer edge of the
longitudinal comes on a raised strake the thickness of the bottom
plating at that part must be allowed for and the lower batten set
out to these points. The mould is then made as in the Fig. p. 233,

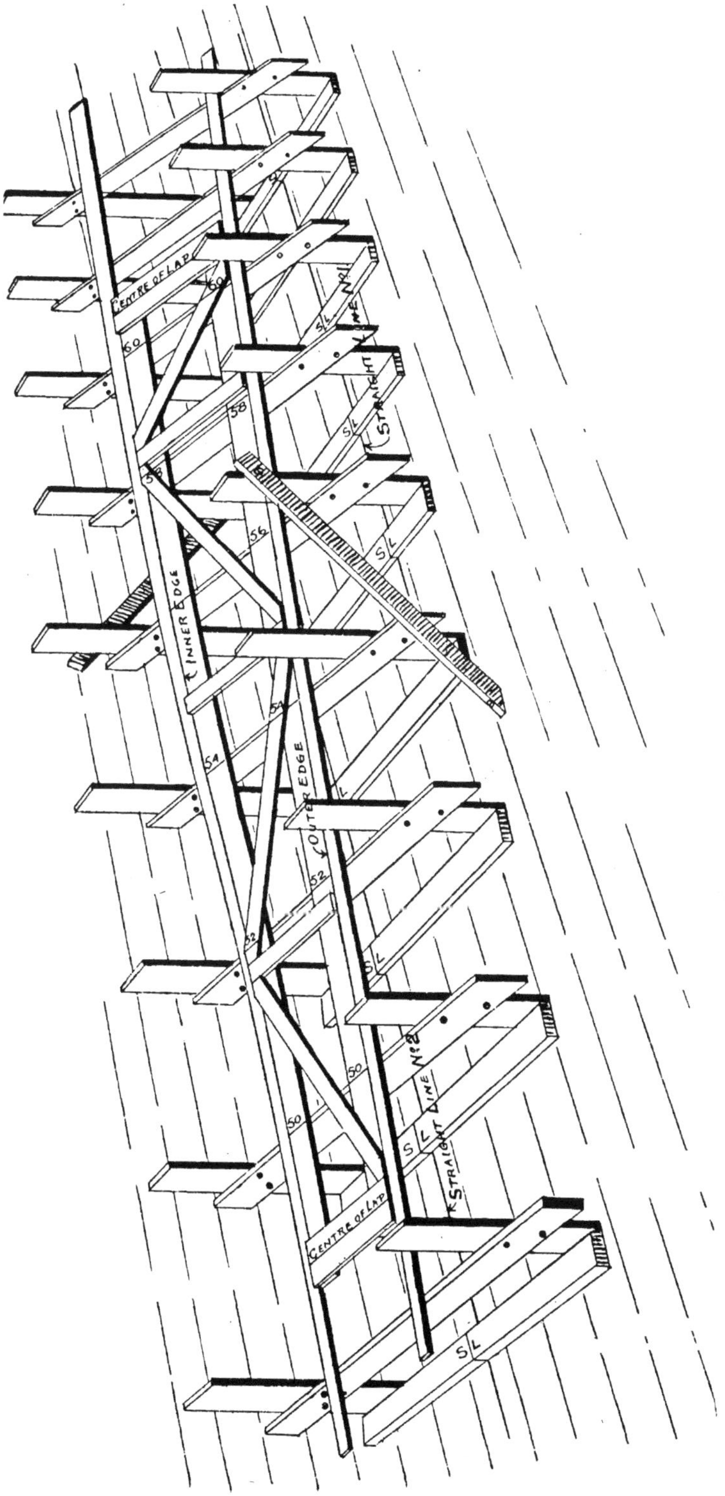
CENTRE OF LAP
60
58
56
INNER EDGE
54
OUTER EDGE
52
50
CENTRE OF LAP
STRAIGHT LINE No. 1
STRAIGHT LINE No. 2
SL

the position of laps and frames being marked on it. A straight line
is also struck across the mould and the bevellings for the angles
at inner and outer edges placed upon it.

Laying off a Longitudinal with Considerable Twist.

Referring to p. 235, Fig. A shows a portion of a longitudinal
with considerable twist.   Produce the rays aa', bb', cc', which will
pass through the same point marked 1.   Proceeding to the ray dd'
it will be found not to pass through the point 1, but intersect the
previous ray in the point 2, through which point the next ray will
also pass.

Thus a series of points marked 1 to 7 in Fig. A are
obtained, through each of which a number of rays will be found to
pass, not necessarily an equal number through each point. This is,
of course, a process of trial.   Next draw a base line as in Fig. C,
and mark off along it the frame stations as 100, 102, etc., and
erect perpendiculars to the base line.   Then in Fig. A place a
batten as indicated along 100, and mark on it the inner and outer
edges of the longitudinal a' and a respectively.   Now swing the
batten round 1 as a centre and mark off in succession bb' and cc'.
The batten will now be swung round 2 as centre, the inner and
outer edges of all rays passing through 2 being marked on it, and
so on, using the points 3, 4, etc., in succession as centres round
which to swing the batten.

The batten will then be marked as shown on Fig. B, the inner
and outer edges of the longitudinal being denoted by 100, 102,
104, etc.

Take any point X on the batten, and place the batten in turn
on the frame stations 100, 102, etc., in Fig. C, and mark off the
inner and outer edges (a_1a_2), (b_1b_2), etc.   In doing this the point X
is always placed on the base line.

To obtain the developed stations draw perpendiculars from
a', b', c' (Fig. A), and set these off on Fig. C, obtaining a curve AB
(exaggerated for clearness), and proceed similarly for the points
a, b, c, obtaining the curve AB' (Fig. C).

Pen a batten to AB, and mark off points as D where AB
crosses the frame lines, and let it spring along the base line,

obtaining points as D', marked X.   Obtain the points for the outer edge similarly.

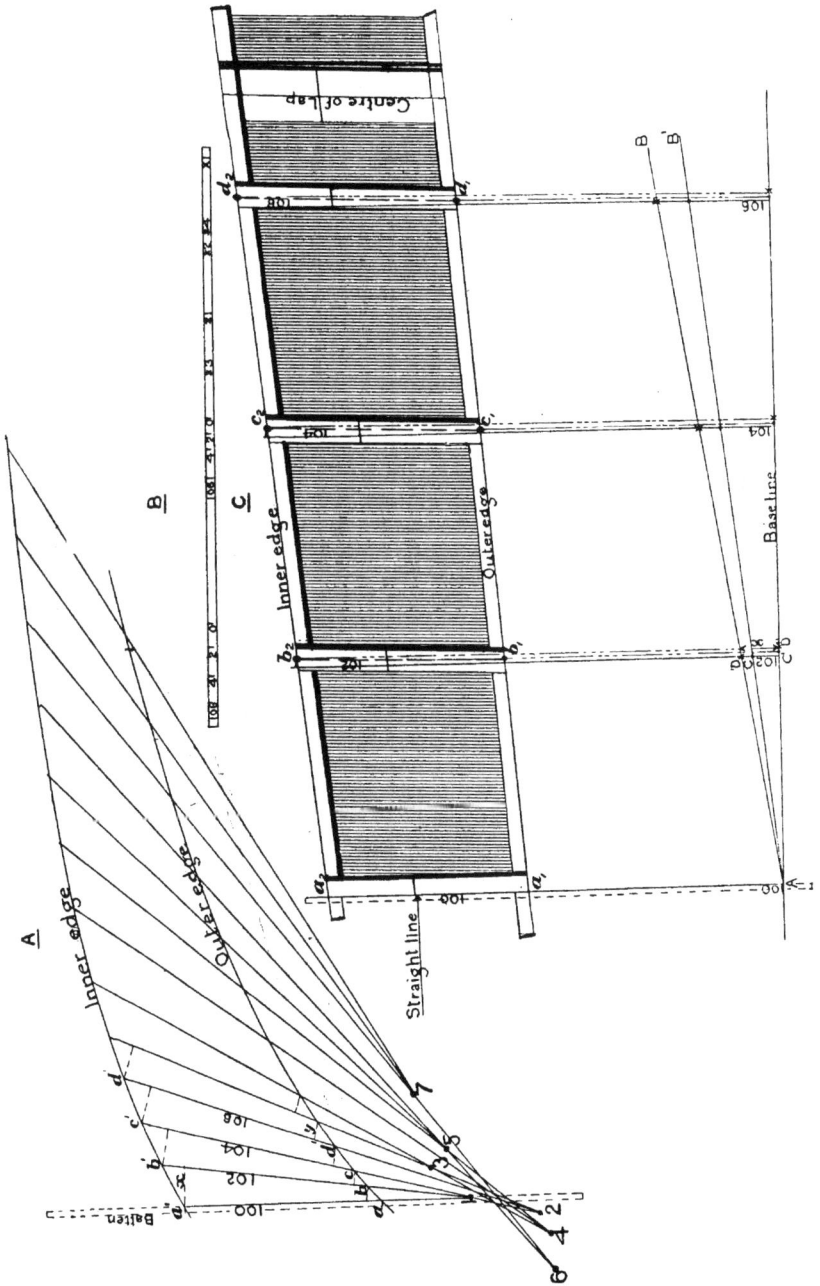

Through D' draw a line parallel to frame line 102 to meet the line drawn through the point of inner edge parallel to base line in b_2, similarly through C' draw a line parallel to 102 frame line to meet that drawn through the outer edge in b_1; then b_1 and b_2 are the outer and inner edges of the longitudinal, and $b_1 b_2$ is the developed position of frame station 102. Proceed similarly with the other frame stations.

MOULDS FOR WATERTIGHT AND OILTIGHT FRAMES.

The moulds are made for the actual frame, and therefore the thickness of longitudinals must be allowed for, and also any compensating liners (see Fig.). The bevellings of the frame angles

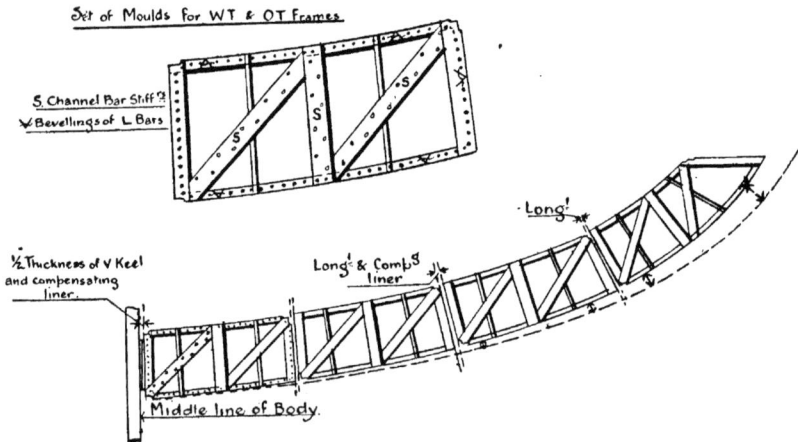

against inner and outer bottoms, and longitudinals, are measured and marked on the moulds as shown in the figure.

The method of obtaining these is given on p. 22.

Battens are placed on the moulds to show the stiffeners, and all holes for these and the boundary angles are marked off. The joggling of the boundary angles over the angle bars of longitudinals are all formed, and the mould when complete gives the exact shape of the frame.

MOULDS FOR DECK STRINGERS.

There are two cases to be considered, viz. ordinary stringer and stringer on slope of protective deck.

Ordinary Stringer. Referring to p. 237, Figs. C and D, the

A Body Plan

Plate edge

Beam end line

1 2 3 5 6

B

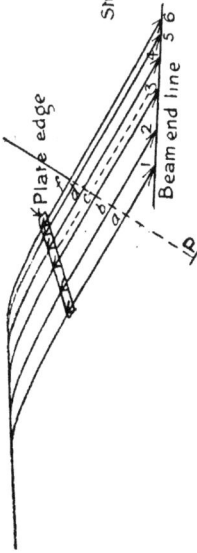

Protective Deck Stringer

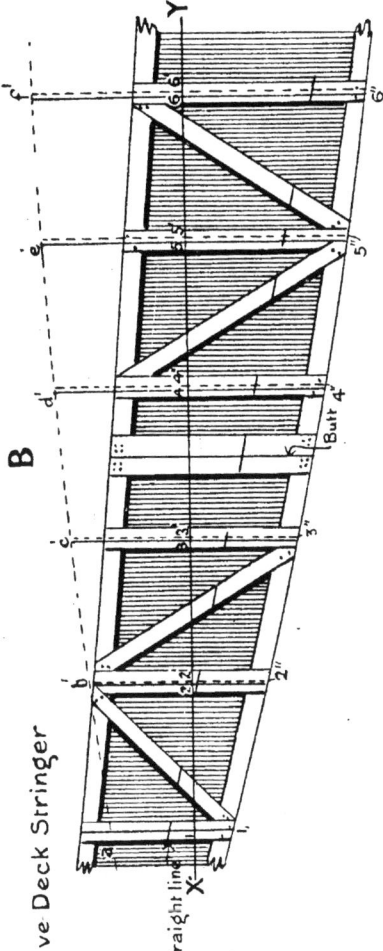

Straight line

Butt 4

C Body Plan

Beam at side

Batten

ML

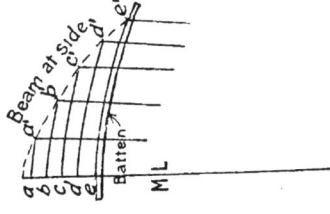

Deck with roundup & Sheer

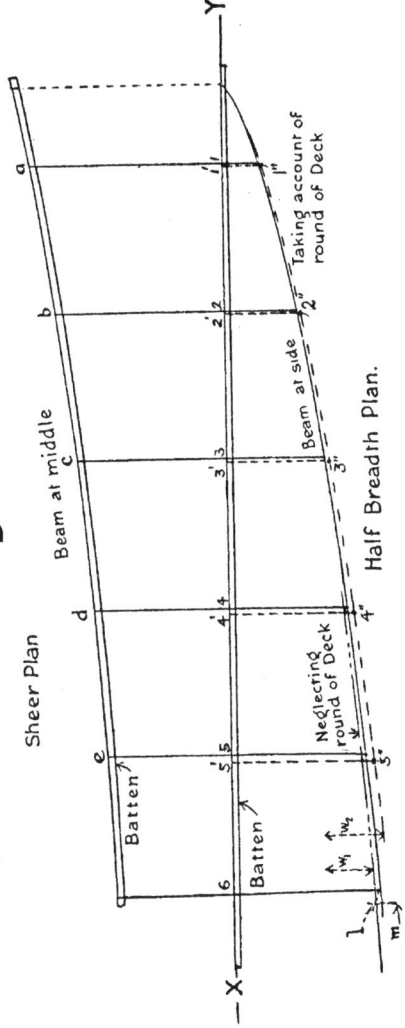

D Sheer Plan

Beam at middle

Batten

Beam at side

Batten

Taking account of round of Deck

Neglecting round of Deck

Half Breadth Plan.

most general case is taken, the deck having considerable sheer and round up. Pen a batten to beam at middle in sheer, and mark on it the frames *a, b, c,* etc. Place batten along base line XY, fix one end, let batten spring straight and transfer marks, obtaining points 1′, 2′, 3′, etc. Square out lines 1′1″, 2′2″, etc., which will give positions of developed frame stations. Through the intersection of beam-end line and frame stations draw lines parallel to XV, to cut developed frame stations. The curve through these points will be the shape of outer edge of stringer, and is the one marked W_1 " neglecting round of deck." The inner edge of stringer, not shown, is copied from plan of deck plating fixed, and mould made to these lines.

To take account of round of deck when stringer plates are very wide, the inner edge is marked off on the body, Fig. C, a batten is penned to the round of deck, and inner and outer edges marked on it. The batten is then laid along the developed frame station in the halfbreadth, and inner and outer edges transferred. The resulting curve is W_2. The positions of developed frames are marked across · the mould, also a straight line for checking distortion.

Protective Deck Stringer. Referring to Fig. A, the plate edge is copied from model of deck and faired in. From any point P a line is drawn at right angles to slope of deck. Measure distances P*a,* P*b,* etc., and set these off from XY, obtaining the curve *a′b′c′d′*. Pen a batten to this curve, marking points *a′, b′, c′,* etc., place it along XY, let it spring straight, transfer these points giving 2′, 3′, 4′, etc. Draw lines through these square to XY (the ticked lines), which will be the developed stations. Measure distances *a*1, *b*2, *c*3, etc., to beam-end line, and also those to plate edge in Fig. A. Set off these above and below XV in Fig. B, and the resulting curves will be inner and outer edges respectively of stringer.

The moulds are made as in Fig. B, the developed stations and a straight line being struck across them.

This method is similar to that described for laying off a longitudinal, but as the lines *a*1, *b*2, etc., are parallel, only one base line is required, the developed stations being parallel.

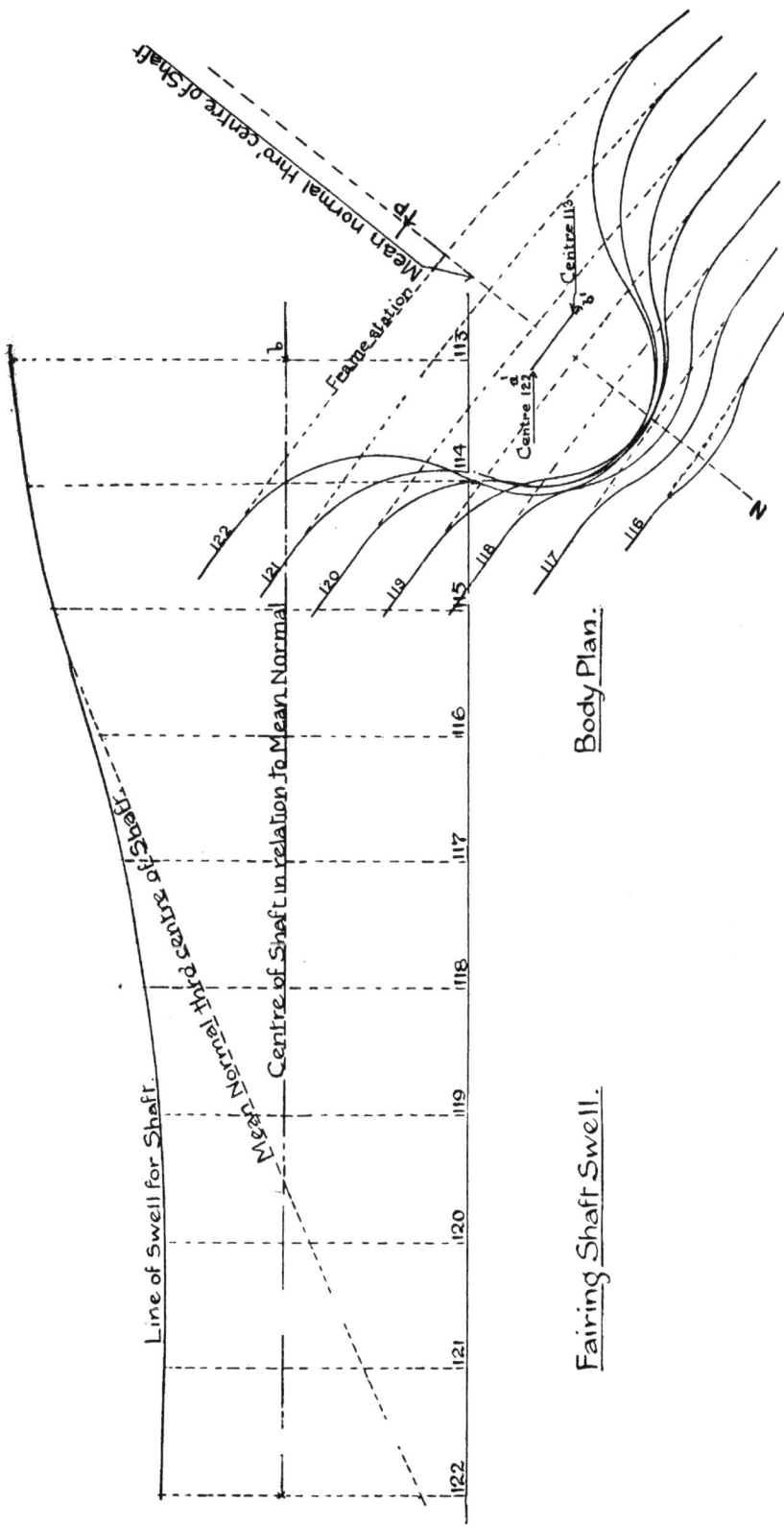

Fairing Shaft Swell.

Body Plan.

Fairing Shaft Swell.

Referring to p. 239, a' and b' represent centres of shaft at stations 113 and 122. Draw a mean normal NP to pass through middle point of $a'b'$. Then the centre of shaft can be assumed to lie in the diagonal plane PN without appreciable error. Lay off the diagonal PN marked on the plan " mean normal through centre of shaft," and also centre of shaft ab. The radius at point a is known and is set off, the line of swell for shaft being faired into the mean normal as shown, when the distances from ab to line of swell give the radii at the stations 115 to 121. Arcs are swept out from the corresponding centres on the line $a'b'$, and connected to the frame lines by easy curves, and then faired by a series of diagonals so as to include the last frame which has no swell. These diagonals are not shown in sketch. A shaft recess is faired in a similar manner.

Moulds for Side Armour.

These are the moulds supplied to the contractors for preparing the armour plates, and the two methods of preparing them are (1) *by laying off*, which is used for the armour over the middle portion of the ship, and (2) *by mocking up*, which is used for the armour at the ends of the ship, where there will be considerable twist.

The armour is generally in tiers, and if in two, the lower edge of upper tier is termed the "joint line," and from this line all the work is carried out.

To ensure the edges being in contact, the joint line is always in a plane perpendicular to the sheer plane, so that every transverse section of the joint is a horizontal straight line. For the middle part of the ship this line will be at a constant height above load water-line, and at the ends incline upwards.

Laying-off Method (p. 241).

The first moulds obtained are the "surface" moulds, which give the area and shape of the outside surface of armour, and are

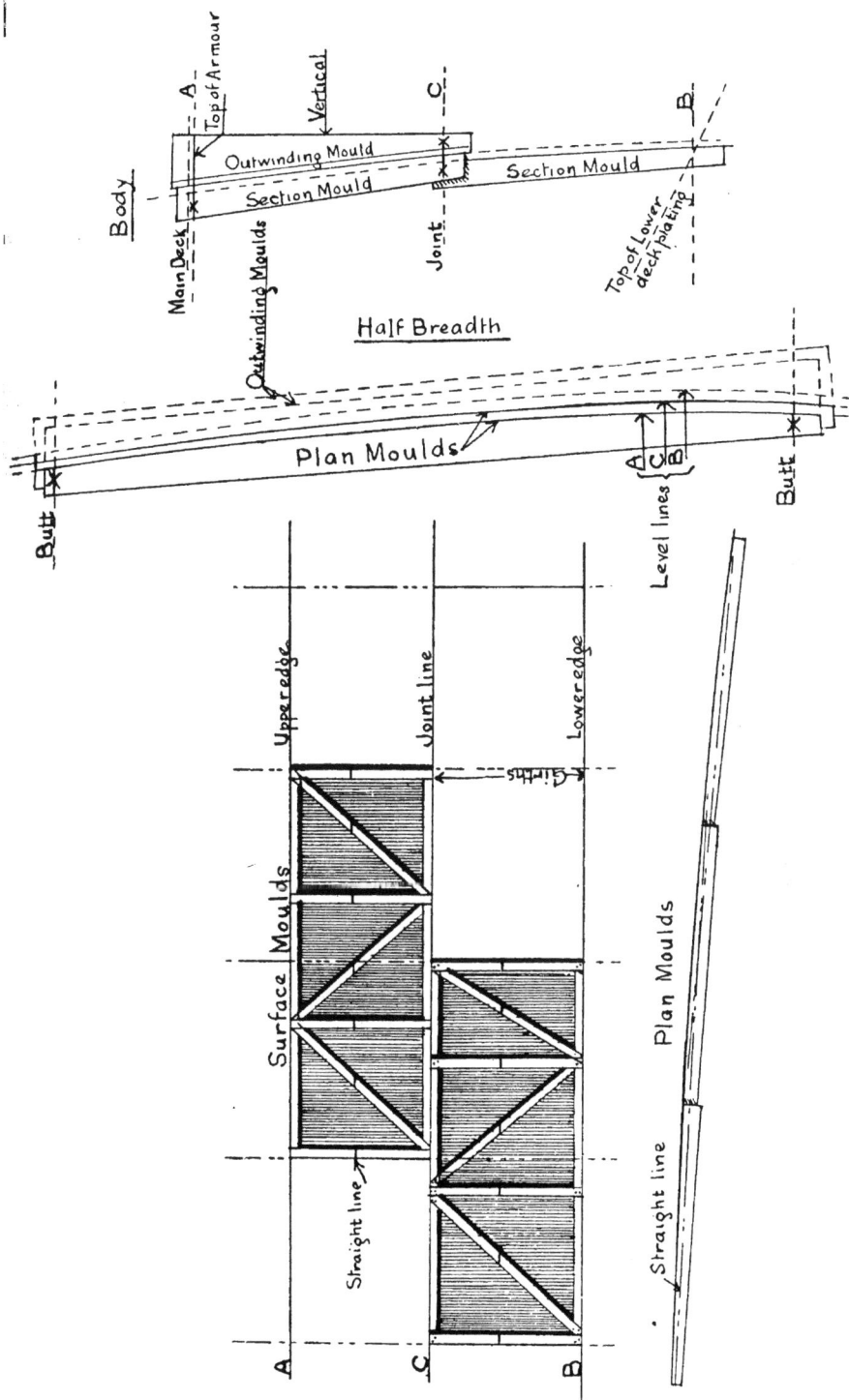

Body

Top of Armour

A

Vertical

C

B

Outwinding Mould

Section Mould

Section Mould

Main Deck

Outwinding Moulds

Joint

Top of Lower deck plating

Half Breadth

Plan Moulds

Butt

Level lines { A C B

Butt

Upper edge

Joint line

Lower edge

Surface Moulds

Girths

Straight line

Plan Moulds

Straight line

A

C

B

made thus : strike a line along the floor, and measure on the body plan from the joint line along the frame line to upper and lower edges of armour at convenient stations, and set these girths off above and below the line on the floor. This will give the development of the surface, but in order to ensure the armour being got in place at the ship, it is necessary to allow some amount of clearance at the upper and lower edges of armour.

It is now necessary to draw on the floor the butts of the armour plates, and these are copied from a drawing previously prepared. " Surface moulds " are now made for each plate, and straight lines are struck across each mould and cut in to serve as a check on the mould.

The " section " moulds are next made, one to each butt of each plate, the outer edge of the mould being the frame line and across it are marked the top and bottom edges of each plate. On the section moulds a line parallel to its outer edge is drawn, which serves as a check should the edge of the mould be damaged.

To each section mould an " outwinding " mould is made such that its inner edge agrees with outer edge of section mould, and its outer edge is vertical, the top and bottom edges of armour being marked across these moulds also.

" Plan " moulds, which give the shape of upper and lower edge of armour, are made to level lines laid off at upper edge, joint, and lower edge of armour respectively. Outwinding moulds are made to these, their outer edges being parallel.

The plan moulds are laid in their correct positions on the floor, and straight lines, to include as many of them as possible, struck across them. By means of these lines the moulds can be correctly laid down at the works.

Nine moulds are made for each plate, viz. two " section " with outwinders, two " plan " with outwinders, and one " surface."

To test the correctness of the armour all the plates are erected at the works to form the complete belt, and moulds called " erecting " moulds are supplied for this purpose. An erecting mould is made at each butt of lower tier, the mould extending the whole depth of belt.

While the armour is erected at the contractor's works it is noted at what points the several plates give a good fair surface, and at such places " well spots " are painted on, so that when the armour is subsequently erected at the ship, it is known that by making these spots coincide the best surface has been obtained.

MOCKING-UP METHOD.

The side of the ship is represented by the temporary frames, which are made as indicated on p. 244. The lower and upper edges are cut to level lines, and the positions of deck beam ends and joint line are marked across each. The outer edges of these frames are made to lines representing the surface of the bottom plating, *i.e.* the thickness of bottom plating is added outside frame line, as the surface of armour has to be flush with bottom plating to enable covering plate to be worked.

Moulds are erected at the stations and correctly pitched to the level lines A and B, bringing the lower corner of each to coincide with B, and plumbing down from points as "A" on the frames to the level line A on the floor.

The lower ends are secured by short pieces of wood to the floor, and before finally securing them it is necessary to sight the joint lines on the frames to see that they are in a plane to take account of any unevenness of the floor. The surface moulds are made as indicated in sketch, and outwinding moulds at each butt, the section moulds being made from these.

Another method of mocking up is illustrated on pp. 245, 246. It differs from the previous method in that the middle line of ship is now taken as horizontal.

Moulds as in Fig. VI. are made at the frame stations, the outer batten U denoting outside surface of armour, and that marked H the outside surface of frames behind armour. The moulds extend beyond the beam-end lines to assumed level lines PP. The edges of armour are marked on them, the one just above the protective deck plating being marked M, and that below the beam-end line of deck to which the side armour is carried by O. The upper and lower edges of plating behind armour are indicated, and a straight line SSS struck across the mould.

Similar moulds are made at the butts of plating behind armour, one such being shown at E (Fig I.).

The moulds shown in Fig. IV. are made to level lines P_1, P_2, which correspond with P,P (Fig. VI.). These "plan" moulds have

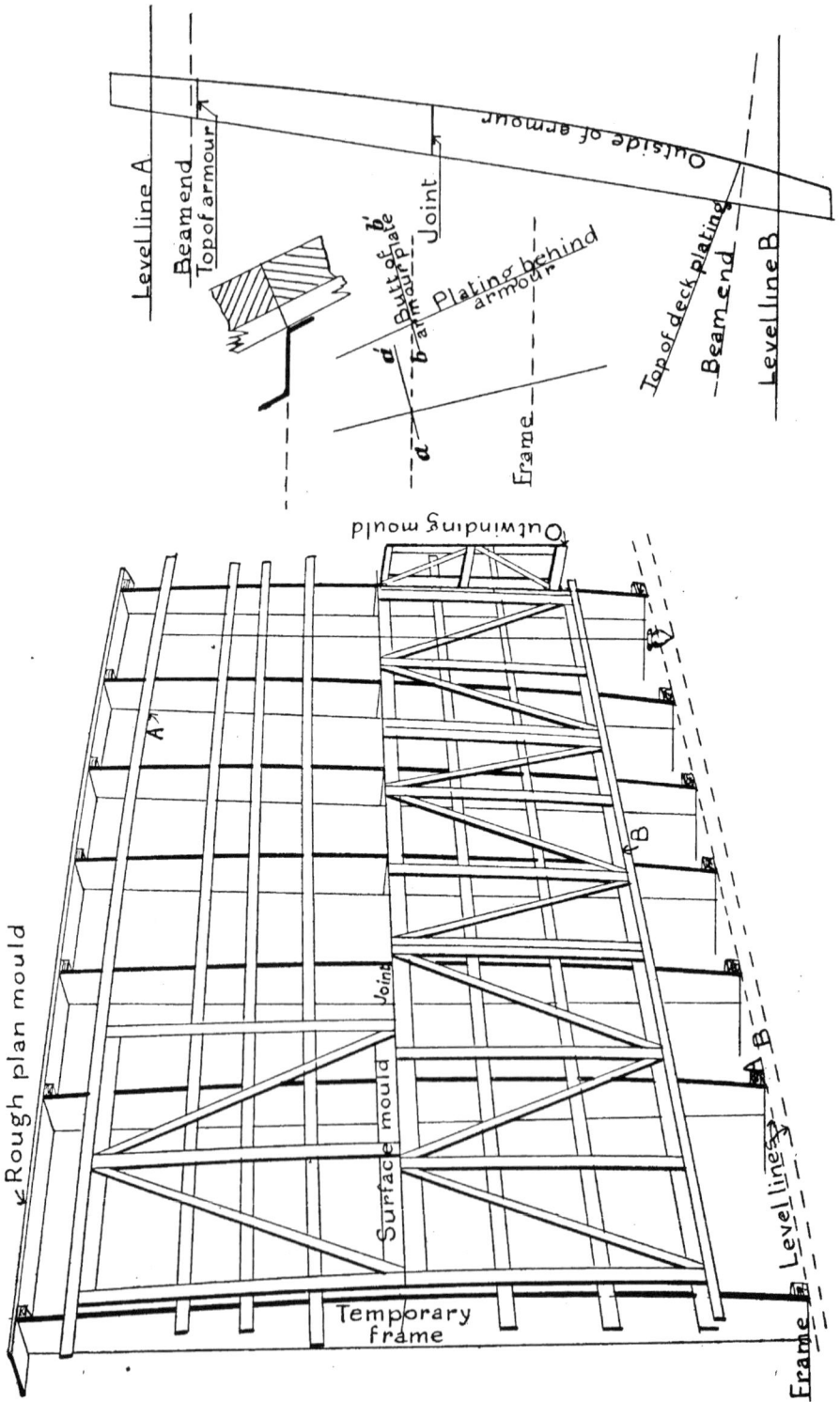

the frame stations 1, 2, 3, etc., marked across them, also a straight line SS, and are then erected as in Fig. I. To avoid building to

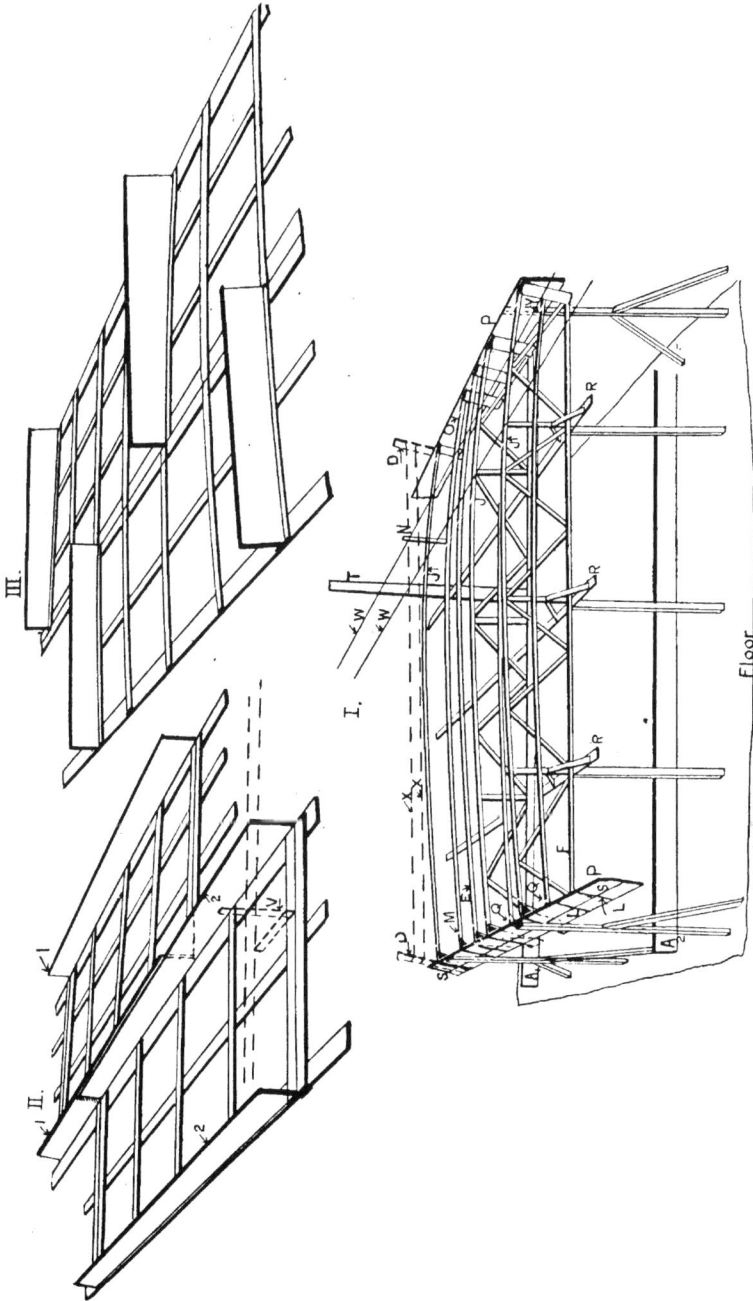

an excessive height a line AA (Fig. IV.) is drawn, which is the base line for all measurements, and the upper edges of the battens A_1A_2, etc., which are in a horizontal plane, are sighted in to represent this line.

The lines q_1p_1, q_2p_2 (Fig. IV.) will be vertical when the plan moulds P_1, P_2 are erected as in Fig. I., and in order to check the correctness of the erection the distances pq_1, pq_2 (Fig. IV.) are required. A wire is stretched along the top of the battens A_1A_2,

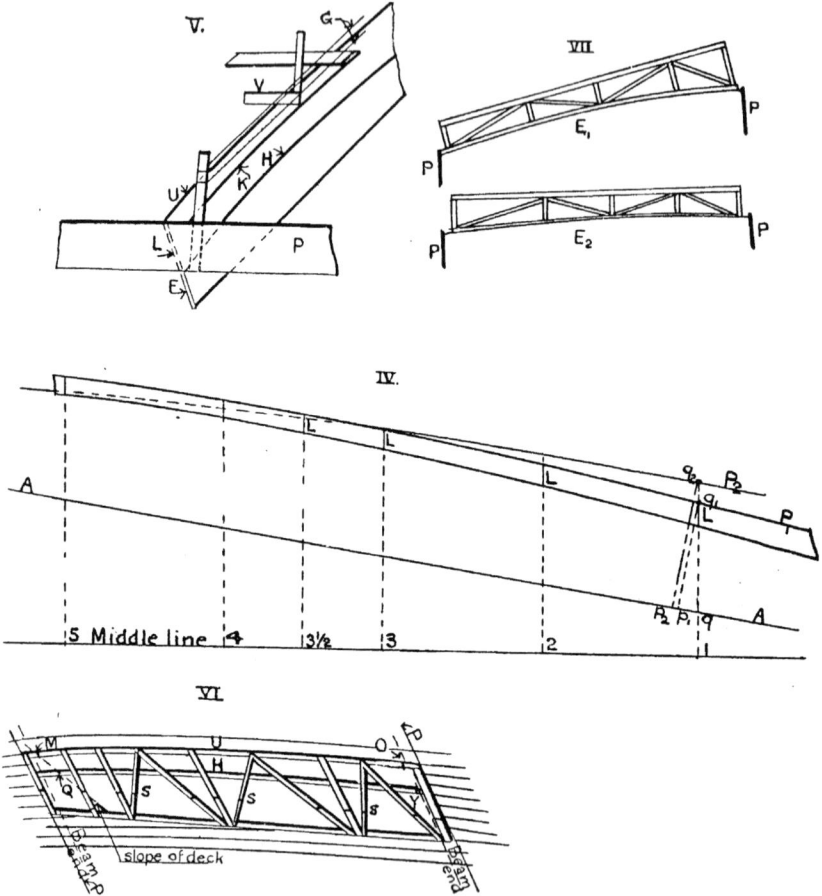

and the distances as p_1q, p_2q set off, and a plumb bob suspended from q_2, q_1 should coincide with the points p_2, p_1 respectively.

The moulds (Fig. VI.) are attached to the plan moulds as in Fig. I., being made to coincide with the line LL on them, and stiffened by ribands R. Before finally fixing them by means of struts, it must be ascertained that there is no twist or sag. The twist is checked by erecting battens DD and outwinding them, and stretching cords XX across which will be the plane to which the moulds must come. The sag is checked by means of straight lines SS (Fig. VI.) marked on the moulds.

The joint line is marked as follows: Vertical battens T (Fig. I.) are erected at intervals and sighted in, and wires WW stretched along the whole length of mocking. By means of a batten N these are projected on to the moulds giving points JJ, representing the joint line.

The surface is now ready for making the surface moulds shown in Figs. II. and III. The butts of the armour plates come at the frames, and must be in a plane normal to the surface. Where the side of the ship is running parallel to its middle line the plane of the butt will be parallel to the plane of the frames, but forward and aft where the outside surface of armour is inclined to the plane of the frames the butt is obtained as follows: In Fig. V., U and H represent surface of armour and frames behind armour as before, F being one of the moulds illustrated in Fig. VI. A batten is then placed with its lowest point agreeing with H to outwind with the tongue of a bevel V, the stock of which rests on the surface mould. Cords GG are stretched across, and by moving the bevel across with the tongue in contact with these cords the butt K is obtained. This butt will be in a plane, which at one point will be normal to surface of armour. It cannot be normal at all points owing to the form of the armour, and a point must therefore be selected at which it shall be made normal to the surface. The plane will not be vertical, but inclined to the horizontal and vertical planes, *i.e.* a double-canted plane.

After the surface moulds are made horizontal outwinders (Fig. I.) and vertical outwinders (Fig. III.) are prepared and sighted as shown, two of each being made for each plate with their edges outwinding, and not necessarily outwinding with the moulds for other plates.

Erecting moulds extending from top to bottom of armour are then made at intervals of about 15 to 20 feet, and to avoid making them large the mocking is divided into sections, the moulds for middle portion constructed as E_2, and those at the ends as E_1 (Fig. VII.).

When the moulds for armour have been made the original moulds are cut down to the batten, and the surface then represents that of the plating behind armour. The upper and lower edges of this plating were marked off from the floor on the original moulds, and thus the exact form of the surface is given, and moulds for plating behind armour can be made from this surface.

HOLES IN ARMOUR.

These comprise the holes in the back of the plate for securing it to the plating behind armour, those along the top surface of *lower tier* for lifting the plate, and the holes for covering plates at upper and lower edges of armour, and other holes as may be necessary, *e.g.* attachment of torpedo boom fittings, soil pipes, etc. It is important that a complete account be given of all holes in armour, as these must be drilled before the plate is hardened.

Brass ferrules are inserted in the holes in back of armour for lifting purposes, to prevent damage to the thread. No holes are necessary for lifting purposes in upper tier, these plates being lifted by means of pieces of plate secured by bolts which screw into holes provided for covering plate. The lifting holes in lower plate are plugged before upper tier is placed in position.

CLOSING PLATES.

In each tier of armour certain plates are chosen for " closers," moulds being made for these as for the other plates, but the plates are not completed until the remaining plates have been erected at the ship, as it may be found that the original moulds give a plate too large or otherwise, and the final moulds to which the closers are completed are therefore made at the ship. The original moulds are useful in permitting a certain amount of work to be done on the closer, it being always understood that there may be slight departures to be made from these, which will be shown by the moulds made at the ship.

DEVELOPMENT OF PLATING BEHIND BARBETTE ARMOUR.

Considering the most general case, that is, when the plating rests on a deck having round and sheer, the development is obtained thus : Referring to figure, draw the plan and elevation of barbette, the beam at middle line and the beam mould, as in the figure, and divide up the circumference into any number of parts, 12, 23, 34, etc.

Take any one of these as 3, and measure the amount of round

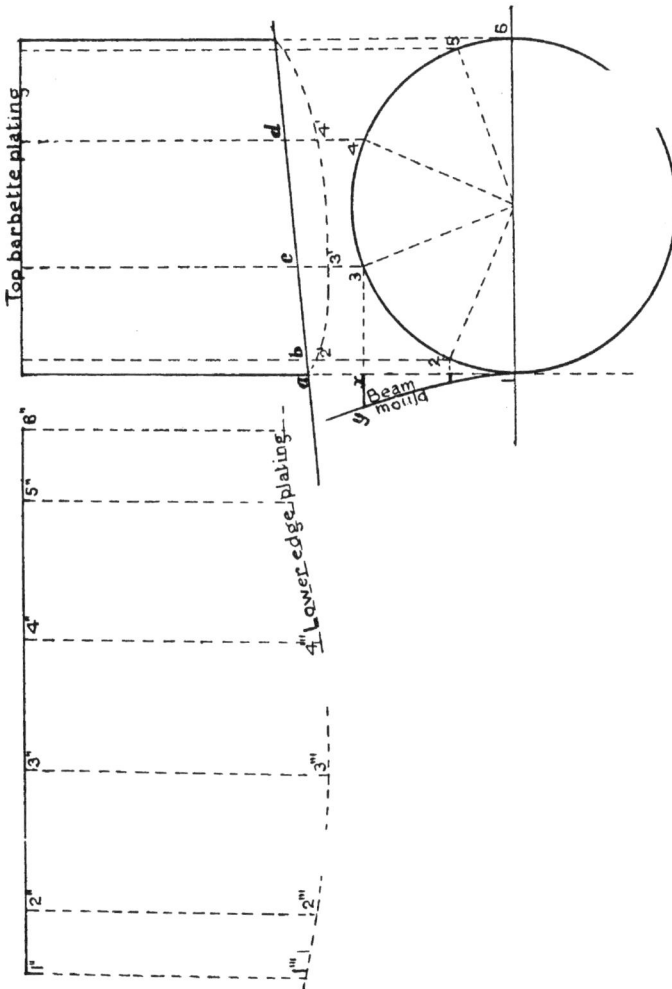

down of deck at this point, which is got by drawing 3*x* parallel to 16, *xy* then being the round down which is set down from *c* so that *c*3′ is equal to *xy*. The points 2′, 3,′ 4′ on will be points on the projected shape of the intersection of the plating with the deck.

Next draw a straight line 1″6″, giving top of plating, and set off 1″2″ = 12, 2″3″ = 23, etc. (the distances 12, 23 being measured along the arc), and draw vertical lines through 1″, 2″, 3″, 4″. Project across the points 2′, 3′, 4′ on to the corresponding lines on the development, and the curve through these points 1″′, 2″′, 3″′ will give the shape of the lower edge of the plating, and the moulds are made to the plates after the butts have been arranged.

In the case of a broadside barbette, the plating rests partly on the flat and partly on the slope of the deck, and it is laid off thus : Draw plan and transverse elevation as above, and to work the problem exactly it would be necessary to draw in the elevation the beam lines at the points 1, 2, 3, 4, and 5, and square these across to the development as before, but it will be sufficient to select those at the points 1, 3, 4, and 5, and square across 1', 2', 3', 4', 5' to 1'', 2'', 3'', 4'', 5''. The curve through these latter points gives true shape of plating.

LAYING-OFF BILGE KEEL.

The trace of the keel is shown on the body plan, and lies in a diagonal plane (Fig. 1, p. 252). The first thing to be done is to lay off the diagonal. Take a base line, XY, and on it mark off the frame stations 1, 2, 3 in the figure and set off 11', 22', 33' . . . equal to P1, P2, P3 . . respectively, and draw the curve 1'2'3' . . .

The depth of the keel normal to the ship's bottom is given, and this is then set off as in Fig. 2, the curve 1''2''3'' being obtained. Next measure off 1f = 1'1'', 2h = 2'2'', 3n = 3'3'', in Fig. 1 the points f, h, n, being then the outer edge of keel at the respective stations. The width of keel is also given, and this is set off, giving the points d, g, m . . . Join df, gh, mn ; these lines will then represent the section of the lower face of the keel at the stations 1, 2, 3 . . . The upper face is obtained in the same way.

In Fig. 1 draw a line Qabc, at right angles to the parallel lines df, gh, mn, produced to cut them in a, b, c . . . From XV set off 1c', 2b', 3a' . . . equal to Qc, Qb, Qa . . . respectively, and draw the curve a'b'c' . . . Pen a batten to this curve, and mark on it the points a', b', c' . . . Place the batten along XY, and transfer these points to it, and the points so obtained will give the positions of the developed stations represented in Fig. 3 by the ticked lines. Now, along these developed stations set off 1d', equal to cd, 1f' equal to cf: 2g' equal to bg, and 2h' equal to bh, and so on for the other stations. Draw the curves d', g', m' . . ., f', h', n' . . ., and these will be the inner and outer edges of the lower plate of the bilge keel. The mould is made to these as in the figure. It is to be observed that the thickness of bottom plating must be allowed for, and this is done by drawing the curve d'g'm' . . . this amount nearer the outer edge. To obtain the position of the bilge keel on the ship's bottom, it is only necessary to know the

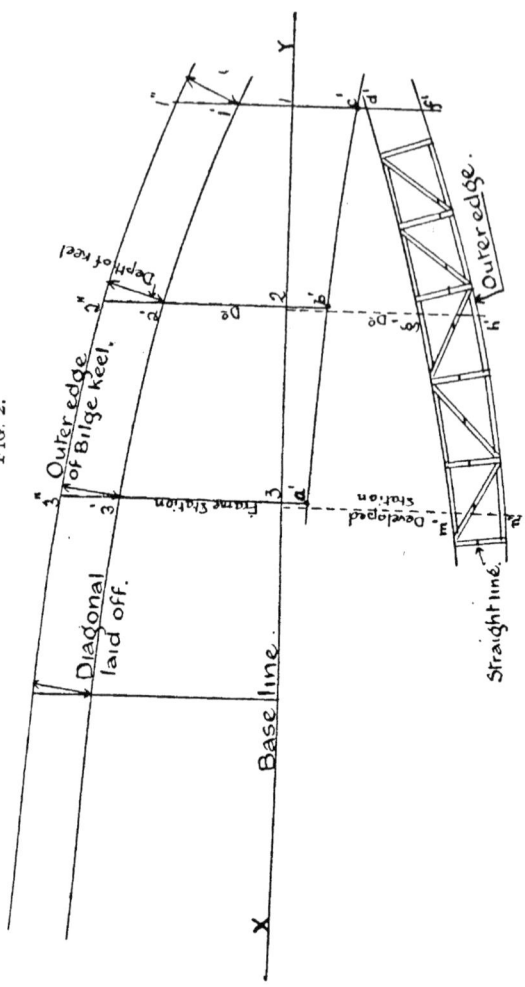

FIG. 1

Trace Diagonal

Frame line

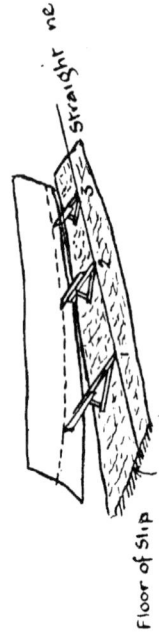

FIG. 2.

Outer edge
of Bilge keel.

Depth of keel

Diagonal
laid off.

Frame station

Base line.

X

Y

FIG. 3.

Straight line

Developed
station

Outer edge.

Straight line.

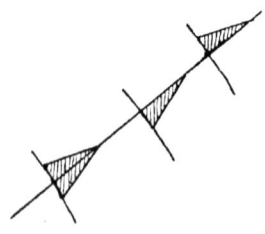

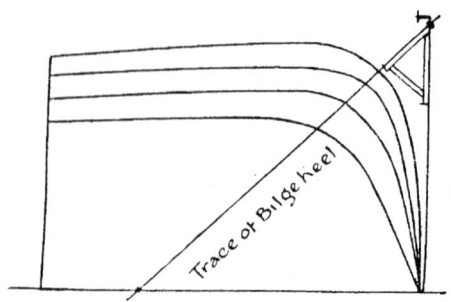

Floor of Slip

LAYING OFF BILGE KEEL

Trace of Bilge keel

FIG. 4.

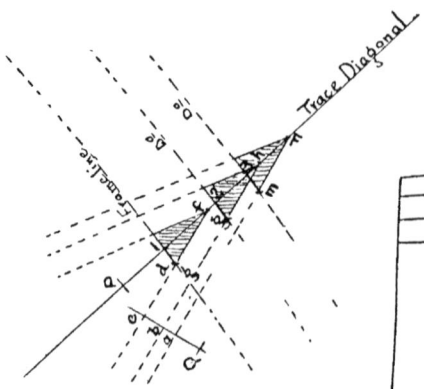

distance ST in Fig. 4, and the angle STO. A mould is made to give this angle, and a straight line is drawn along the floor of the slip, the corner of the moulds being placed at this line, and then by means of straight-edges the trace is sighted in on the bottom as indicated in Fig. 5. The sighting in is done in lengths at a time, looking over the tops of the first and third moulds and under the centre mould.

In Fig. 4 three methods of arranging the plates are shown.

INTERSECTION OF HAWSE PIPE WITH SIDE AND DECK OF SHIP.

The hawse pipe is assumed to be cylindrical, the central line being given and the diameter of the pipe.

Referring to p. 254, ab, $a'b'$ is the centre line of the pipe. First find the curve in which the cylinder cuts the sheer plane thus : Set up $b'c' = bc$ and join $a'c'$, then $a'c'$ is the angle between the centre line of pipe and sheer plane. Set off $a'f' =$ radius of pipe, and through f' draw $f'a'$ parallel to $a'c'$, then $a'd'$ is the major axis of the ellipse in which the cylinder cuts the sheer, the minor axis being $a'f'$.

To obtain a point in the side of ship, take a single-canted plane q_1q_2 parallel to the centre line of pipe as in the plan. Project q_1 to cut ellipse in q_4 and q_5, and draw the lines q_5q_7, q_4q_6 through these points parallel to centre line of pipe in sheer. These lines represent the intersection of the single cant with the cylinder.

Next obtain curve of intersection of single cant with ship's side by projecting up the points in which the single cant cuts the level lines in plan to the corresponding levels in the sheer. The curves are marked "intersection of side" in the sheer plan. Then the intersections of the lines q_5q_7, q_4q_6, with these curves give points in the curve of intersection of cylinder with ship's side.

To obtain intersection with deck, draw beam curve as in plan and measure round down X at any buttock. The buttock and single cant intersect in a vertical line, and from the point in which this intersects beam at middle set down the distance X, then the curve through q_3 and points thus obtained is the curve of intersection of the single cant with the deck. The points in which q_4q_6 and q_5q_7 intersect the curves made by the single cants are points on the intersection of cylinder with the deck.

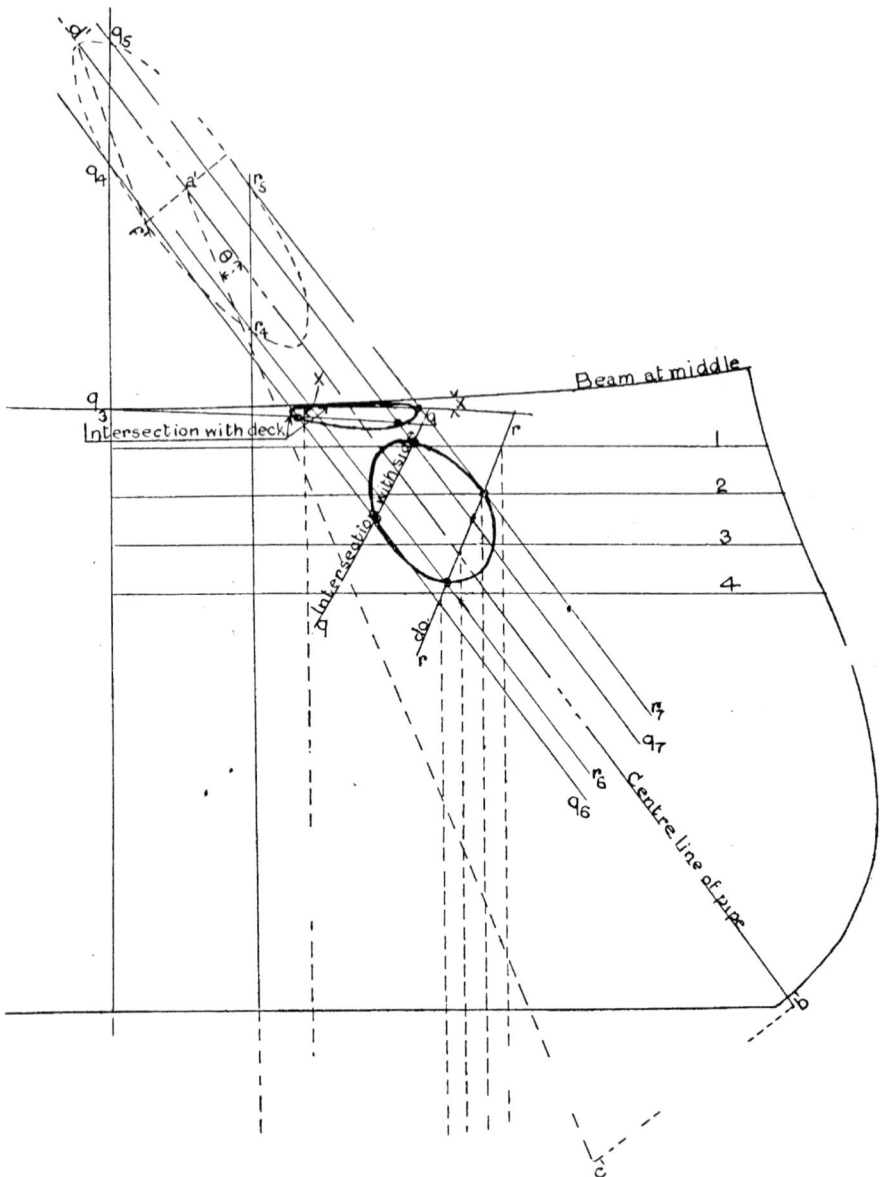

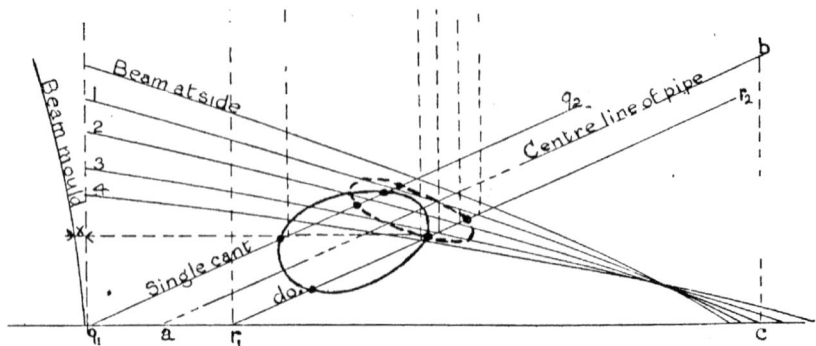

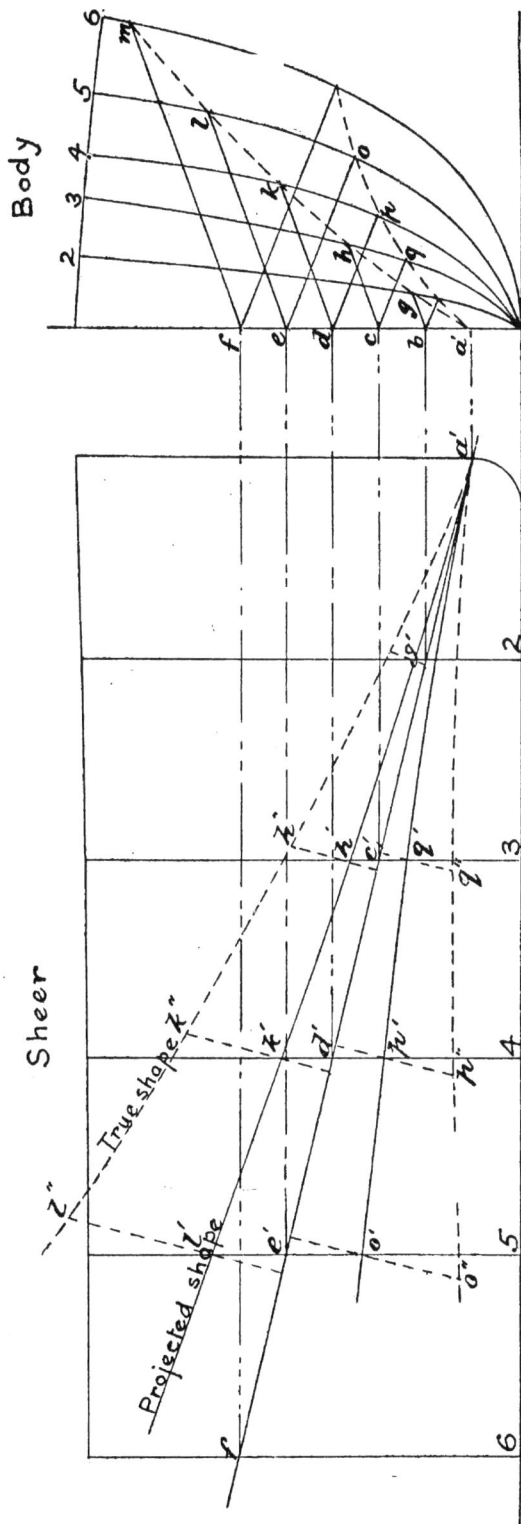

To find Projected and True Shapes of Water Plane in a Ship having List and Trim.

Referring to p. 255, $a'e'$ denotes trim, the drafts at ends being known. From any point c' square across to c in body and draw ch to represent the list. Square across from h to h' in sheer, then h' is a point on projected shape. To obtain the true shape the water plane is rabbated round $a'e'$ as axis, and through $g'h'$, etc., the points of intersection of square stations 2, 3, 4, etc., with the projected shape draw perpendiculars to $a'e'$ and produce them. Then measure $c'h'' = ch$, $d'k'' = dk$, etc., and the curve through these points give the line and shape.

To obtain the other side through points as c draw cq at same inclination to horizontal as ch and proceed as before.

To find Shape of Heel Plates of a Raking Mast on a Deck.

(1) When the deck is flat.

Draw the diameter 1, 5, and the semicircle 1, 2, 3, 4, 5, divided into any number of parts, 12, 23, 34, 45. Through these points draw lines parallel to the centre line of mast, then they will represent generators of the mast (Fig. A, p. 257).

Girth 12, 23, etc., round the semicircle, and set off these distances along a straight line, obtaining the points $2'$, $3'$, $4'$, $5'$, and set down the distances $2'b'$, $3'c'$, $4'd'$, $5'e$, equal to be, cf, dg, $5h$ respectively; then the curve passing through $1'b'c'd'e'$ is the developed shape of one-half the circumference, the other half being similar.

(2) When the deck has round and sheer (Fig. B, p. 257).

Both the round and sheer are much exaggerated in the figure to make the method clear.

Obtain generators as before and then find the round of beam at the distances of these from the middle, thus make ($1a' = 2a$), then $a'b'$ is round; hence set down vertically a distance from top of deck $= a'b'$ to cut $2a$ produced in b, then b is a point in projected shape of curve of intersection of mast with deck.

To develop, obtain $2'3'4'5'$ as before and measure $2'c' = ab$, $3'd' = gd$, $4'e' = cf$ . . ., and the curve $1c'd'e'f'$ is shape of lower edge of one-half the heel plates, the other half being similar.

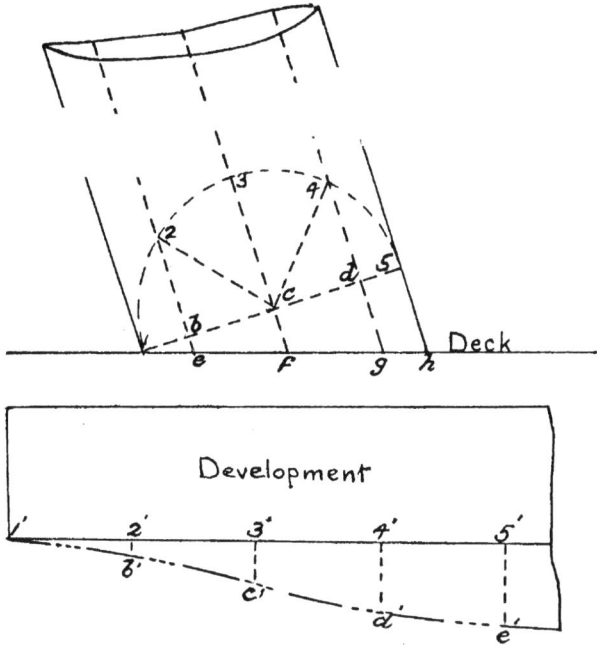

FIG. A.

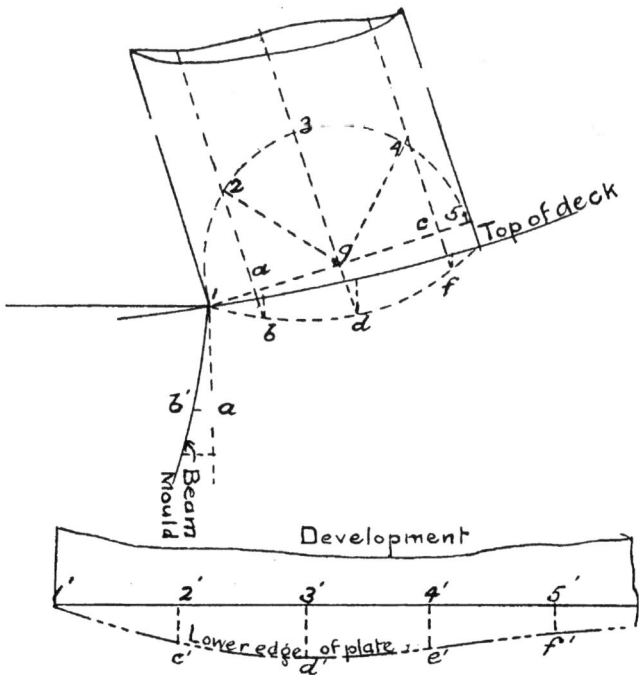

FIG. B.

To find the Curve in which a Strut is cut by a Mast.

The plan and elevation are given of mast and strut, the elevation for convenience being on a vertical plane through the centres of mast and strut.

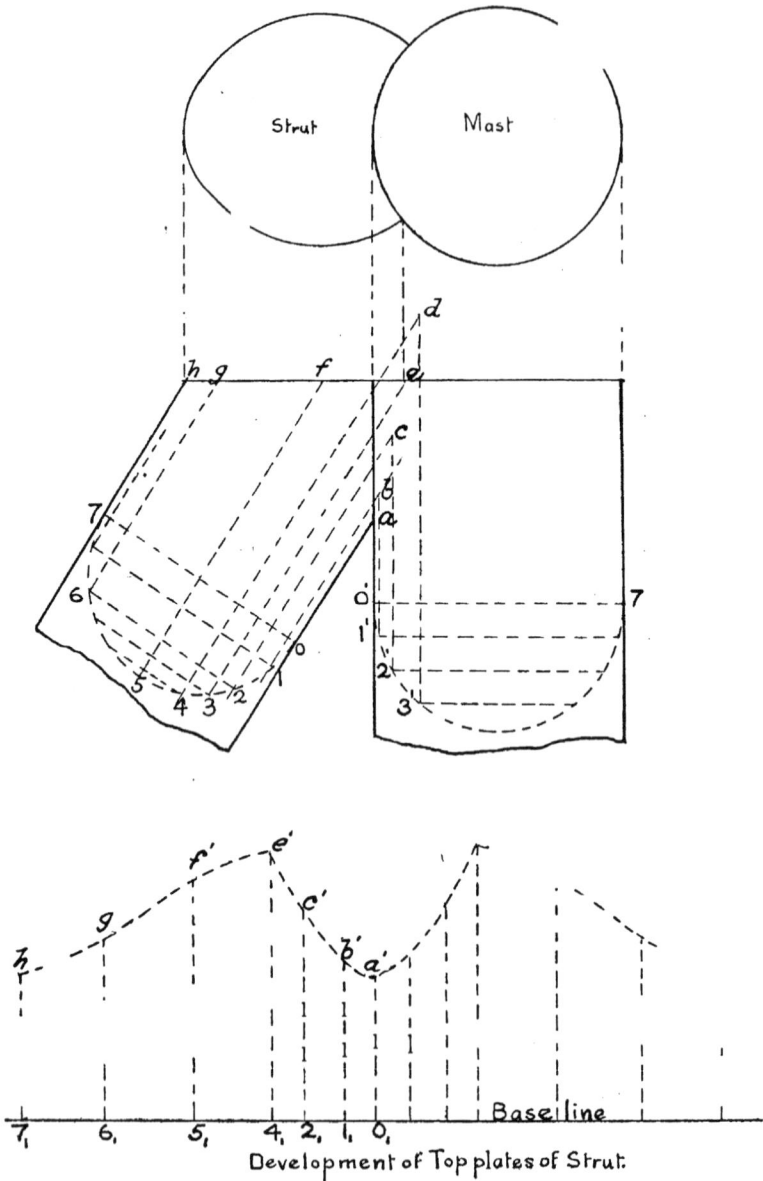

Development of Top plates of Strut.

The method adopted is to cut both by planes parallel to centre lines of the mast and strut. These planes intersect mast and strut respectively in straight lines inclined to each other, and their intersection will give a point in desired curve.

Draw 0567 and $0'2'3'7'$, the semicircular sections of strut and mast respectively, and draw lines parallel to 07 and $0'7'$ the same distance apart. These will represent the planes referred to above, and line drawn through their intersections with the semicircle parallel to centre line of mast and strut give the lines of inter-section. Through 1 draw $1b$ parallel to centre line of strut, and through $1'$ draw $1'b$ parallel to centre line of mast, and the point where they intersect will give a point on the required curve. Other points are obtained in the same way, giving the curve *abced*, which is the projected shape of the curve in which the mast cuts the strut. To develop this, take a straight line and set off $0_1 1_1$, $0_1 2_1$, $0_1 4_1$, $0_1 5_1$, etc., equal to 01, 02, 04, etc., and erect per-pendiculars $0_1 a'_1$, $1_1 b'$, etc. Measure off $0_1 a' = 0 a_1$, $1_1 b' = 1 b$, etc., and so obtain the curve $a'b'c'd'e'f'g'h'$, which will be the developed shape of one-half the top plating of strut, the other half being similar.

TAKING OFF PLANK.

In a sheathed ship the offsets on the displacement sheet are given to the outside of the sheathing, and to obtain the frame lines it is necessary to " take off the plank," as it is called.

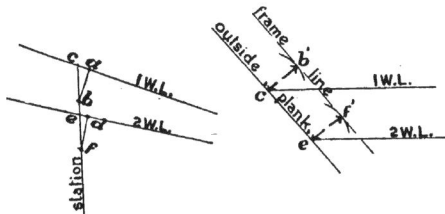

Approximate Method. Take any water-line laid off to the outside of planking, and set off normal to this at the displace-ment stations the thickness of plank plus plating, *ab*, to cut the displacement station in a point *b*. With a pair of compasses measure off the distance *bc*, and then refer to the body plan at the same water line and displacement station, and with centre c' describe an arc. The frame line will be the line touching all these arcs $b'f'$.

This method is sufficiently accurate for all ordinary cases.

PATTERN FOR STEM CASTING.

A flight mould is made, the upper surface representing the middle line of the ship, and across it are marked the positions of the water and level lines and frames. Section moulds are made at each of these positions and attached to the flight moulds, but to allow for shrinkage in the casting they are placed just above the level and water lines, it being assumed that the casting will contract towards the thickest part, that is the ram, or foremost extremity of casting.

The section moulds are made to the shape of the inside of the casting, and battens are worked round them as in the sketch, to build up the pattern. These battens are glued and screwed together, the· fastenings in each portion being marked so that those of the subsequent ones shall be clear of them. Saw cuts are made in the battens so that they can conform to the curvature. The outside surface is trimmed to section moulds, and the rabbets for plating, and armour if any, formed. The pattern is made in two pieces, dowelled together, and, to enable the correctness to be checked, outwinders are made and applied as shown in the sketch (p. 261).

At deck lugs and webs, the contraction of the metal tends to pull the two sides of the casting together, causing hollow portions, and to avoid this, wedge-shaped pieces are attached to the pattern, these being about $\frac{3}{16}$ of an inch at their thickest part. One such is shown in way of deck lug L in the sketch.

Sharp edges on the casting, as at plate rabbet, will show a series of cracks which must be chipped, and in order to leave the full amount of good metal, pieces P are worked on the pattern. When the defective portion is chipped off, the full amount of metal remains.

PATTERN FOR A SHAFT BRACKET.

The centre of shaft is already marked in on the body plan, and its position, corresponding to the fore and after ends of the casting, are picked up, these being marked A and B respectively (p. 262).

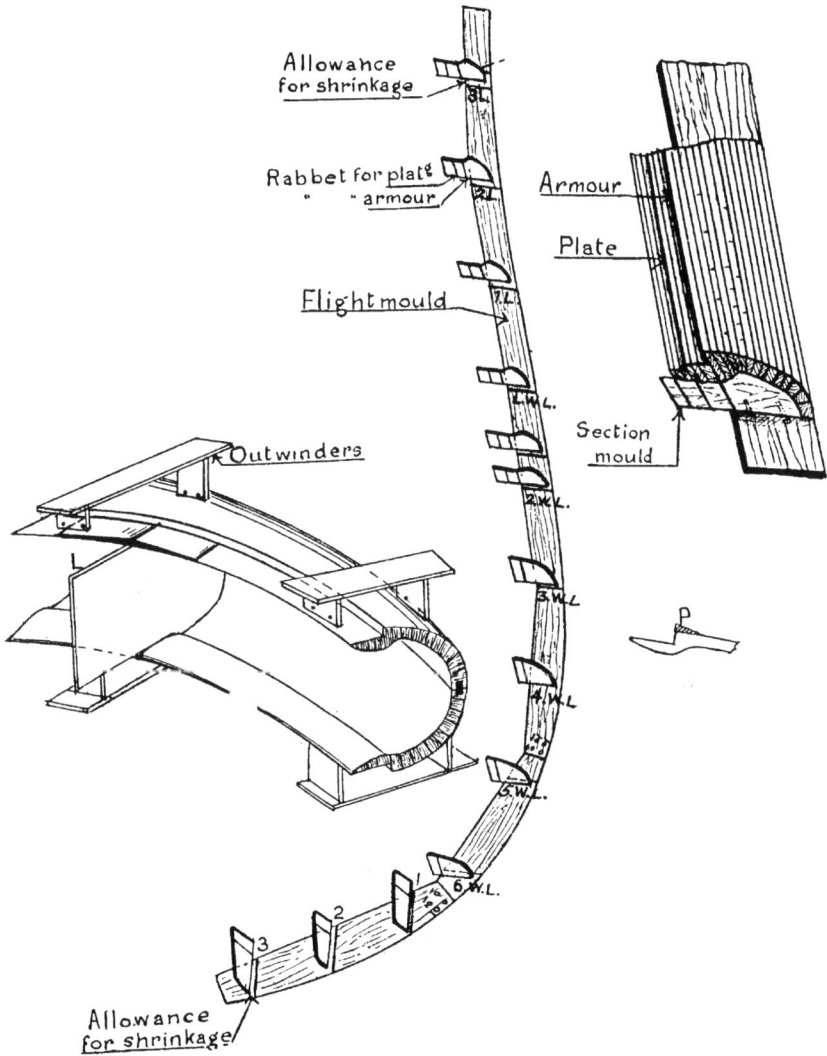

PATTERN FOR A STEM CASTING.

A circular mould is made at each end, using the two centres in turn, the diameter being equal to that of the boss of casting, and to each such mould two battens are secured which represent the arms of the casting and their thickness, as in the figure.

Convenient points, P and Q, are selected on the floor, and these

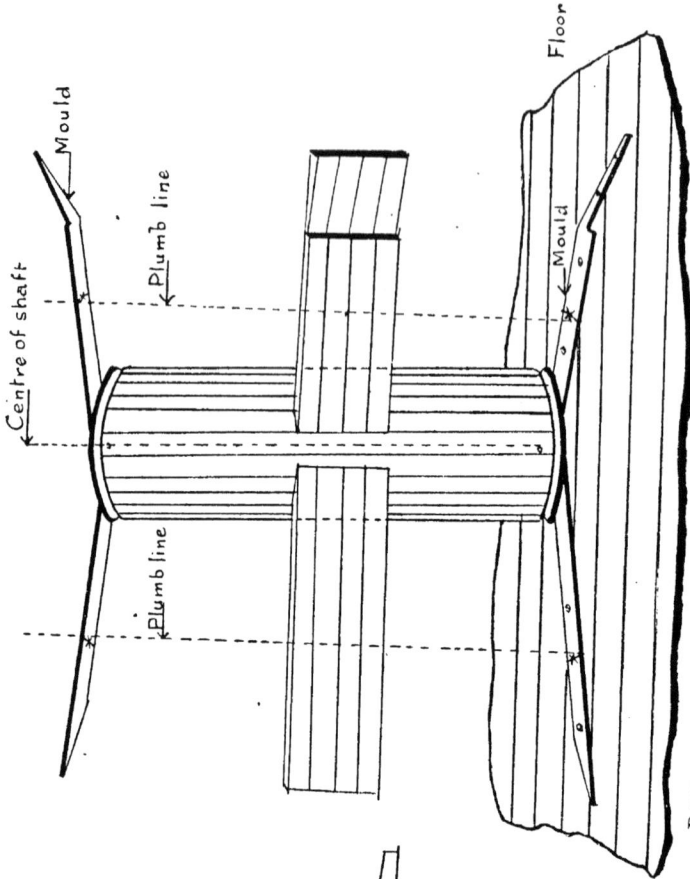

PATTERN FOR A SHAFT BRACKET.

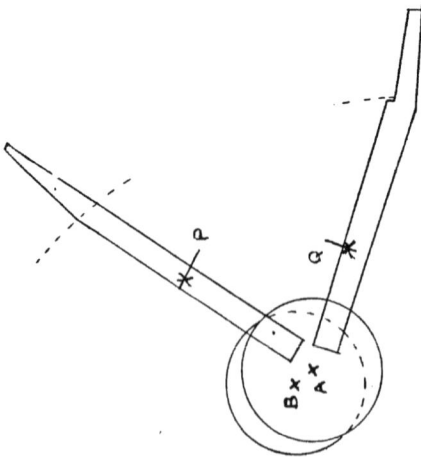

marked across the arms of both moulds, the arms being in gonal planes.

These moulds are now erected as shown in figure, the points PP and QQ on the moulds being plumbed down ; this is, in effect, making the diagonal plane of the arms a vertical plane. The lower mould is secured to the floor, and the upper by battens, etc. (not shown for the sake of clearness). It will now be clear that the centres of the circular pieces of the two moulds give the centre of the shaft, which will be the axis of the barrel of the casting, and will be inclined to the vertical. The vertical distance apart of the moulds will be slightly more than the length of the barrel, to enable the latter to be got in and out.

The barrel is built up as shown on solid ends, with the centres marked, to enable them to be pitched correctly by aid of the moulds.

The arms are now built up of pieces glued and screwed together, and these are fitted to the barrel, the inner ends passing through the barrel and connecting inside. The next process is to form the upper and lower palms, the sections of which are given by the two moulds, and they are formed to agree with these, which is done by a straight-edge, carefully plumbed. The pattern will then be as in the figure, and pieces are now fitted round the arms where they enter the barrel, these being afterwards shaped as Fig. A.

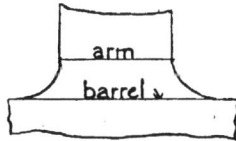

FIG. A.

The arms are finished to a pear-shaped section as in Fig. B.

FIG. B.

The ends of the barrel are formed with rabbets for attachment of shaft casing, and the inside surface gulleted to receive bearing bushes as Fig. C.

FIG. C.

A certain amount is allowed for machining on the palms, as it may happen these will have a slight twist, and also on the inside and ends of the barrel to allow for boring out after the brackets are in position.

The moulds for the brackets can subsequently be used for erecting purposes, as described on p. 72.

LAUNCHING CURVES.

These have already been referred to in dealing with the question of position of ship and height of blocks, and the method of obtaining them will now be described (pp. 264–270).

The ship is assumed to slide down the slip at the slope of the tangent to groundways ; in other words, the camber of the ways is neglected in making the calculations. On the longitudinal section of slip draw in the height of tide at one hour before high water, and find what travel of the ship corresponds to any particular increase of draft as measured on after perpendicular. In the case worked out (see p. 267) the tide line gave a draft of 4 feet on AP for no travel, the draft and corresponding travel being given in columns (1), (2), p. 268. The method then consists in drawing horizontal lines on the profile of the ship on the slip, and finding the buoyancy and C.B. up to each. To do this, take the frame lines at each displacement station which are curves of sectional areas at the stations, and give the cross-sectional area up to the various water-lines. Thus the volume up to any line can be obtained by Simpson's rule, the sectional areas being known.

These curves are conveniently dealt with by making them into a sort of "body plan," as in Fig. A, p. 265. It is necessary to mark at 1 and 21 the points at which a known water-line cuts these stations, and any other water-line can at once be obtained. To obtain the displacement and centre of buoyancy up to any water-line, measure the ordinates of the curves at each station where it is cut by the water-line, and put them through Simpson's rule in the ordinary way, the centre of buoyancy being got by multiplying by the levers in the ordinary way.

To avoid the confusion which might arise through having to draw a set of lines for each water plane on the body, it is convenient to have a piece of tracing paper and draw on it parallel lines to represent the displacement stations as in the figure. The distance apart of these lines will, of course, depend on the inclina-

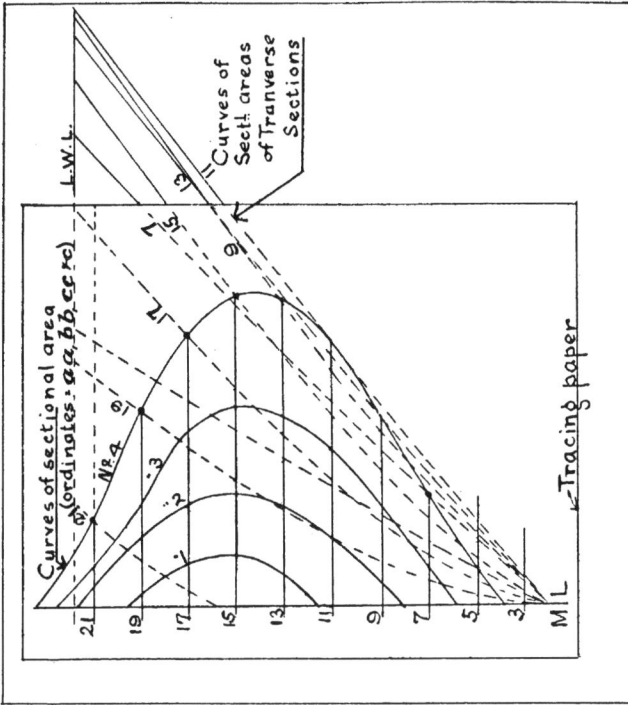

F G . A.

Launching curves.

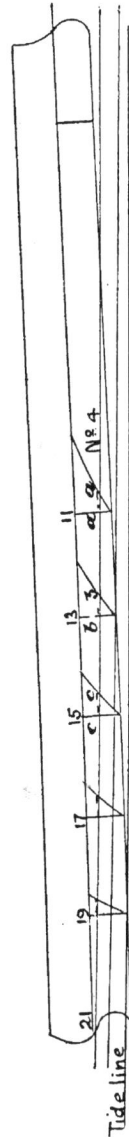

Fig. B.

tion of the L.W.L. of the ship at rest on the blocks with water-lines, and is easily measured.

This tracing is placed on the body plan with the middle lines coinciding, and 21 station is placed in succession at the points where the various water-lines cut the middle line of the body. Thus in the figure the ordinates of the curve, marked No. 4, give the sectional areas below the deepest water-line taken, and its area will be proportional to the buoyancy.

By this method it can be seen if the curves of sectional area are fair, and as the curve for each water-line is distinct, there is little chance of mistakes in reading off.

The areas of these curves being proportional to the buoyancy, a convenient check can readily be made on the calculations by means of a planimeter.

The actual work is given on pp. 267, 268.

The buoyancy B and moment of buoyancy M_B are given in columns (5) and (8), and are plotted on a base of travel of ship.

The weight and position of C.G. of hull is estimated, and the lines showing the weight W and moment of weight about the fore poppets M_W will evidently be straight lines parallel to the base line.

It is now necessary to find out if tipping will occur, and for this the moments of buoyancy and weight respectively about after end of ways must be obtained. As similar curves for moments about fore poppets have been obtained the latter can be deduced from them, as the distance between after end of way and fore poppet is known. To ensure that tipping shall not occur the moment of buoyancy about after ends of ways must be greater than the moments of weight about after end of ways, *i.e.* the former curve must lie above the latter when once C.G. of ship has passed over end of ways. If the stern lifts before the C.G. has passed end of ways there is, of course, no risk of tipping. It has only to be taken into account if ground ways are so short that stern has not lifted before C.G. passes end of ways.

When the moments of buoyancy and weight about fore poppet are equal, the stern will lift and the pressure on fore poppets will be given by intercept between weight and buoyancy curves.

The remaining columns on p. 268 deal with the question of pressure on cradle, etc., throughout the motion of the ship, and the curve obtained from them, p. 269, shows how the intensity of pressure varies up to time of stern lifting.

The total pressure on the cradle is the difference between the weight of ship and cradle and the buoyancy (W − B) in column (6). The ship is taken at different distances down the slip, the length of cradle, *i.e.* bilgeways, still on the ways (column (3)) varying

|CALCULATION OF BUOYANCY AND POSITION OF CENTRE OF BUOYANCY.

23 FEET DRAUGHT AT 21. (Travel = 297·3 feet.)

No. Ord.	S.M.	Ord.	Function dispt.	Lever.	
21	1	2·45	2·45	0	0·0
19	4	4·5	18·0	1	18·0
17	2	7·65	15·3		30·6
15	4	9·55	38·2		114·6
13	2	9·3	18·6		74·4
11	4	7·6	30·4		152·0
9	2	4·6	9·2		55·2
7	4	1·7	6·8		47·6
5	$\frac{3}{2}$	0·3	0·45		3·6
4	2	0·2	0·4	$\frac{1}{2}$	3·4
3	$\frac{1}{2}$	0·0	0·0		0·0
			139·8		499·4

$$B = \frac{50 \times 100 \times 139 \cdot 8}{3 \times 35} = 6657 \text{ tons}$$

$$\text{C.B. before 21} = \frac{499 \cdot 4}{139 \cdot 8} \times 50 = 178 \cdot 4 \text{ feet}$$

21 FEET DRAUGHT AT 21. (Travel = 266 feet.)

No. Ord.	S.M.	Ord.	Function dispt.	Lever.	
21	1	2·25	2·25	0	0·0
19	4	3·63	14·52	1	14·52
17	2	6·3	12·6	2	25·2
15	4	8·1	32·4	3	97·2
13	2	7·65	15·3	4	61·2
11	4	6·0	24·0	5	120·0
9	2	3·25	6·5	6	39·0
7	4	0·8	3·2	7	22·4
5	1	0·0	0·0	8	0·0
			110·77		379·52

$$B = \frac{50 \times 100 \times 110 \cdot 8}{3 \times 35} = 5276 \text{ tons}$$

$$\text{C.B. before 21} = \frac{379 \cdot 5}{110 \cdot 8} \times 50 = 171 \text{ feet}$$

19 FEET DRAUGHT AT 21. (Travel = 234·7 feet.)

No. Ord.	S.M.	Ord.	Function dispt.	Lever.	
21	1	1·85	1·85	0	0·0
19	4	2·85	11·4	1	11·4
17	2	5·22	10·44	2	20·88
15	4	6·5	26·0	3	78·0
13	2	6·15	12·3	4	49·2
11	4	4·27	17·08	5	85·4
9	$\frac{3}{2}$	2·0	3·0	6	18·0
8	2	0·9	-1·8	7	12·6
7	$\frac{1}{2}$	0·2	0·1	8	0·8
			83·97		276·28

$$B = \frac{50 \times 100 \times 83 \cdot 97}{3 \times 35} = 3998 \text{ tons}$$

$$\text{C.B. before 21} = \frac{276 \cdot 28}{83 \cdot 97} \times 50 = 164 \cdot 5 \text{ feet}$$

17 FEET DRAUGHT AT 21. (Travel = 203·4 feet.)

No. Ord.	S.M.	Ord.	Function dispt.	Lever.	
21	1	1·4	1·4	0	0·0
19	4	2·25	9·0	1	9·0
17	2	4·1	8·2	2	16·4
15	4	5·2	20·8	3	62·4
13	2	4·5	9·0	4	36·0
11	4	2·75	11·0	5	55·0
9	1	0·75	0·75	6	4·5
			60·15		183·3

$$B = \frac{50 \times 100 \times 60 \cdot 15}{3 \times 35} = 2864 \text{ tons}$$

$$\text{C.B. before 21} = \frac{183 \cdot 3}{60 \cdot 15} \times 50 = 152 \cdot 3 \text{ feet}$$

15 FEET DRAUGHT AT 21. (Travel = 172·1 feet.)

No. Ord.	S.M.	Ord.	Function dispt.	Lever.	
21	1	0·5	0·5	0	0·0
19	4	1·7	6·8	1	6·8
17	2	3·0	6·0	2	12·0
15	4	3·7	14·8	3	44·4
13	2	2·95	5·9	4	23·6
11	4	1·25	5·0	5	25·0
9	1	0·0	0·0	6	0·0
			39·0		111·8

$$B = \frac{50 \times 100 \times 39}{3 \times 35} = 1857 \text{ tons}$$

$$\text{C.B. before 21} = \frac{111 \cdot 8}{39} \times 50 = 143 \cdot 3 \text{ feet}$$

13 FEET DRAUGHT AT 21. (Travel = 140·8 feet.)

No. Ord.	S.M.	Ord.	Function dispt.	Lever.	
21	1	0·0	0·0	0	0·0
19	4	1·26	5·04	1	5·04
17	2	2·15	4·3	2	8·6
15	4	2·38	9·52	3	28·56
13	$\frac{3}{2}$	1·5	2·25	4	9·0
12	2	0·75	1·5	$4\frac{1}{2}$	6·75
11	$\frac{1}{2}$	0·0	0·0	5	0·0
			22·61		57·95

$$B = \frac{50 \times 100 \times 22 \cdot 61}{3 \times 35} = 1076 \text{ tons}$$

$$\text{C.B. before 21} = \frac{57 \cdot 95}{39} \times 50 = 74 \cdot 2 \text{ feet.}$$

CALCULATIONS FOR LAUNCHING **D**, E t.

Draught at after perpendicular. 21 stat. (1)	Corresponding travel of cradle. (2)	Length of cradle on ways. L (3)	Weight of ship. W (4)	Buoyancy, B (5)	$W-B$ (6)	Pressure (p) $\frac{W-B}{L}$ (7)	Moment of buoyancy about fore poppet, M_B (8)	Moment of weight about fore poppet, M_w (9)	M_w-M_B (10)	Centre of pressure from fore poppet, $\frac{M_w-M_B}{W-B}=y$ (11)	$x=(L-y)$ (12)	$\frac{6px}{L}$ (13)	$2p$ (14)	$a=\frac{6px}{L}-2p$ (15)
feet	feet	feet	tons	tons	tons	tons	ft-tons	ft-tons	ft-tons		feet			feet
4	0·0	376·0	7930	0	7930	21·2	0	1,427,400	1,427,400	180·0	198·0	66·2	42·4	23·8
11	109·5	376·0	7930	554	7376	19·6	168,000	1,427,400	1,259,400	170·7	205·2	64·1	39·2	24·9
13	140·8	376·0	7930	1075	6854	18·25	307,500	1,427,400	1,119,900	163·3	212·6	61·9	36·5	25·4
15	172·1	348·9	7930	1855	6073	17·3	506,100	1,427,400	921,300	151·7	197·2	58·6	34·6	24·0
17	203·4	317·6	7930	2858	5066	15·95	747,100	1,427,400	680,300	134·2	183·5	55·2	31·90	23·3
19	234·7	286·3	7930	3975	3932	13·73	993,300	1,427,400	434,100	110·4	176·7	50·8	27·46	23·34
21	266·0	255·0	7930	5275	2654	10·4	1,281,000	1,427,400	146,400	55·1	199·9	48·9	20·8	28·1
23	297·3	213·7	7930	6670	1273	5·9	1,571,000	1,427,400	143,600	—	—	—	—	—

in corresponding manner, and the average pressure is simply $\dfrac{W - B}{L}$. It is assumed that the pressure on the cradle will vary uniformly, so that it is necessary to slope this pressure line to bring

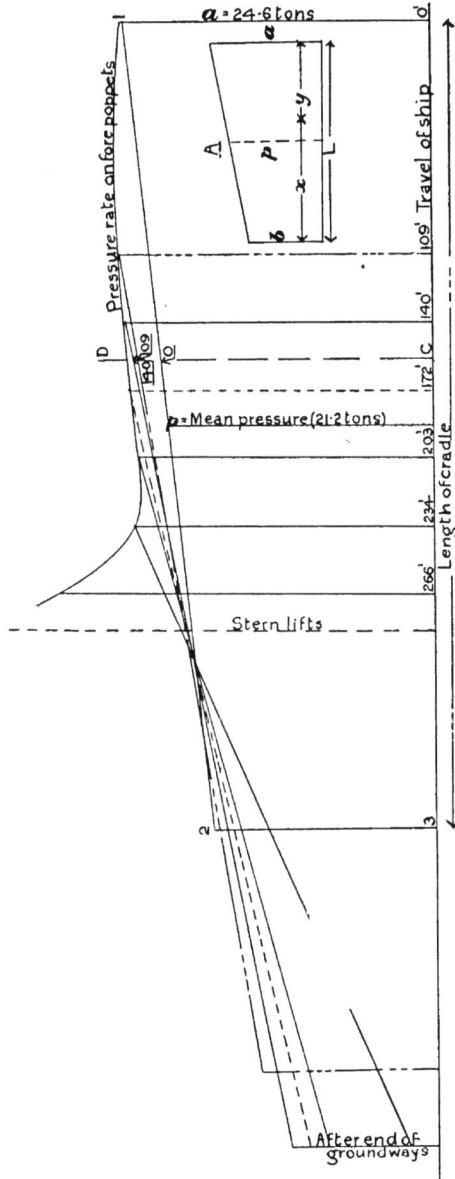

the centre of pressure at the distance $\dfrac{M_B - M_W}{W - B}$ from fore poppet.

From column (8) we obtain moment of buoyancy and from

column (9) moment of weight, column (10) giving the difference
$M_B - M_W$. Column (6) gives $\overline{W - B}$, and so the distance of centre
of pressure on cradle from fore poppet is at once obtained and
is shown in column (11). The length of base is length of cradle on
ways, so that mean ordinate p is got at once and set up as in inset
diagram in Fig. on p. 269. From formula in column (15) the end
ordinate a can be got and pressure line drawn in line through the
tops of a and p.

These pressure lines are constructed for ship in different
positions down the slip, and then a curve is passed through all the
points corresponding to fore poppet, the result being a curve giving
pressure rates on fore poppet for all positions of ship.

If now a cross curve be plotted at any section as D in the figure,
its ordinates will show how the pressure rate varies at this point,
throughout the motion of the ship down the slip.

DRAFT MARKS.

These are the marks on the ship's bottom which give the
draft of the ship at the point where they are placed. They are

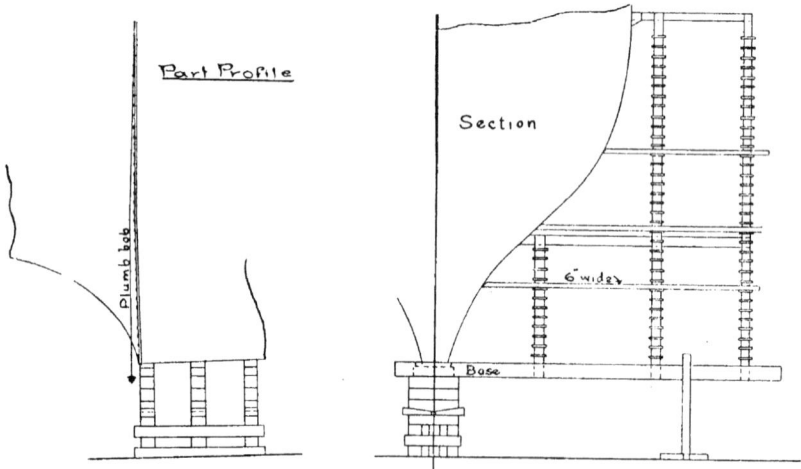

SETTING OFF DRAUGHT MARKS.

set up from the keel line, or keel line produced, and square to
the load-water line, and generally at the points of cut up forward
and aft. The draughts are indicated by Roman figures 6 inches
deep, vertically the bottom of the figure giving the number of feet
from the •keel line. Where the bottom of the ship departs from

the vertical, the actual width of the figures as measured on the plating will be more than 6 inches.

Standards are erected at the position where the marks are to be placed with cross battens 12 inches apart, as indicated on p. 270. These standards rest on a levelled base, touching the keel and square across the ship. The standards are next placed square to the L.W.L., which can be done by plumbing or by a declivity base.

It will be necessary to have two or three of these standards at the after end owing to the form of the bottom, while two only will be required forward.

To mark the top and bottom of the figures on the bottom plating a batten 6 inches wide, with one end tapered to a point, is used.

This batten is placed on the cross pieces and slid in till the point touches the bottom ; this will be the bottom of the figure. It is then turned over, resting still on the same cross pieces, and the point will then show the top of the figure. These points are indented in and the upper and lower edges of the figure drawn parallel to the L.W.L. by means of a small declivity batten and spirit level.

After the figures have been marked off they are centre-punched in.

In the case of torpedo boats and destroyers three sets of draught marks are set up—

(*a*) A set of marks on the after propeller brackets, measured from a line drawn parallel to the straight line of keel produced and touching underside of the lowest propeller sweep.

(*b*) A set of marks forward from the keel ; the position in a fore and aft direction to be at the touching point of a tangent line drawn to touch lowest propeller sweep aft and the keel forward.

(*c*) A set of marks at the elbow or lowest point of keel measured from the keel. Above these marks on the upper deck a gun-metal plate is fixed and engraved " For docking purposes only."

Paint or Copper Line. A line is drawn parallel to the L.W.L. and about 12 inches above it. The spring or sheer of the line is set off at the ends of the ship above the parallel line, the amount being about 18 inches forward and 6 inches aft. Divide the level line into a number of equal intervals and erect ordinates. The heights to be set up on these ordinates are obtained by the formulæ, viz. $\left\{ h \times \left(\dfrac{n}{m} \right)^2 \right\}$, where h is the amount of spring above parallel line, m the number of equal intervals between amidships and forward or after ends, and n the number of ordinate. Assume ten ordinates used from amidships and 18 inches spring, at a third ordinate from amidships the spring would be equal to $(\frac{3}{10})^2 \times 18$ inches $= 1\cdot62$ inch. The curve is faired in the sheer by girthing the water-line and setting off on the expanded ordinates the amount of spring at each when a batten is made to pass as nearly as possible through these points consistent with fairness.

The line so obtained is transferred to the ship by setting up bases and levelling and sighting them.

INCLINING EXPERIMENTS.

These are for determining the position of the centre of gravity of a ship. The theory of the method adopted is that if a weight be shifted across a ship this will alter the centre of gravity and hence produce a list of the ship to bring the C.B. and C.G. in the same vertical line. The shift of the C.G. can be found when that of the weight is known thus—

Let a weight W be shifted across the ship, causing her to heel θ degrees, then G moves to G_1, and B to B_1; then θ being small, M is the meta-centre obtained from the formula—

$$BM = \frac{I}{V}; \text{ also } GG_1 = GM \cdot \tan \theta;$$

now
$$wd = W \cdot GG_1 = W \cdot GM \cdot \tan \theta$$

$$GM = \frac{wd}{W \cdot \tan \theta}$$

Now M is known from formula $BM - \dfrac{I}{V}$

$$\therefore \text{ G can be found at once.}$$

The success of an inclining experiment depends largely on the precaution taken in preparing the ship and taking the readings. To obtain the best result the ship should be as complete as possible, but it is not always possible to arrange for the inclining to be done at this stage, and so a complete account has to be taken of all tool chests, lumber and other gear to be removed, of all weights to be added, and of gear on board which has not been put into its final position.

All boilers, tanks, etc., which contain water, oil, or oil fuel, must be full or empty, all double bottom compartments opened up and examined. Any coal on board must be carefully measured, and for greater accuracy in estimating, trimmed level. All weights, as boats, derricks, anchors, etc., likely to move when the ship heels must be lashed. All boats examined for water.

The ship should be as upright as possible and not have excessive trim. These can be corrected if necessary by filling double compartments the capacity of which is known and allowed for.

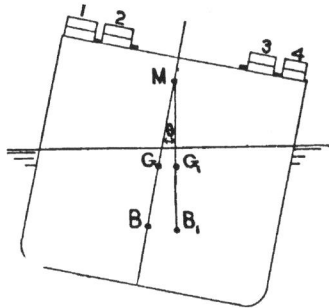

The inclining ballast is in 1 cwt. pieces, and is placed on the upper deck in rows parallel to the middle line to obtain as great a shift as possible. For a battleship 100 to 150 tons of ballast will be required, arranged into four rows, two each side of the ship, as in above diagram. The ballast is placed on battens to prevent damage to the deck.

Two plumb bobs are generally used, 15 to 20 feet in length, placed one at each end of the ship.

To carry out the experiment the ship must be placed head to wind clear of jetty, with all hawsers slack or, better still, in a basin or moored to a buoy in midstream. If inclined alongside a jetty, inclining should be done at slack tide.

Proceed next to check spread and quantity of ballast, place all men at middle line of ship, and then read drafts forward and aft, port and starboard ; this will show if ship has any list and also amount of trim.

Now shift ballast as follows :—

1st shift.	Top tier of 1 on to top tier of 3	Reading
	Top tier of 2 on to top tier of 4	
2nd shift.	Remainder of 1 on top of 3	Reading
	Remainder of 2 on top of 4	
3rd shift.	Replace ballast to bring ship upright and check zeros.	
4th shift.	Top tier 4 on top of 2	Reading
	Top tier 3 on top of 1	
5th shift.	Remainder of 4 on top of 2	Reading
	Remainder of 3 on top of 1	

Thus four sets of readings will be obtained, and the one to be taken will be the mean of the four.

After the inclining has been carried out the drafts should be taken again to see if any water has got into the ship during the inclining, or any weights shifted. It might happen that a valve has not been properly closed and water enter in a compartment, and this could be discovered from the draft, investigated and allowed for in the subsequent calculations.

When inclining in the stream or alongside a jetty the water is not always still and the drafts will not be easy to read, but if a glass tube be used the still water surface can be readily seen, and the draft corresponding to it measured.

The pendulums are placed in hatchways, and must be securely fixed at the upper end, with no possibility of the string touching anything when ship has extreme list in both directions. This should be tested before commencing operations. It will be found that though the ship is apparently perfectly steady the plumb-bob will make oscillations of more or less amount about a certain mean position, and it is this mean position which must be found and marked on the base to obtain the tangent of the angle of inclination.

In small boats the pigs of ballast should not be placed on top of one another, as if the boat lists the ballast may slide off the deck. Also no one should be on board when taking readings, but the pendulum be rigged at bow and stern and readings taken from a boat. As these boats oscillate under the slightest disturbance of the water the inclining should be done in a sheltered spot.

The formula for finding the metacentre being only true for

small angles, a preliminary calculation is made to find the amount of ballast which will give the ship an inclination of about 3 degrees from the vertical. The number of men required for shifting ballast is about one man per ton of ballast.

CIRCLE TURNING TRIALS.

These are trials carried out to determine the steering qualities of a new ship.

Three sets of " circles " are obtained : (1) At full speed. (2) At twelve knots. (3) One engine ahead and one astern at 12 knots rudder assisting, in all cases with the rudder at 35 degrees.

The method of determining the curves is as follows :—

A buoy or any easily distinguishable floating object is thrown overboard from the ship, which then proceeds about two miles, when a second buoy is thrown over. The reason for having two buoys is that after the ship has circled round one buoy, there will be time for her to recover her speed before circling round the second.

The ship then steams back to the first buoy, and by means of instruments at bow and stern the angles between the lines join-ing the buoy and instruments and the middle line of the ship are found, giving a series of triangles with the buoy as apex. The readings are taken every time the ship turns four points (45 degrees) from the start as measured by the ship's compass, and thus to complete each circle nine readings will be necessary. In order that the readings may be taken simultaneously, some signal is necessary from the bridge, where the angle through which ship is turning is read, and the observers at the instruments forward and aft. This signal is conveniently made by means of the syren. The times of putting helm hard over and subsequently through each four points is also taken, and observers stationed in the steering compartments take these times and check angle of helm shown by the indicator with results obtained on the bridge. The instrument for measuring angles is simply a telescope mounted on a vertical axis, and also being free to move round a horizontal axis with a pointer passing over a graduated arc, as shown on p. 276.

At the first signal, readings are taken and helm put hard over simultaneously. In recording the readings it is necessary to

distinguish between cases when buoy is on the forward or after side
of observers, and it is convenient to do so by recording whether
eye-piece of telescope is forward or aft of the line through the
centre of graduated arc at right angles to middle line of ship.
Thus in the figure the reading would be 12 degrees *before* for
forward instrument, and 47 degrees *abaft* for after instrument.
Draw all the triangles on tracing paper with a line as common
base representing to scale the distance apart of the instruments,
giving a figure which really shows the motion of the flag relatively

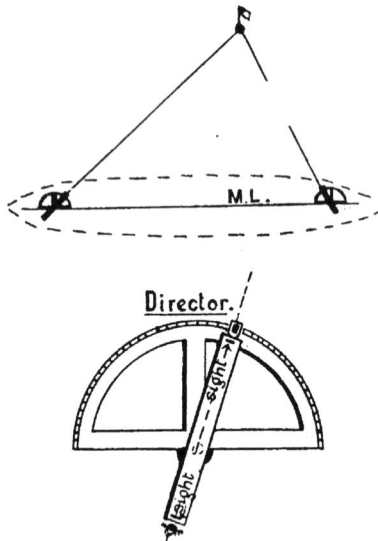

to the ship, the ship considered as at rest; and it can be seen
from this if there are any doubtful results before the circle itself
is plotted. To plot the circle, draw through a point lines at 45
degrees, the apex of the triangle being placed at the point which
will represent the buoy, its base being at right angles to that
marked 1 (p. 277); and then take second triangle and place apex
at spot with base at right angle to 2, and so on until the nine different positions of the ship have been marked off. The turning
circle is the curve passing through the middle points of the lines.

The information obtained from the curves is—

(1) **The Advance.** That is, the distance the ship moves forward
in turning through eight points (90 degrees) from first putting the
helm over.

(2) **The Tactical Diameter.** That is, the distance the ship has

moved parallel to herself by the time she has turned through sixteen points (180 degrees).

(3) The time taken in turning through each four points.

If the whole circle is not required, the ship can be circled round the two buoys figure eight fashion, which will give sufficient information to allow of (1) and (2) being obtained.

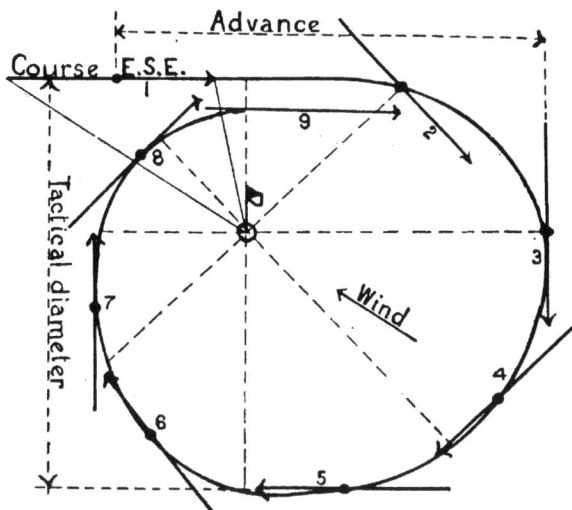

TESTING FITTINGS.

The link plates on mast and fittings on derrick itself are tested to the amounts shown on the sketches, all these parts being tested in the cable-testing machine before they are fitted in place, and also incidentally when the derrick is tested.

BOAT DAVITS.

Tested to twice the working load, and, as each davit takes half the weight of boat suspended from it, the test load will be equal to the weight of the boat with all its equipment and men.

BOAT SLINGS AND LIFTING PLATES.

The tests for these are given on the sketches of boats.

Capstan Gear.

This is tested at the makers' works, and also after fitting on board. The following is a sample of the test for a battleship :—

(1) Speed of hauling in slack cable is 40 feet per minute.

(2) Heaving and veering with two anchors and cables laid out : (*a*) heave one or both anchors ; (*b*) veer one or both anchors ; (*c*) heave one anchor and veer the other—all at 25 feet per minute.

At the makers' works the capstans were required to lift a load of 37 tons at the rate of 25 feet a minute.

Steering Gear.

This is tested on the trials of the ship, the requirements being (1) that the steering engines shall be capable of moving the rudder from hard over to hard over in both directions in   minutes (the steering engines are connected up in turn for this test) ; and (2) that with the hand gear only, at a speed of   knots, it shall be possible to move the rudder from hard over to hard over in a reasonable time ; (3) that when going astern at   knots the rudder can be put hard over to hard over in   minutes (this astern speed will depend on the full speed of the ship). This latter test is generally the most severe test of all, as the centre of pressure on the rudder is at a much greater distance from the axis of the rudder when going astern than when going ahead, causing a greater moment to come on the rudder head. When going ahead the rudder is more or less balanced, *i.e.* the centre of pressure is somewhere near the axis of the rudder.

Tests for Steel Derricks.

(*a*) Lift and lower a load of[1]   tons carried at the point of suspension through the full lift at $1\frac{1}{2}$ feet per second.

(*b*) Lift top and lower a load of   tons through full travel of derrick, speed of end of derrick not less than   feet per minute, any portion of derrick.

(*c*) Lift top and lower   tons, load moved through full range

---

[1] The test load depends on the weight of the heaviest boat to be lifted by the derrick. •

due to each motion, and derrick swung across the ship, to test all the fittings attached to the mast.

(*d*) To hold at a short distance from the ground, in any selected position of the derrick, and to lower a load of   tons carried at the point of suspension to test the strength of all parts, to be carried out on both sides of ship.

(*e*) Lower the maximum load of   tons at the greatest rate permitted by the gear.

CHAIN CABLE.

Size of cable.	Breaking strain.	Proof strength.	Weight of 100 fathoms.	Size of cable.	Breaking strain.	Proof strength.	Weight of 100 fathoms.
ins.	tons	tons	cwts.	ins.	tons	tons	cwts.
$3\frac{1}{2}$	264·6	176·4	588·0	$1\frac{7}{8}$	94·9	63·3	168·15
$3\frac{1}{4}$	242·4	161·6	507·0	$1\frac{3}{4}$	82·7	55·1	147·0
3	218·7	145·8	432·0	$1\frac{5}{8}$	71·3	47·5	126·15
$2\frac{3}{4}$	194·0	129·3	363·0	$1\frac{1}{2}$	60·8	40·5	108·0
$2\frac{11}{16}$	187·8	125·2	346·15	$1\frac{3}{8}$	51·0	34·0	90·15
$2\frac{9}{16}$	177·3	118·2	315·0	$1\frac{1}{4}$	42·2	28·1	75·0
$2\frac{1}{2}$	168·8	112·5	300·0	$1\frac{1}{8}$	34·1	22·8	60·15
$2\frac{3}{8}$	152·3	101·5	270·15	1	27·0	18·0	48·0
$2\frac{1}{4}$	136·9	91·1	243·0	$\frac{7}{8}$	20·6	13·8	36·15
$2\frac{1}{8}$	121·9	81·3	216·15	$\frac{3}{4}$	15·2	10·1	27·0
2	108·0	72·0	192·0				

The breaking strain must not be less than $1\frac{1}{2}$ times the proof strain.

The space required for stowage of chain cable is $35d^2$ cubic feet per 100 fathoms, *d* being diameter of cable iron in inches. This is for close stowage; but for rough stowage about $\frac{1}{3}$ should be added. Weight of 100 fathoms stud link cable $= 2\cdot4d^2$ tons (*d* = dia. cable in inches).

Cable clench is proved to the full breaking strength of the cable.

CHAIN RIGGING.

Size of chain.	Breaking strength.	Proof strain.	Weight per fathom.	Size of chain.	Breaking strength.	Proof strain	Weight per fathom.
ins.	tons	tons	qrs. lbs.	ins.	tons	tons	qrs. lbs.
$1\frac{3}{8}$	50·9	22·6	3 24	$\frac{9}{16}$	8·4	3·8	0 21
$1\frac{1}{4}$	42·0	18·8	3 8	$\frac{1}{2}$	6·8	3·0	0 17
$1\frac{1}{8}$	34·3	15·3	2 17	$\frac{7}{16}$	5·0	2·2	0 $13\frac{1}{2}$
1	27·0	12·0	2 5	$\frac{3}{8}$	3·6	1·6	0 $9\frac{1}{2}$
$\frac{15}{16}$	23·6	10·5	1 25	$\frac{5}{16}$	2·5	1·1	0 $6\frac{3}{4}$
$\frac{7}{8}$	20·5	9·1	1 20	$\frac{9}{32}$	2·0	0·87	0 $5\frac{3}{4}$
$\frac{13}{16}$	17·7	7·9	1 11-	$\frac{1}{4}$	1·7	0·75	0 $4\frac{3}{4}$
$\frac{3}{4}$	15·2	6·8	1 8	$\frac{3}{16}$	0·9	0·4	0 3
$\frac{11}{16}$	12·7	5·6	1 2	$\frac{1}{8}$	0·4	0·19	0 2
$\frac{5}{8}$	10·4	4·6	0 25				

Slips, eyes, etc., used with this chain are tested to twice the proof strain of the chain.

STEEL WIRE ROPES.

Flexible steel wire rope for hawsers and running rigging.			Steel wire rope for standing rigging.		
Size of rope.	Weight per fathom.	Breaking strain not less than	Size of rope.	Weight per fathom.	Breaking strain not less than
ins.	lbs	tons.	ins.	lbs.	tons.
8	53	170	$4\frac{1}{2}$	19	51
$6\frac{1}{2}$	35	112·7	4	$15\frac{1}{2}$	40
6	31	96·6	$3\frac{1}{2}$	$11\frac{1}{2}$	32
$5\frac{1}{2}$	28	81·6	3	8	24
5	23	67·8	$2\frac{1}{2}$	6	16
$4\frac{1}{2}$	15	44·8	2	4	10
4	12	35·6	$1\frac{3}{4}$	3	8
$3\frac{1}{2}$	9	27·6	$1\frac{1}{2}$	2	6
3	7	19·5			
$2\frac{1}{2}$	$4\frac{1}{2}$	13·5			
2	$2\frac{3}{4}$	8·0			
$1\frac{3}{4}$	2	6·3			
$1\frac{1}{2}$	$1\frac{3}{4}$	4·6			
$1\frac{1}{4}$	$1\frac{1}{4}$	3·3			
1	$\frac{3}{4}$	2·0			

RIDING BITTS.

Diameter of bitts.	Diameter of cable iron.
ins	ins.
16	$1\frac{1}{2}$
18	$1\frac{5}{8}$
20	$1\frac{3}{4}$
22	$1\frac{7}{8}$
24	2
26	$2\frac{1}{8}$
28	$2\frac{1}{4}$

For steel-wire hawsers, bitts should not be less than 4 times the circumference of the hawser.

SHEAVES FOR STEEL-WIRE ROPE.

Circumference of rope.	Diameter of sheave.
ins.	ins.
3	14
$3\frac{1}{4}$	14
$3\frac{1}{2}$	16
4	18

DOCKING SHIPS.

The first operation is to ascertain if the dock is long and broad enough, which is done by placing the profile of the ship on a

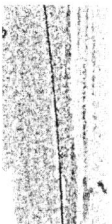

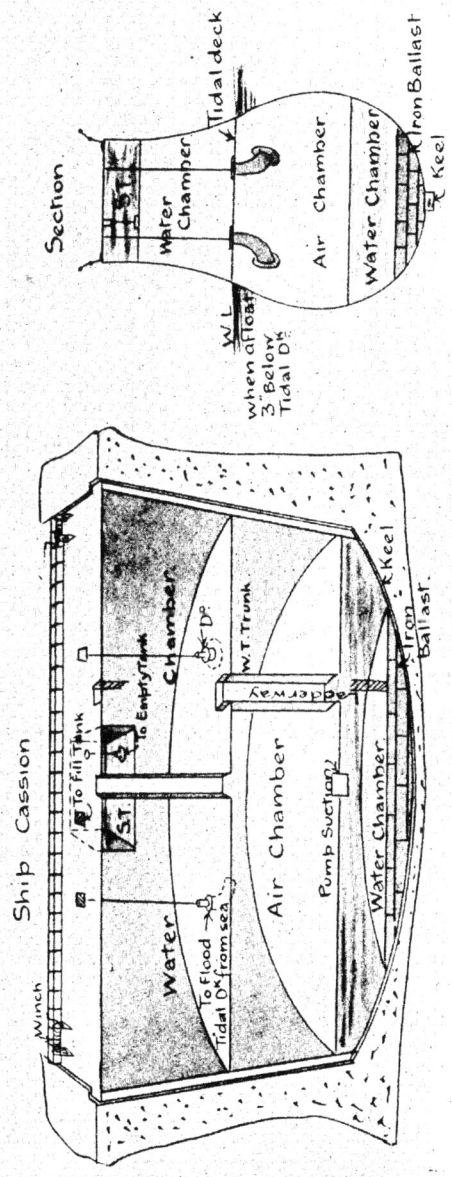

longitudinal elevation of the dock, and noting if the ship is clear at the ram, and that the stern, or stern walk if there is one, is also clear of the caisson. The bilge of the caisson projects some distance into the dock, and must be taken account of when considering the length of ship that can be docked. Next take a section of the dock and one of the ship at its fullest part, and show on the section of the ship all projections on the side, as torpedo net shelves, gun sponsons, bilge keels, docking keels, etc., and see if these clear the dock steps when the keel rests on the blocks (Fig. A, p. 283).

It has now to be determined whether the ship will clear the dock entrance and pass over the sill. On the sketch of dock entrance the datum line is shown, this being the low water ordinary spring tide (L.W.O.S.T. line), and the depth of the water on the day is obtained by setting up from this the rise of tide given in the Tide Table. To ensure getting the caisson in place before tide turns, the ship is brought into the dock about an hour before high water, and so the height to set up will be about one foot less than the rise of tide. On the section of the ship draw the draught line, and place this to coincide with the tide line, and it will then be easy to see if the projections on her side will clear. For ships of great beam in which there is no rise of floor, it will be found that the turn of the bilge will determine the clearance between the sill and bottom, as indicated in Fig. B, p. 283. The clearance will depend on height of the tide and draught of ship, as the slope or batter of the dock walls at the entrance lessens the clearance the lower the tide. It is also necessary to note the *maximum* draught of the ship, to see if this will permit of her passing over the sill and blocks, as, should the maximum draught be at the bow and the blocks have a declivity, the bow may touch the blocks before the ship has been got right into the dock. The trim, if excessive, can be lessened by filling double-bottom compartments, which brings her more on an even keel, but of course also increases the mean draught ; the latter being very much less than the former, it need not be considered.

Assuming the ship will go into the dock it is necessary to obtain the lengths of the shores. First select the positions at which these are to be placed, which will be at or near watertight frame and bulkheads, decks, longitudinals, etc. The upper tier or tiers of shores termed the breast shores, will be nearly horizontal, sloping slightly upwards from the dock side to the ship, and the number will depend on the length of the ship, being about 15 to 20

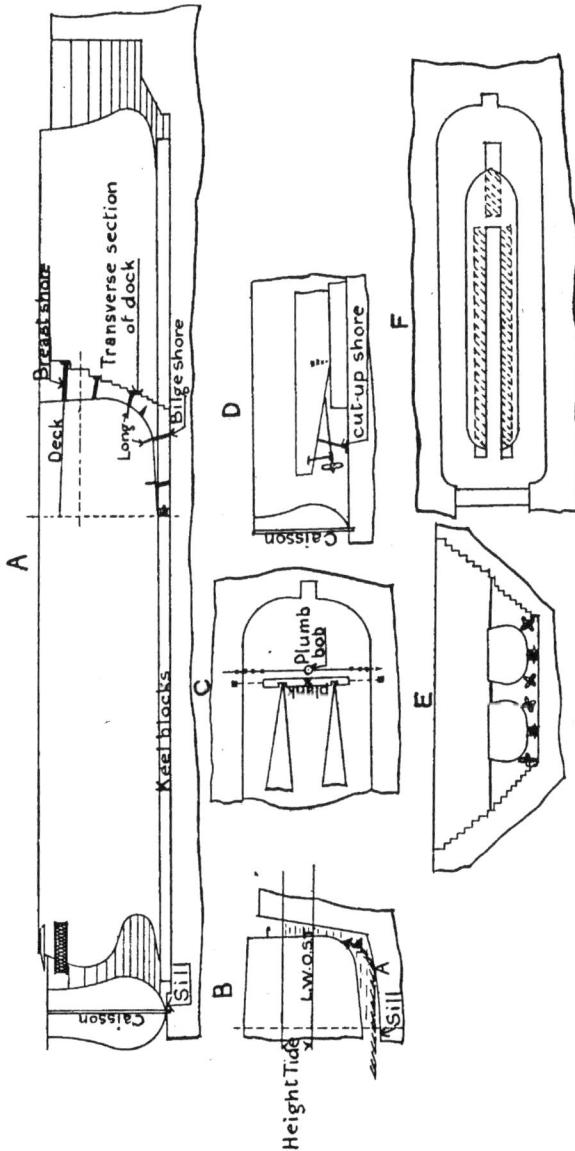

each side for a long ship. If the section of the ship be placed on that of the dock it can be seen on which altar the heels of the breast shores will come, and at the same time where the heads will come against the ship's side. A plan of the ship is made at this position, and put on a plan of the dock, and the lengths of the breast shores measured, and are cut about 6 inches short to allow for packing pieces and wedges. At the ends of the ship the shores will stand square to the ship's side, and allowance is made for this when measuring lengths. The breast shores must not be placed at positions so far forward as to incline them at too large an angle to the dock sides, or there will be difficulty in securing their heels, unless these should happen to come at the curved portion of the dock, and the bevel of the head of the shore must not be excessive, or the shores will not grip the ship's side. The position at which the bow and stern will come is marked on the coping of the dock, and plumb-bobs placed at these marks after the ship is docked.

For future reference a record is kept of the lengths of the shores for each ship in the docks in which she can be placed.

The docking blocks are examined to see that their upper surface is straight, and that they are all secured. The blocks are spaced about 2 feet 6 inches apart from centre to centre, and secured by means of chains to eye-bolts in the floor of the dock. The upper surface of the blocks is horizontal in some docks, and at a declivity of from one to two feet in others. All the shores having been prepared, lines are placed across at intervals to prevent the hawsers used for hauling the ship into the dock from sinking to the bottom when slack, and tripping the blocks when again made taut. These lines are termed "tripping lines," or " swifters."

The dock is flooded, the caisson. floated, and hauled clear of the dock entrance. The ship's rudder is locked amidships, and all watertight doors opened, as it may be difficult to open these when the ship is settled on the blocks.

It should carefully be noted when the ship comes in that there are no ropes, etc., hanging over the sides of which the ends are not visible, as these may go so far down below the water as to trip the blocks. The ship is guided along the centre of the dock by means of plumb-bobs at bow and stern, held by men on both sides of the dock, and it is important to check these bobs to see that the knots in the lines are at equal distances from the bob. This is done after the lines have been wetted, as if only portions of the lines are wetted there may be a difference between the two parts, and the bob when hanging will not point over the centre of the dock. The ship is kept in position by steel hawsers at bow and stern, to which jigger tackles are attached to permit of the ship being kept central as the surface of the water falls. Before the caisson is replaced, the groove is examined by the diver, to see it is all clear. It is then sunk, and, after the ship's bottom has been examined by divers, pumping is commenced. To enable it to be seen when the ship takes the blocks, the water-line is marked at bow and stern (sewing marks) while she is afloat, and directly it is seen one end is touching, or "sews," that end is breast shored. When the ship is taking all along her length, the whole of the remaining breast shores are shored up simultaneously. The opportunity is taken to scrub the bottom as the water falls in the dock and the remaining tiers of shores are got in place. As the ship will be insulated when she rests on the blocks, it is necessary to provide electrical continuity to avoid risks to men when working in the dock should she be struck by lightning ; this is conveniently done by means of a steel wire hawser from a bollard passing into the water outside. In some cases, where the ship is being docked only for testing gun-sights or other work not necessitating pumping the dock dry, breast shoring only is necessary, and the dock pumped till water level is a few feet below water-line. The water affords support to the ship, but it must be observed that level of water does not rise through leakage or otherwise, or there will be danger of ship lifting and disturbing all the shores.

If unacquainted with a dock which has already been flooded, it is necessary to send down the diver, to see that all the blocks are correct.

The preceding remarks to the case of docking an ordinary ship, and a few special cases, will now be considered.

Torpedo Boats and Destroyers. In many of these boats the propeller is some distance below the line of keel, and it is necessary to see that the blocks are high enough for the propeller blades to clear the floor of the dock. These boats are often docked abreast ; and those selected must be floating at the same draught, and be about the same depth, so that they shall take the blocks simultaneously, and that the breast shores between them shall be approximately horizontal. Further, it must be seen that their propellers are clear of the steps at the side of the dock, and that there is sufficient distance clear of the caisson to allow of withdrawing the shaft. To ensure the boats being centrally over the blocks, an arrangement is made use of as shown in Fig. C, p. 283. Shores are usually placed under the after cut-up, as there is considerable overhang in these boats.

Dredgers and Hoppers. In these craft there is a central well, so that middle-line blocks can only be placed at the bow and stern, and for the remaining portion two rows of side blocks must be provided, as shown in Fig. F, p. 283.

Ships with Excessive List or Trim. If this cannot be corrected by filling compartments, great care must be taken to hold the ship directly she touches, which is done by getting the breast shores immediately over the point at which she touches at once in position, but not shoring them, simply keeping them ready to hold the ship, should she show any tendency to list. To prevent the risk of these shores catching the ship as water surface falls, sufficient clearance should be allowed between end of shore and ship's side. This clearance is in addition to the amount always allowed for the packing-pieces and wedges at the heel of all shores. If a ship has excessive list, it will not be easy to tell if she is centrally on the blocks by means of plumb-bobs, and it is best in such case to send down a diver to make sure.

Ships with a Considerable Length of the Bottom Plating at the Middle Line damaged. It will be clear that no blocks can be placed at the damaged part, and it is therefore first necessary to send down a diver to ascertain the extent and area over which the damage has been sustained, and arrange the blocks accordingly.

Ships with Docking Keels. The docking keels are parallel to the middle line of the ship, and distant from it about 24 feet. The lower surface of the keels is generally at the same level as the flat keel; but in some ships it is above this, and it is necessary to find out from the docking plan the correct particulars. The blocks for the side-docking keels are built up to the correct height, so that the ship shall take the three sets of blocks simultaneously.

UNDOCKING.

Before flooding the docks it must be ascertained that all openings in the ship's bottom, such as valves, have been closed. In ordinary ships the bilge shores are removed before undocking ; but if any of these are left, they are lashed together and hauled clear when the ship floats.

In the case of shores under an overhanging position at bow or stern, these must be taken away before flooding the dock, as, if the ship does not lift off the blocks along her whole length simultaneously, then these shores may damage the bottom plating : this applies particularly to torpedo boats and destroyers. It is important that a record be kept of all weights placed on board or shifted while in dock, so that it can be determined what her draughts will be when undocked. In the case of a ship of deep draught undocking on a low tide this point may be of importance.

All new ships, after completion, are sighted for breakage on the first two occasions of docking, armoured ships both transversely and longitudinally, and unarmoured the latter only.

CAISSONS.

These are of two types, viz. ship or floating, and sliding.

Floating Caisson. This is a ship-shaped vessel with sloping ends, corresponding to the batter of the dock entrance, strongly built of steel, and divided up internally into chambers as shown on p. 281. The lowest chamber contains iron ballast at the bottom part, the remaining space, up to the top of the watertight trunk, being always kept filled with water. The middle chamber is an air-chamber, a pump being provided for clearing any leakage. The upper chamber can be flooded by means of valves on the tidal deck, and just below the deck is a sinking tank, provided with a valve in the bottom.

When the sinking tank is empty, the tidal deck clear of water and valves closed, the caisson is so designed as to float with this deck about 3 inches above the water. Filling the sinking tank brings this deck below water, and on opening the flood valves the caisson sinks rapidly. When it is desired to lift the caisson the valve in the sinking tank is opened, and the water falls on to the tidal deck and through the flood valves into dock. The flood valves are then closed, and when the water rises the caisson lifts out of the groove.

Staffs placed on the dock side show if the caisson is sinking level, any adjustments being made by means of the winches.

In some caissons the upper water-chamber is divided into several compartments, each with its own flood valves. This method has the advantage that the trim of the caisson is more under control when being sunk in the groove, as the compartments can be flooded as necessary to keep the caisson horizontal.

COMPOSITIONS, PAINTING, ETC.

All confined spaces as double-bottom spaces, except those for storage of oil fuel and reserve feed water, are coated with oxide of iron, the oil fuel and reserve feed water spaces being thoroughly cleaned and coated with mineral oil.

Red lead is not used for confined spaces owing to the poisonous effect it has on men working in such spaces, but it can be used for all other spaces.

All living spaces are painted, the underside of decks being cork cemented to absorb moisture, the cork being afterwards coated with paint.

Fresh water storage tanks are coated with rosbonite.

The compositions applied to the bottom of a ship below the water-line are protective and anti-fouling, the protective being applied first and then the anti-fouling. In the neighbourhood of the water-line a special composition known as "boot-topping" is applied, which will withstand the constant changes from wind to water that this part of the ship experiences. It is a protective composition, not anti-fouling. Above the boot-topping ordinary grey paint is used.

Protective composition can be applied and left for any length of time exposed to the air, but anti-fouling must only be applied just prior to the ship being undocked, or in a new ship, just before launching.

ZINC PROTECTORS.

The method of fitting zinc protectors to floor plates and bulk-head stiffeners is shown below. These protectors are fitted in reserve feed, feed, overflow feed, drain and hydraulic tanks, and also on the inner bottom in the engine and boiler rooms to arrest corrosion, and are left bright and unpainted. On the shell of the ship zinc protectors are fitted at all openings, *e.g.* sea cocks, torpedo tubes, engine-room inlets, and also in the vicinity of the propellers and around the rudder head and to the inside of the manhole covers in shipbuilders' tubes. The fastenings for these protectors should not pierce the bottom plating.

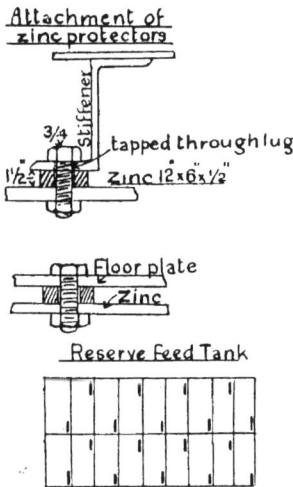

GLOBE PNEUMATIC HAMMERS.

Referring to p. 290, live air enters the hole G upon the throttle lever F being actuated. It then flows along the hole H and port I to the rear of the piston, and drives the latter forward into contact with the snap. Live air constantly enters the interior of the valve box through the restricted passage J in the cover, and into the space in the box between the cover and the valve. The air flowing into this space passes through the slot K and hole M into the cylinder, and thence through the port N to exhaust. The pressure on the rearward end of the valve will therefore be wholly relieved during a certain interval of time.

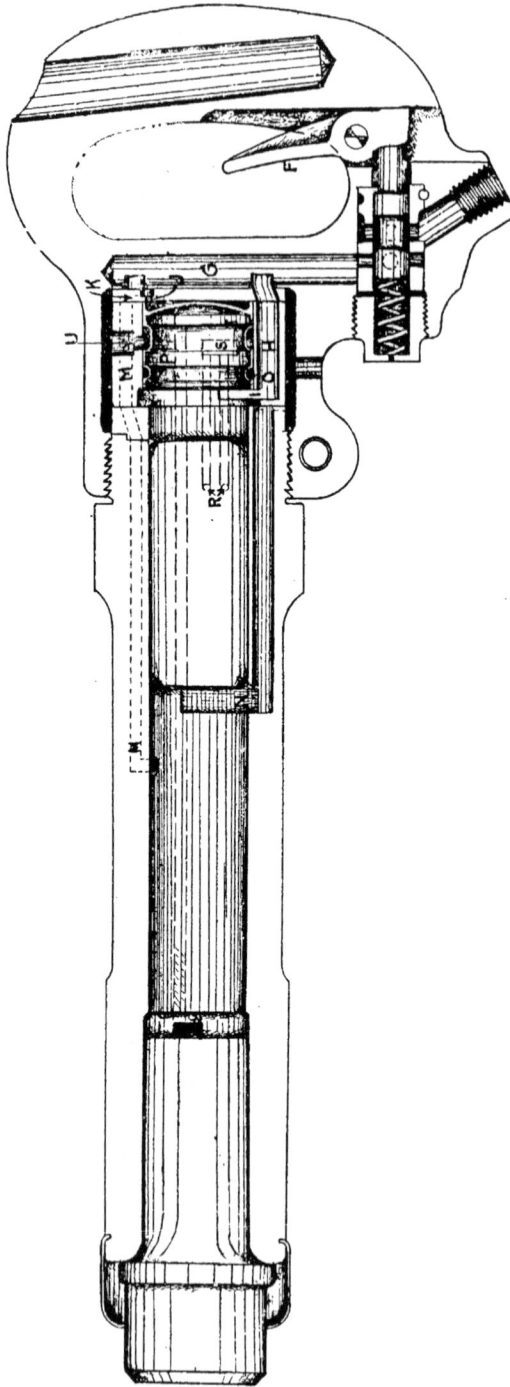

Globe Pneumatic Hammer.

When the piston in its forward movement closes the hole M in the cylinder the pressure in space begins to accumulate, and by the time that the rear of the piston uncovers the port N and relieves the pressure driving the piston forward, the pressure in the space is high enough to move the valve to the forward limit of its stroke.

This movement of the valve places the groove O in its external diameter not only into communication with the inlet I, but also into register with the hole P, which leads into longitudinal holes in the cylinder, which opens into the latter at Q adjacent to the shank of the snap. Live air now acts on the front end of the piston and drives it backward, the air which was used to drive the piston forward having passed to exhaust through the port N. When this port is covered by the piston on its further backward movement, the air continues to exhaust through the hole R, slot S, and groove T of the valve, and thence through the port U to atmosphere, it being noted that the groove T places the ports S and U into communication with each other when the valve is in its forward position. As soon as the holes R are covered by the piston the further backward movement of the latter by its kinetic energy compresses the air which is trapped between the piston and the valve to a high pressure, thus overcoming the live air pressure on the back of the valve which is now free to escape through the passages K and M. As soon as this occurs the valve at once moves rearwardly and a fresh cycle of operations occurs. The piston enters the valve in its backward travel and is cushioned thereon.

BOYER PNEUMATIC HAMMER.

A section of a caulking and chipping hammer is shown on p. 292. The action is as follows: Air enters passage A and passes into space *e*, thence through passage X to space E, and the latter area being greater than the former the valve V remains in the position shown, the piston being assumed at the front of the hammer as shown. The air then passes round annular space E through S and R to front of piston driving it back, the air at the back of the piston exhausting through the holes L in valve (see also Fig. IV.) and the passage H to the atmosphere.

When the fronts P and Q are uncovered the face E of the valve is put in communication with the atmosphere, and the greater pressure on face e of valve forces the valve forwards, closing the holes·L.

The air at the back of the piston then acts as a cushion, and the live air now enters the interior of the valve and back of the piston as indicated by the arrow in Fig. III., the passage 2 being closed. When the annular space near the backward end of the piston comes opposite M, communication is made between the interior of the valve and the space E by means of the passage T, the valve V being then forced backwards.

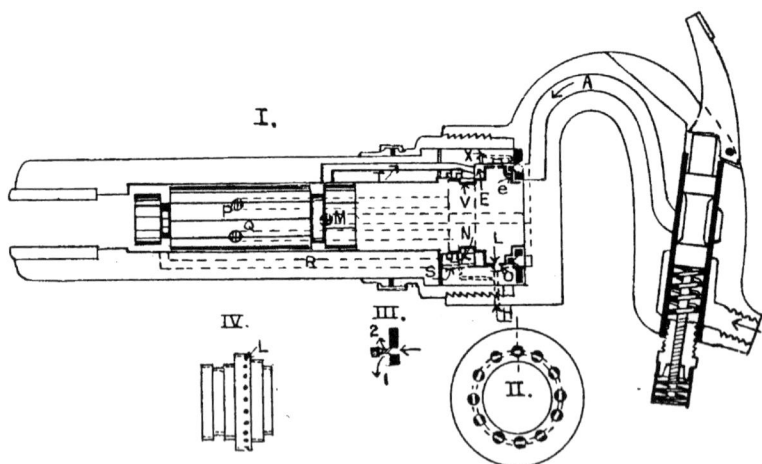

BOYER PNEUMATIC HAMMER.

LITTLE GIANT PNEUMATIC DRILL.

In this drill there are two oscillating cylinders contained in an aluminium case. The pistons drive the crank shaft, to the end of which is fixed a small pinion geared to the socket for the drill as shown on p. 293.

The air supply is through the curved handle, which is arched to fit over the cylinder bearings. The supply ports in this handle have small thimble valves, N, in them, which are kept pressed against the faced-off sides of the cylinder (see sketch) by the air acting on a small shoulder inside the valve.

The air ports in the cylinder correspond to those in the valves N, and the machine can be reversed by raising or lowering the curved arm, as the case may be, which action closes the supply to one side of the piston and opens it to the other.

This property of reversibility is useful, as the machine can be used for boring holes in wood with an auger.

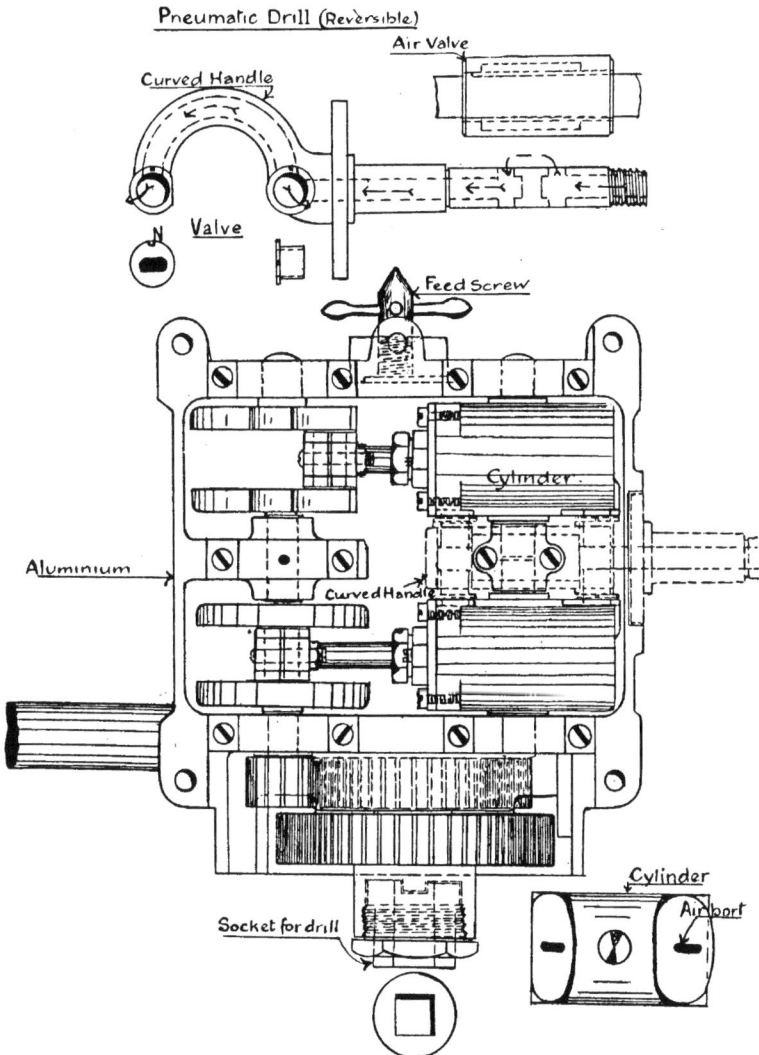

Pneumatic Drill (Reversible)

The action of the air valve is clearly shown by the arrows.

There are other modifications of this drill, such as "corner" drills, for drilling holes in awkward corners where the ordinary type cannot be used.

COUNTERSINKING MACHINE.

This is illustrated below, and consists of a long radial arm on which travels the frame carrying the spindle for the tool and the lever for giving the feed. The feed is obtained by pressing down the lever which acts against a spring, and on releasing the pressure the spindle is raised.

To ensure the same countersink for every hole in a plate there is a fitting consisting of a small roller carried at the end of a vertical arm, and which is attached to the tool spindle. This gauge is set so that the small roller bears on the plate when the countersink is of the required amount.

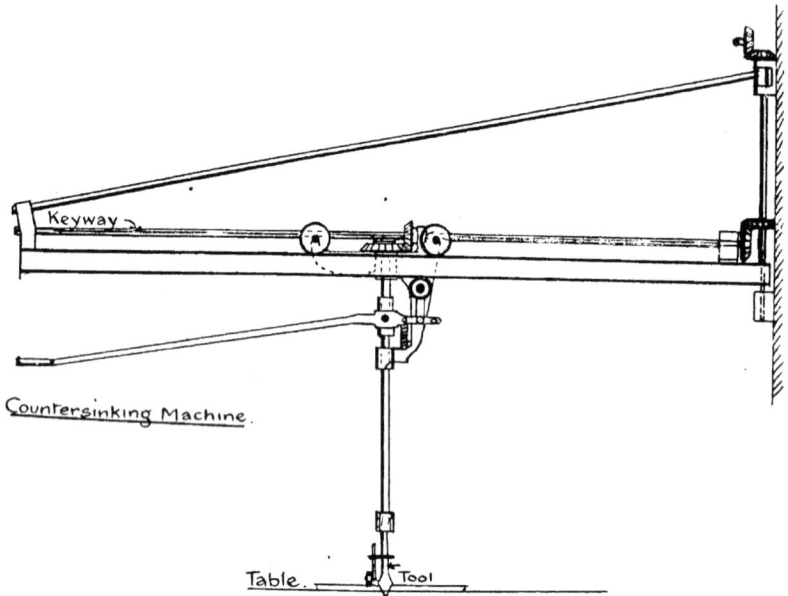

Countersinking Machine.

BOYER PNEUMATIC DRILL.

In this machine there are three oscillating cylinders mounted in a frame which is free to revolve (Fig. I.). At the lower part of the frame is a spigot, which passes through a hole in the division plate M, and has on its end a small pinion P (Fig. III.) gearing with wheels W, W, which engage with a rack formed on the inside of the lower portion of the machine. The wheels W, W are carried on a casting free to revolve, into the lower part of which fits the drill socket. The crank C is fixed to the top of the upper portion of machine, and hence the frame carrying the cylinders revolves. As the frame revolves the cylinders oscillate through a small

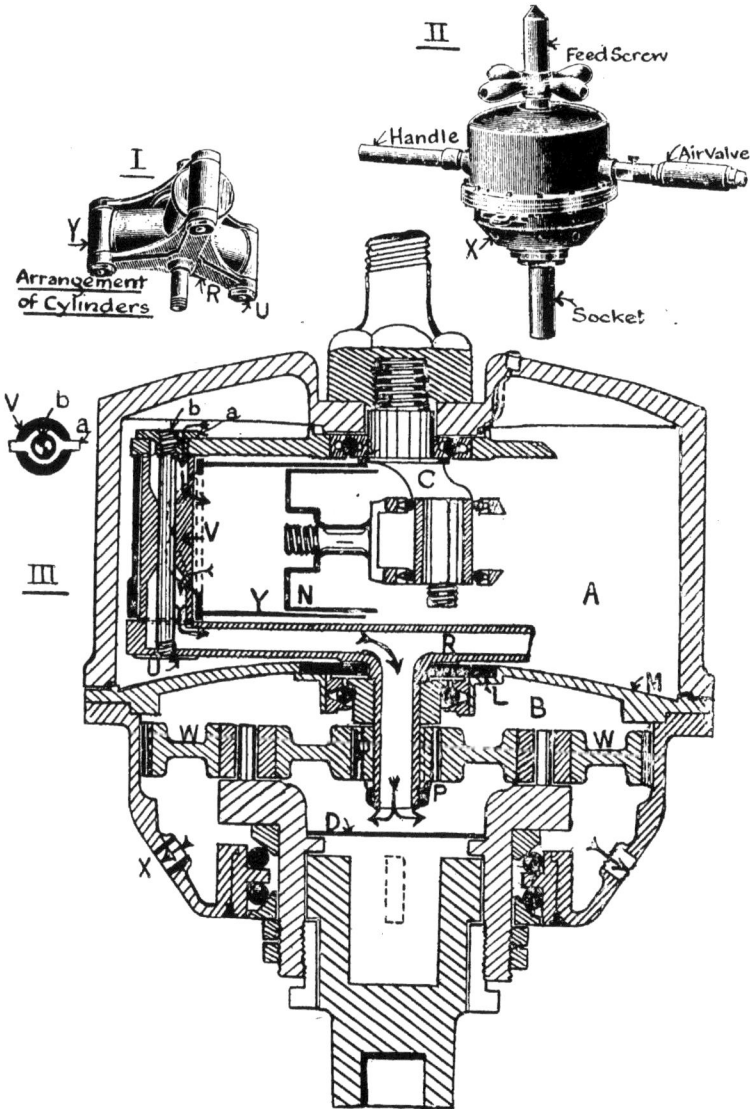

BOYER PNEUMATIC DRILL.

angle, this motion exposing a slit in the end of the cylinder to the supply and exhaust ports alternately in the valve V, which is so formed as to act as supply and exhaust valve (see Fig. III.). The valves V are hollow, and the air passes through them and into the cylinders Y through the narrow slits in them, and exhausts through these slits again on the return stroke of the piston N ; thence along the hollow arm of the revolving framework R to the atmosphere through the ports X, as indicated by the arrows.

The joint where spigot carrying piston P passes through division plate M is made airtight by means of leather washer L.

The framework carrying the cylinders runs at a high number of revolutions, and the gearing down is by means of the train of wheels P, W, W and rack on inside of lower part of machine.

SPAR MAKING.

Sketches of spars are given on pp. 297, 298, and the detail process of trimming yards and booms is given on p. 299, Figs. A and B.

Yard. Before lining-off the log is examined for knots, shakes, and nails, and then secured with dogs (p. 299). Strike a centre line along it and plumb down this at the ends, and divide up length into eight equal parts. Set off the widths at the various places obtained, as in (1), Fig. A, and plumb down the width at the ends. Before starting work, erect battens at the ends with edges vertical, as, if the log moves at all while being chopped, it will be incorrect to test the faces that have been chopped fair by means of a plumb-bob. Hence, this should be done by outwinding a straight-edged batten with the end battens.

When two faces have been chopped fair, turn log over, strike centre line, and mark off widths on these and chop faces fair. The log will then be of square section throughout ; and the next process is to mark off yard arm and square portion as in (3), Fig. A. Remove wood as in figure, marking yard eight-square except at yard arms, and then adze out to sixteen-square. The process is completed with draw-knife and planes. The proportions for a yard are sometimes given in the fractional form, and are approximately $1, \frac{40}{41}, \frac{9}{10}, \frac{3}{4}, \frac{1}{2}$, the unit being the size at the slings.

Boom. The process is similar to a yard, except for the lining-off. The greatest diameter of a boom is two-thirds from the heel or end against ship's side ; and the diameter at each end is the same as given in Fig. B, p. 299.

SECT THRO'SHEAVE

TOP MAST

LIGHTNING CONDUCTOR

15 P.B. SHEAVE

STEEL FLD 5×3

PAWL RACK

8 P.B. SHEAVE

STEEL FLD 2½×2

46'-6"

2'-0"

STEEL FUNNEL

P.B. SHEAVE

9"

6" O/D

SERVICE
SIGNAL HALYARDS
CONNECTING STRIP TO
LIGHTNING CONDUCTORS

LIGHTNING CONDUCTOR

STEEL FUNNEL

TOP GALLANT MAST

FORGED COPPER SPINDLE

6'-6"

FOR YARD

23'-2"

1'-10"

8'-8"

PLAN OF
UPPER CAP

JACOBS LADDER

DRESSING LINE

T.G. MAST
HEEL HOOP

SIGNAL JACKSTAY

8'-8"

D°

DRESSING

PLAN OF
LOWER CAP

8'-8"

VANE

LIGHTNING COND?

FORGED COPPER SPINDLE

FLASHING LANTERN

FORE TOPGALLANT MAST HEAD

FOR W/T YARD LIFTS

1½" TRUCK

FOR JACOBS LADDER

HINGED GALLOWS

OUTRIGGER FOR SIGNAL HALYARDS

ELEVATION

2"

YARD FOR AERIAL WIRES

ELEVATION

ON MAIN MAST
YARD LIFT

HALYARDS FOR AERIAL WIRES

4" DIA"

ON FORE MAST
YARD LIFT

PLAN   D°
YARD SLING

HALYARD FOR AERIAL WIRES

7" DIA°

18'-0"

UPPER SIGNAL YARD

ELEVATION

EYES FOR
JACK STAY

PLAN

EYE FOR YARD LIFT

SPARE

⅝" EYE FOR SIGNAL HALYARDS

¾" EYE FOR YARD LIFT

⅝" EYE FOR YARD BRACE

4" DIA"

2'-0"

LOWER SIGNAL YARD

ELEVATION

½" SECURING PIN

1" DRILLED HOLE
FOR SLING

½" DIA°

LEATHER

BAND 4"×½"

18'-0"

EYES FOR JACKSTAY

PLAN

⅜" EYE FOR SIGNAL HALYARDS

¾" EYE FOR YARD LIFT

6" DIA°

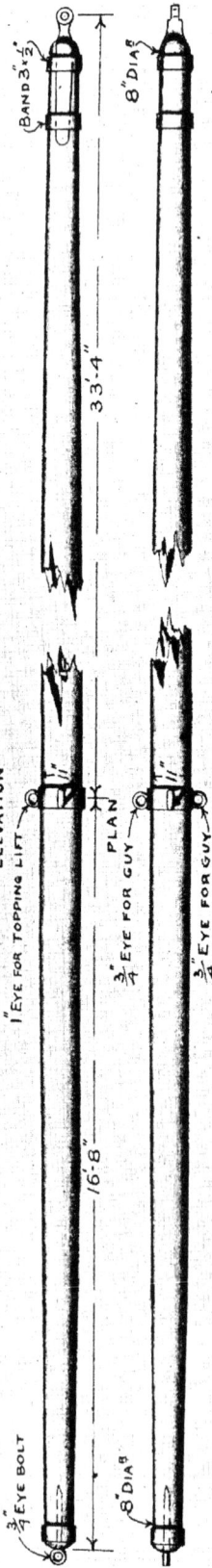

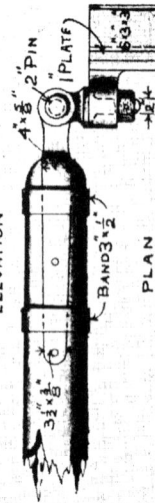

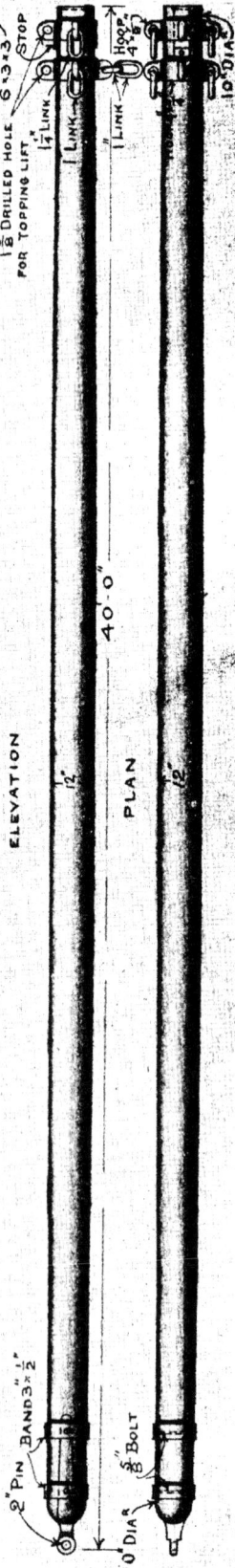

SWINGING BOOMS

ELEVATION

BAND 3"×½"

8" DIA

33'-4"

1" EYE FOR TOPPING LIFT

PLAN

¾" EYE FOR GUY

¾" EYE FOR GUY

16'-8"

¾" EYE BOLT

8" DIA.B

¾" EYE BOLT

8" DIA.B

HEEL FITTING FOR SWINGING BOOM

ELEVATION

4"×⅝" 2" PIN

1" PLATE

BAND 3"×½"

3½"×1⅛"

PLAN

5⅞" BOLT

⅛" DRILLED HOLE FOR TOPPING LIFT

SWINGING BOOMS 2 Nº

ELEVATION

BAND 3"×½"

6" DIA

20'-0"

1" EYE FOR TOPPING LIFT

¾" EYE FOR GUY

¾" EYE BOLT

DERRICKS FOR COALING ETC: 4 Nº

ELEVATION

"STOP

1" LINK

1¼" LINK

1" LINK

1¼" LINK

HOOP 4½"

10" DIA

40'-0"

12"

PLAN

12"

2" PIN BAND 3"×½"

0" DIA

⅝" BOLT

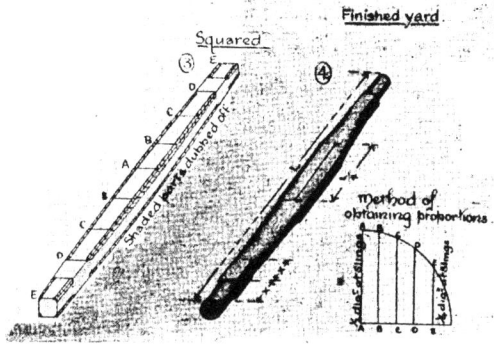

FIG. A.

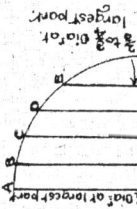

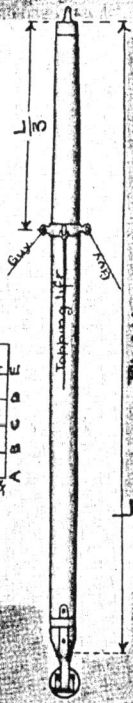

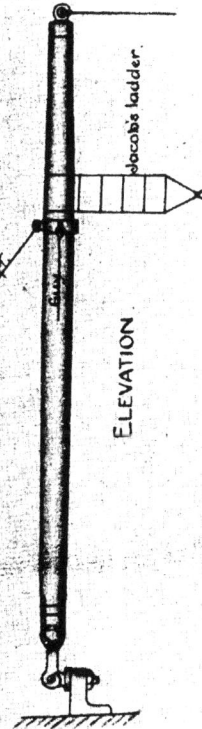

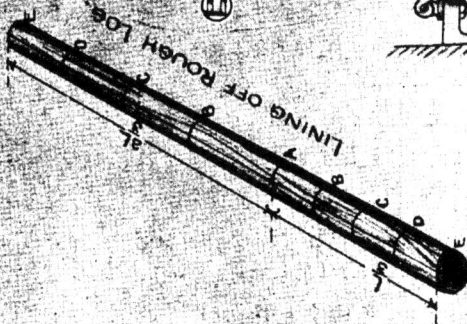

FIG. B.

Divide up the lengths, each side of greatest diameter, into four equal parts respectively. Set out the widths obtained, as in Fig. B, p. 299, at these places, and proceed as before.

Topmast. The proportions of this spar are given on p. 297. The variations in the form of this spar at different places should be noticed. The weight of the topmast is taken by a fid, made either from solid bar of rectangular section or two channel bars riveted back to back. Heeling is fitted, as shown, on three sides of the topmast. A lightning conductor is fitted to topmasts, and consists of a copper strip secured by rag-pointed copper nails. At the sheave holes the copper is carried round both sides of the hole, and secured by screws. The various lengths of the conductor are connected by a thin butt-strap of copper placed underneath the thicker strips. The conductor is carried over the truck, a screw socket being placed at the top of the mast for the heel of the copper spindle carrying the vane.

Oar Making.

Oars are made of pine and ash. The process of making is illustrated by diagrams on p. 301.

The planks from which oars are made are sawn about 7 inches by $1\frac{3}{4}$ inch, and slightly longer than the finished oar. The rough outline is marked on this, and sawn out as in Fig. I., the form of the blade being given by a mould kept for the purpose.

The handle is next formed, and the piece is then turned on its edge, and blade marked as in Fig. II., keeping the curved edge of the mould $\frac{1}{2}$ inch clear of the sides of wood, and a parallel curve is then drawn $\frac{3}{8}$ inch in from this.

The shaded portions are now removed by means of an adze, as in Figs. II. and III. The rough outline of the blade has now been obtained, and it only remains to shape it.

To do this, remove the wood as in section A to the ticked lines with a draw-knife, and hollow out with a plane to the full lines. At the throat of the oar a chamfer is taken off from the curve C to the middle of the blade. To prevent the blades splitting, copper bands are worked at the tips, as shown. The proportions for a 17-feet oar are shown in Fig. IV.

Method of making an Oar.

Deal marked off   Fig I

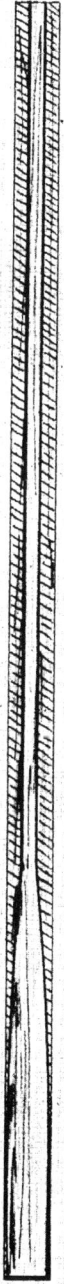

Applying the mould.   Fig II

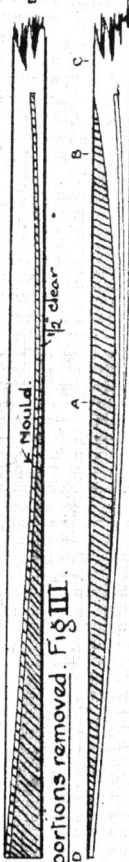

← Mould.

½ clear

Shaded portions removed.   Fig III

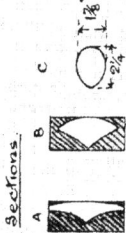

A

B

C

D

Sections

D   A   B   C

← 2¼ →
↑1⅜

Oar finished   Fig IV

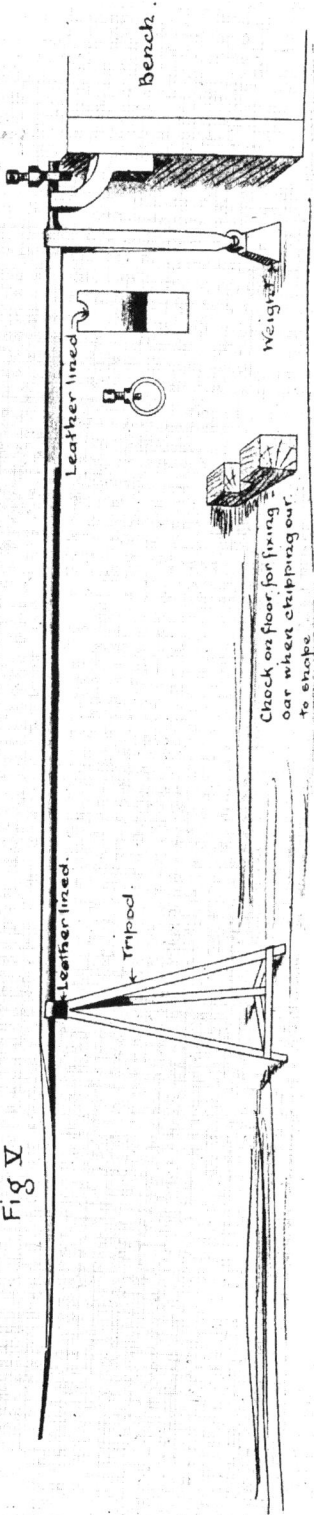

← 5'-0" →   ← 8'-0" →   ← 3'-1" →   ← 11" →

← 5⅝" →

Fig V

← Leather lined.

← Tripod.

Leather lined.

Chock on floor for fixing
oar when chipping oar
to shape.

Weight.

Bench.

Boat Building.

A short description of this for a clinker-built boat will be given, and is illustrated on p. 303.

(1) Prepare a good foundation by means of a piece of timber about 12 inches by 3 inches section, secured with cleats.

(2) Fasten cleats on both sides of keel-piece, and secure it by means of wedges driven on both sides of the keel until it becomes straight. This is checked by means of a line held along the centre line marked on the keel. The keel should have a slight camber.

(3) Plumb stem and stern posts, and secure by means of stays. The stem and stern posts are scarfed to the keel, or " halved," as it is termed, the tongues of the scarfs in the keel being formed on opposite sides.

(4) Prepare garboard, and fit after end first. Then make a template for fore end from a touch mark on the keel, transferred to the garboard before nailing. Fit stopwaters before fastening off garboard. The shape of the garboard should be noted, as it differs from all the other planks.

(5) Prepare and fit section moulds, commencing at the midship section, and connect them by means of the sheer pole, stayed as shown.

(6) Set off planks on moulds thus: mark on each mould and on stem, stern post, and transom the number of planks, making them of least width at the sharpest turns, *i.e.* round bilge and transom, and fair in the lines. Fix plank, roughly cut to shape, with the correct amount of land. Then transfer the marks to it from moulds, stem and stern posts.

Trim plank, and fasten along lower edge and at ends and through floors. The planks are secured at the ends by chisel-pointed nails to the apron (forward) and transom and sternpost (aft).

To fit a Thwart. The thwarts rest on the rising, and the length is taken of the longer edge of the thwart, which will be on the foreside of the thwart in the after portion, and *vice versa*. The bevellings of the ends of the thwarts against the sides are next taken in a horizontal direction, and also vertically if necessary. The thwarts are secured to the sides by means of knees, which

must be scored out to fit over the lands of the planks.   It is best
to make a mould for the knees.

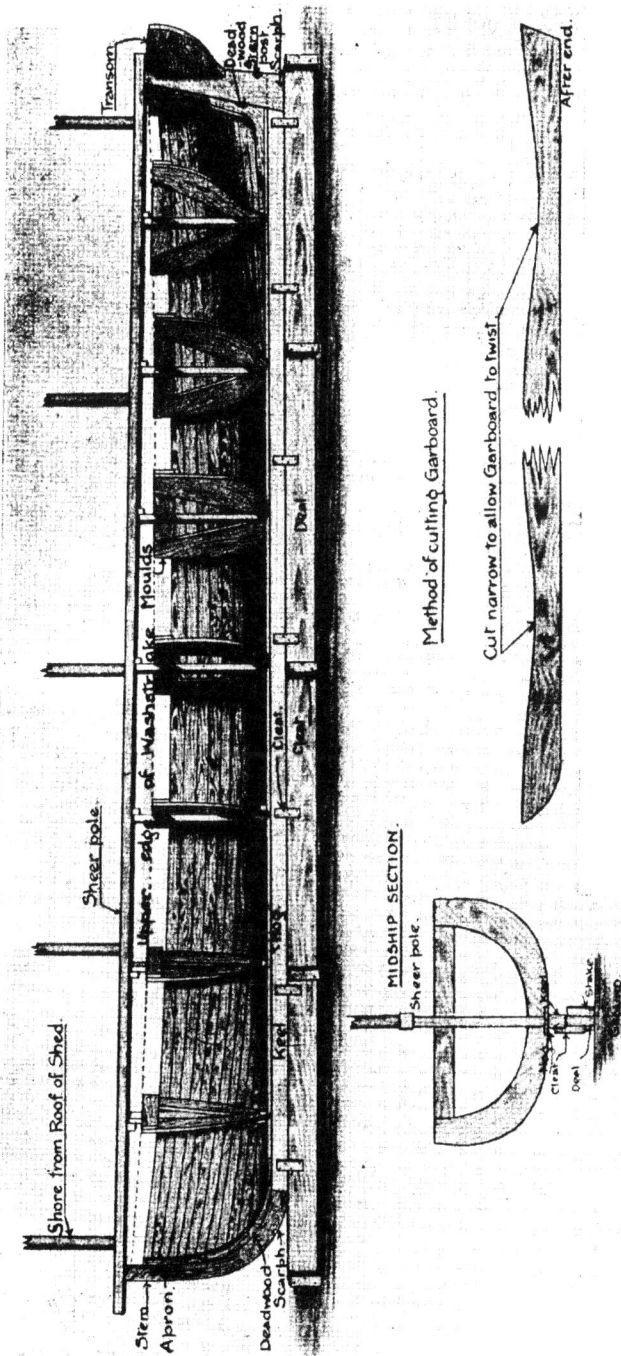

Method of cutting Garboard.

Cut narrow to allow Garboard to twist

After end.

MIDSHIP SECTION.

To fit a Breast-hook (see diagram below).   The oak crook is laid

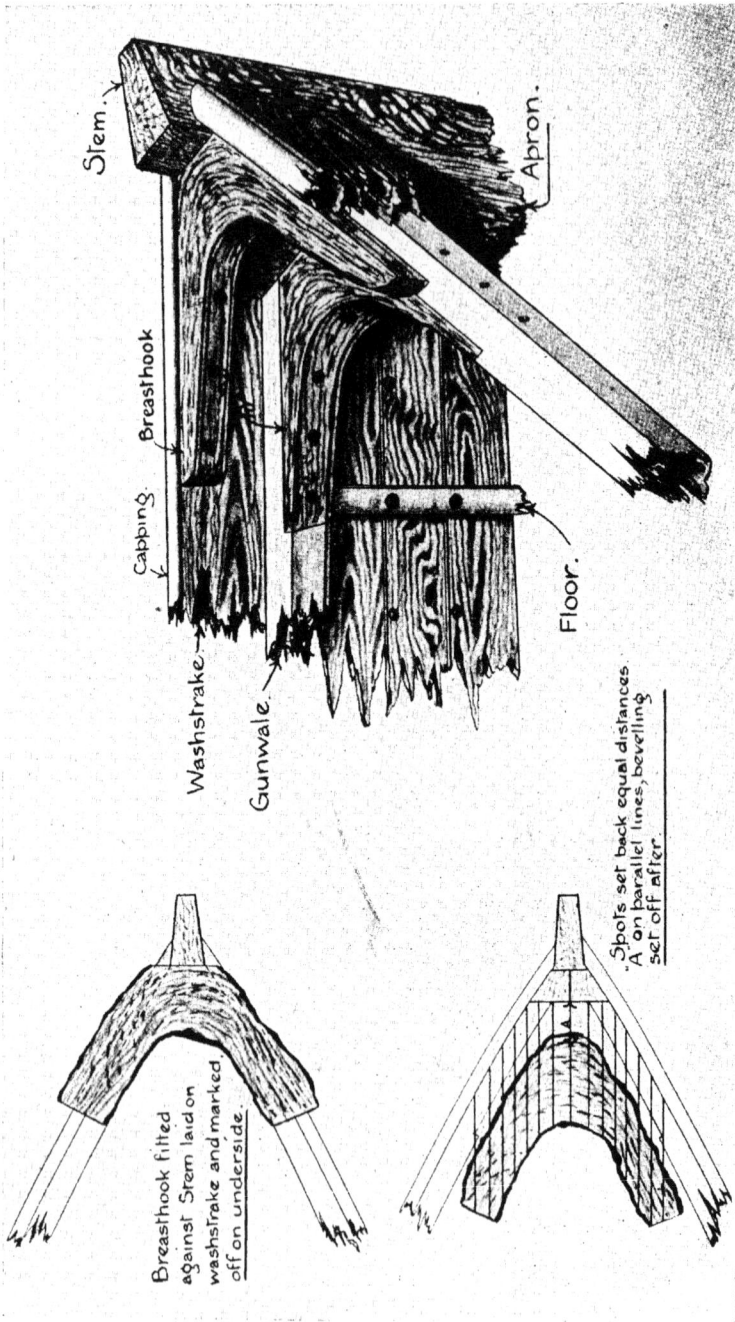

Stem.

Apron.

Breasthook

Cabbing.

Washstrake.

Gunwale.

Floor.

Spots set back equal distances.
A on parallel lines, bevelling
set off after.

Breasthook fitted
against Stem laid on
washstrake and marked
off on underside.

on gunwale, marked off on the under side, and sawn out roughly
to shape.   It is then closely fitted as follows : draw parallel lines in

a fore and aft direction, and measure the amount it is slack, and
set this off from edge of gunwale along the parallel lines and trim

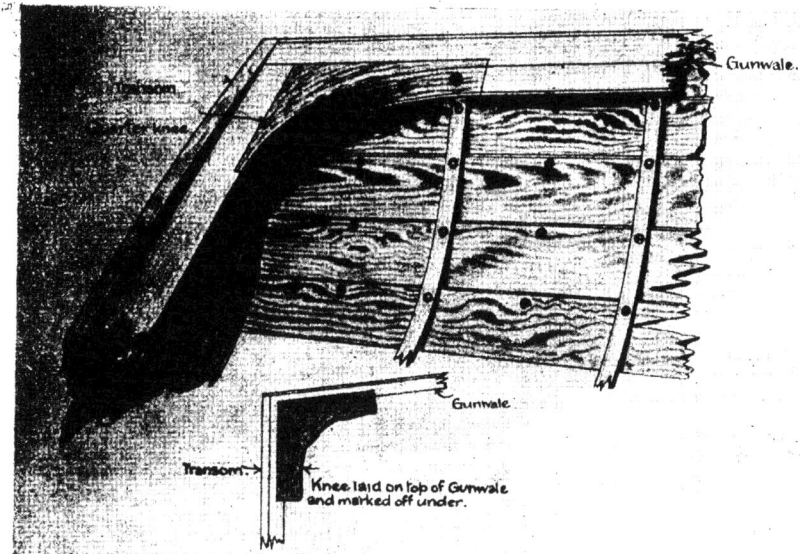

Fig. A.

to these spots. The bevellings against gunwale and apron are
then measured and breast-hook trimmed. Another method is to

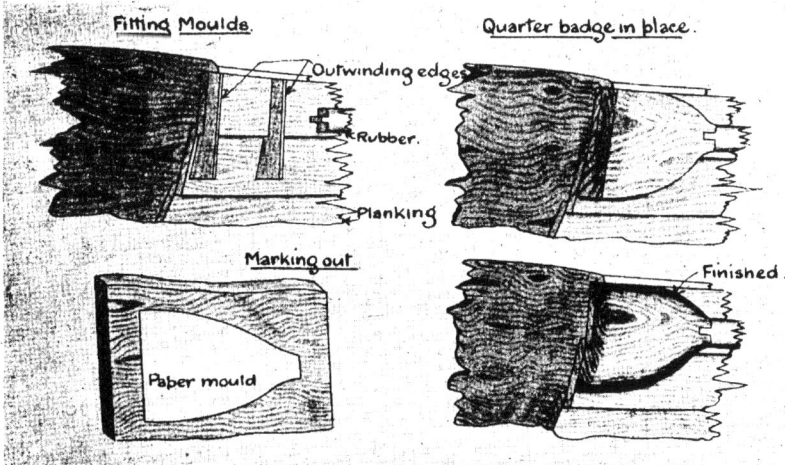

Fig. B.

make a mould to the gunwale and apron. Quarter knees are also
fitted by the above methods (Fig. A, above).

Lifting Plates. The bolts for securing the lifting plates are

X

arranged zig-zag fashion, and in the case of a tapered keel must be driven diagonally, as they would pierce the sides of the keel if driven plumb.

To fit a Quarter-badge or Fashion-piece. Referring to Fig. B, p. 305, a paper mould is made, and the timber roughly cut to form. Section moulds are made to the planking, the edges of the moulds out-winding; reverse moulds made to the section moulds, and the piece trimmed to these. The quarter-badge is then carefully fitted in place, being tenoned into the rubber, and finally trimmed off as shown, " finished."

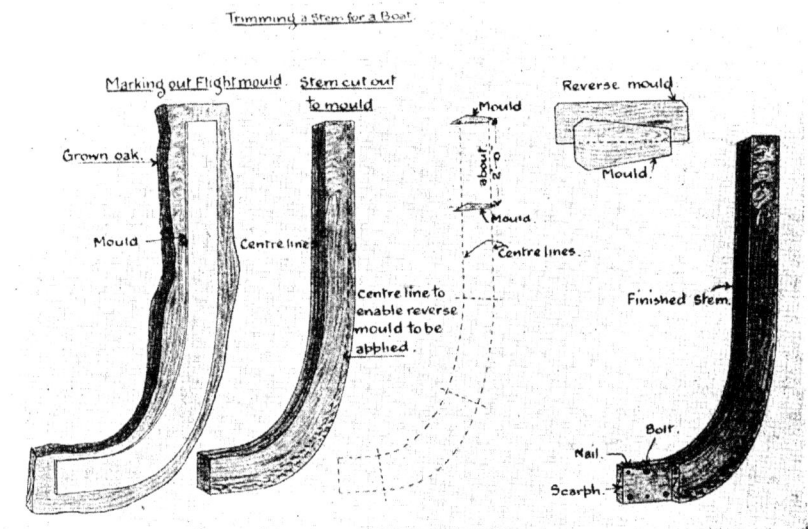

Trimming a Stem for a Boat.

To fit a Stem. A flight mould, giving profile of stem, is applied to the oak pieces, and cut roughly to form as shown. Section moulds are made about 2 feet apart, and reverse moulds made to these. A centre line is struck on fore and after edges of oak piece, and the latter trimmed to the shape of the reverse moulds. The stem is tapered as shown in section mould, this taper being called the bearding of the stem. The scarf at the lower end for attachment to the keel is formed as shown, and the after edges of the stem are chamfered off for the ends of the planking.

When a new stem is fitted to an old boat the opening out of the planks causes their ends to be damaged, and these portions have to be cut off; and when making the section moulds this

must be borne in mind, or the planks will be short of the stem.
The scarf fastenings are nails and clench-bolts as shown, and

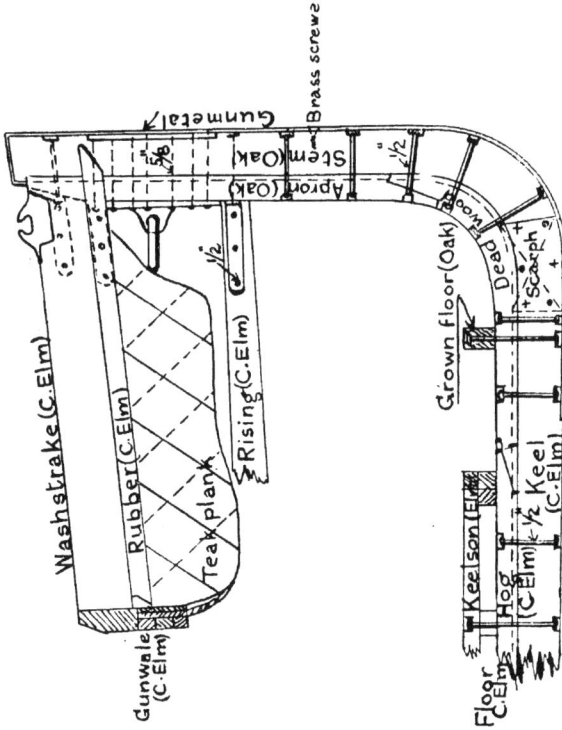

FIG. A. STEM OF 42-FEET LAUNCH.

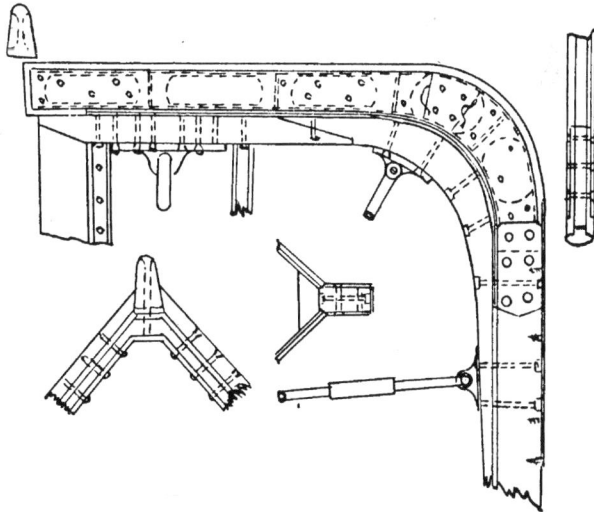

FIG. B.— STEM OF 42-FEET LAUNCH.

they are driven from opposite sides. The holes in keel and stem
are not directly opposite, but arranged so that the driving in of the
fastenings closes the scarf.

The details of fastenings, etc., of stem and stern posts of 42 feet launch are given by Figs. A, on pp. 307, 308.

A metal stem piece for a 42-feet launch is illustrated, Fig. B, p. 307, this being fitted when wood is not obtainable.

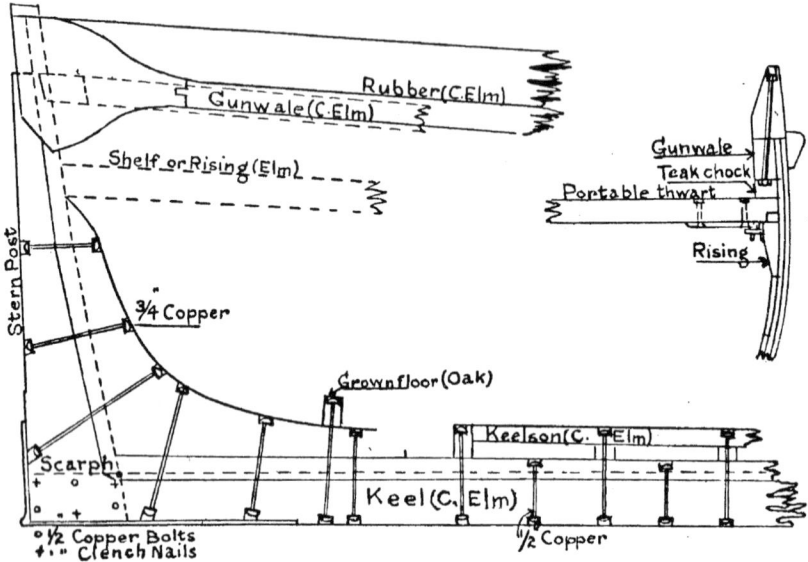

FIG. A.—STERN OF 42-FEET LAUNCH.

The method of working the plank in a diagonal and clinker-built boat respectively is given on pp. 308, 309.

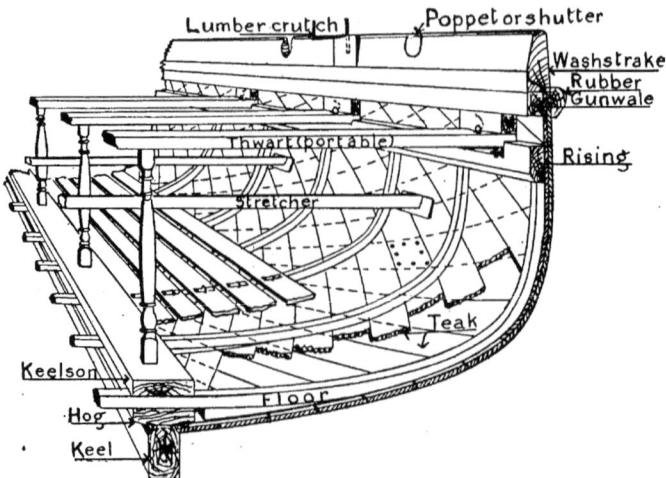

FIG. B.—LAUNCH.

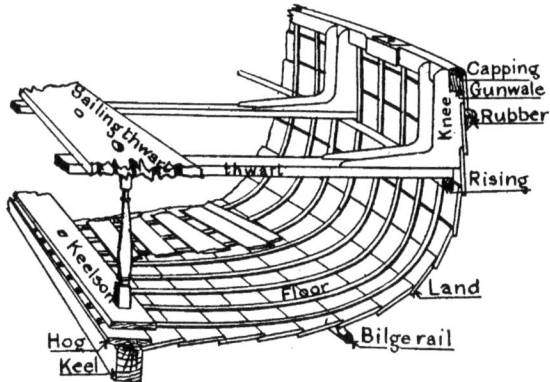

CUTTER.

The construction of typical boats is given on pp. 310, 311, 312, 313, 314

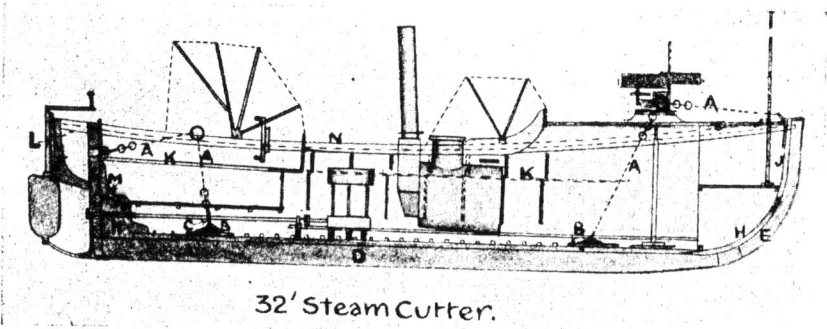

32' Steam Cutter.

A. Slings.
B. Link or sling plates.
C. Shackle (bow-shaped to clear shaft).
D. Keel (C. elm).
E. Stem (E. oak).
F. Aperture knee.
G. Bodypost (E. oak).

H. Fore deadwood (E. oak).
    After deadwood (E. elm).
J. Apron (E. elm).
K. Rising.
L. Transom.
M. Deadwood knee (E. oak).
N. Gunwale.

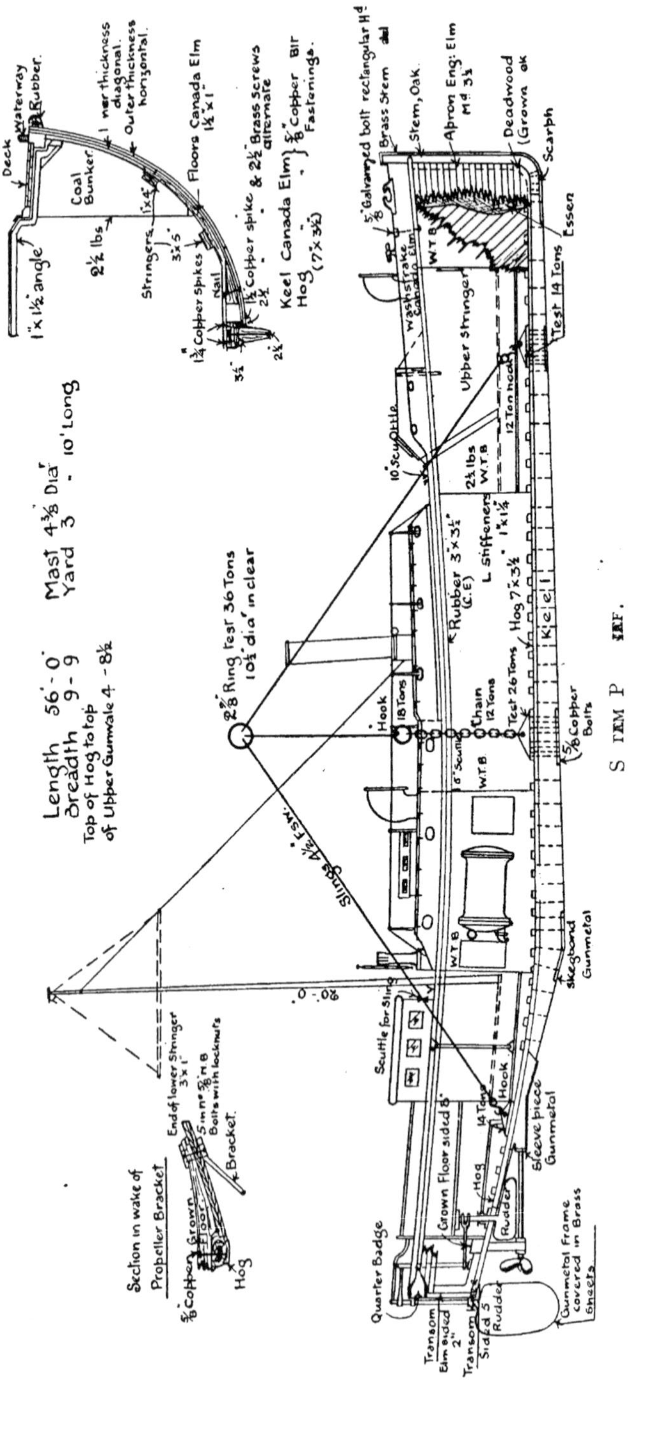
Length 56'-0"
Breadth 9-9
Top of Hog to top of Upper Gunwale 4-8½
Mast 4⅜ Dia
Yard 3 - 10' Long
Section in wake of Propeller Bracket
⅝ Copper
Grown Floor
End of lower Stringer 3"x1"
5 in no ⅝ M.B Bolts with locknuts
Bracket
Hog
Deck
Waterway
Rubber
1"x1½" angle
Coal Bunker
2½ lbs
Inner thickness diagonal.
Outer thickness horizontal.
Stringers
1"x4"
3"x5"
1¾ Copper Spikes
Nail
Floors Canada Elm 1½"x1"
1½ Copper spike & 2½ Brass Screws alternate
Keel Canada Elm
Hog (7"x3½)
⅝ Copper Bit Fastenings.
2⅞ Ring test 36 Tons 10½" dia in clear
Slings 4½" FSW.
80'-0"
Scuttle for Sling
Quarter Badge
Transom Elm sided 2"
Transom Knee Sided 5
Rudder
Grown Floor sided 8"
Hog
14 Tons
Hook
Rudder
Gunmetal Frame covered in Brass Sheets
Sleeve piece Gunmetal
Skegband Gunmetal
W.T.B
Hook
18 Tons
16" Scuttle
Chain 12 Tons
Test 26 Tons
⅝ Copper Bolts
Rubber 3"x3½" (C.E)
L Stiffeners 1"x1¼"
Hog 7"x3½
Keel
2½ lbs W.T.B
10" Scuttle
12 Ton hook
Test 14 Tons
Upper Stringer
Washstrake Canada Elm
⅝" Galvanized bolt rectangular Hd
Brass Stem
Stem, Oak.
Apron Eng: Elm Md 3½
Deadwood (Grown oak
Scarph
Essen

S TEAM P IEF.

Section

Keelson

Floor

Hog

Keel

clench nails
(between Floors)

water
course

silver

5/8 clench bolts
(between Floors)

5/8 Keel

Tiller

Sole pi...

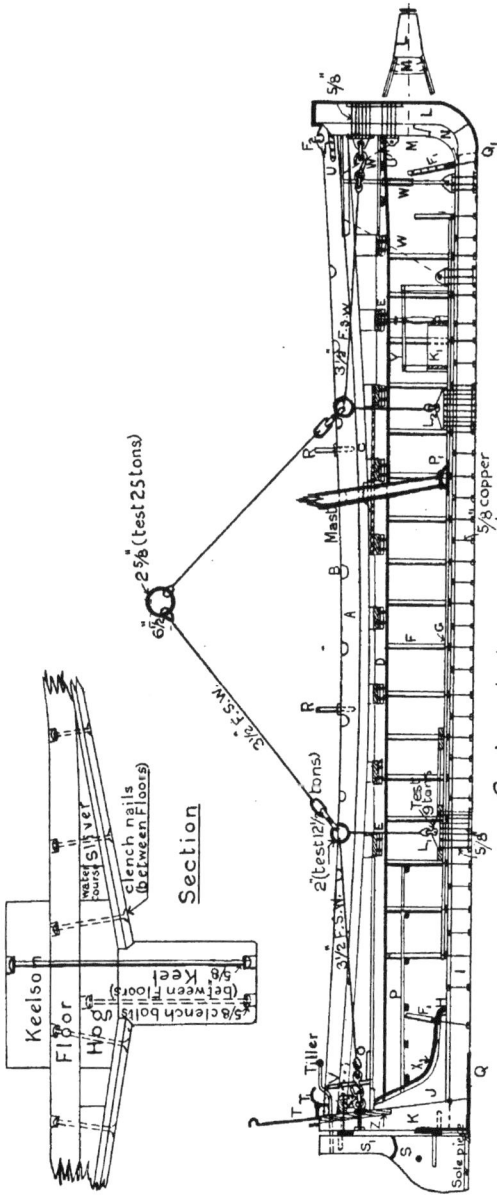

Sectional elevation 42′ Launch.

5/8 copper

2 5/8″ (test 25 tons)

6½″

3½″ F.S.W.

2″ (test 12½ tons)

3½″ F.S.W.

5/8

Test
gear

Mast

A. Solids or ... (C. elm).
B. Shutters or ... pets.
C. Gunwale (C. elm).
D. Rising (C. elm).
E. ... thwarts ... mast where they are of ...
F. ... floors

F2. Fairlead.
G. Keelson (C. elm).
H. Hog (C. elm).
I. Keel (C. elm).
J. ... deadwood (oak
K. ... locker.
K1. ... pipe.
K2. Stem (oak).
L. Stem (oak).
L1. Lifting or sling plates.

M. ... (E. elm).
N. Fore ... (oak
O. Sternsheets.
P. Seat.
P1. Mast ...
Q. Skeg ...
Q. Stem band.
R. Lumber irons or crutches.
S. Rudder.
S1. Rudder

T. Awning stanchion.
U. Pump ...
U. Steel ...
W. Bottle screws for supporting gun stand.
X. ... ipe of ...
Y. Platform for working gun.
Z. ... E. elm.

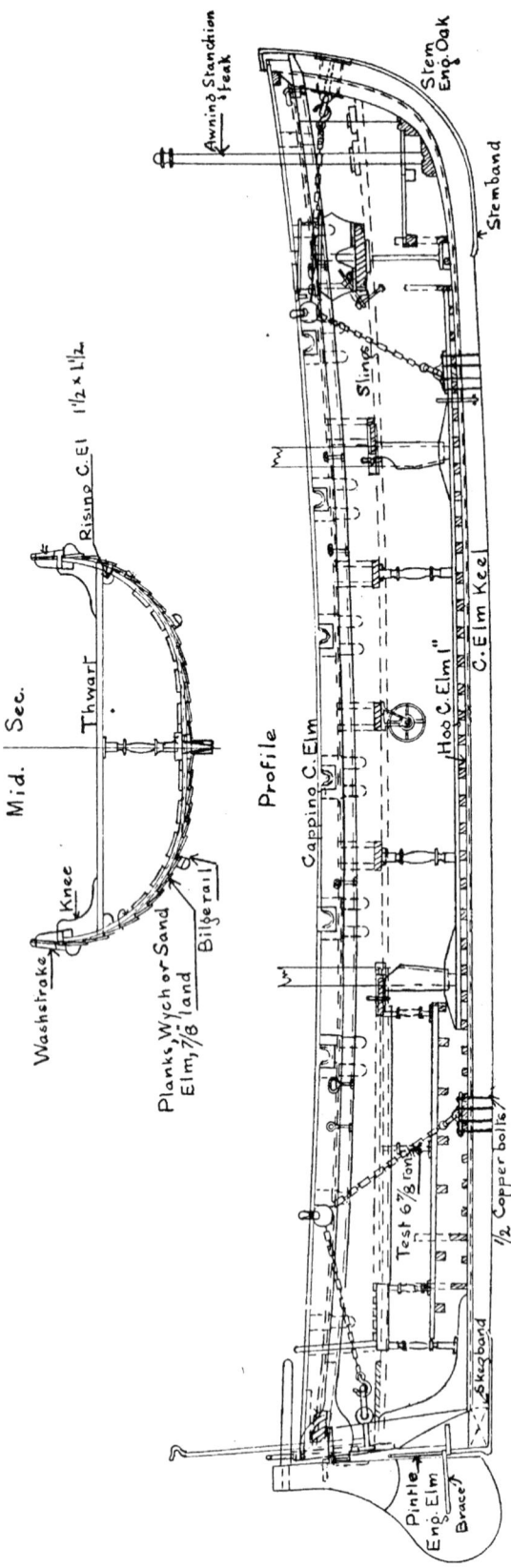

Mid. Sec.

Rising C. El 1½ × 1½

Thwart

Knee

Washstrake

Planks, Wych or Sand
Elm, ⅞" land

Bilge rail

Profile

Awning Stanchion
Teak

Stem
Eng. Oak

Stemband

Slings

Capping C. Elm

Hog C. Elm 1"

C. Elm Keel

Test 6⅞ tons

Slings—Test 4¾ tons (cutters)
" 3 tons (gigs and galleys)

½ Copper bolts

Skegband

Pintle
Eng. Elm
Brace

CUTTER.

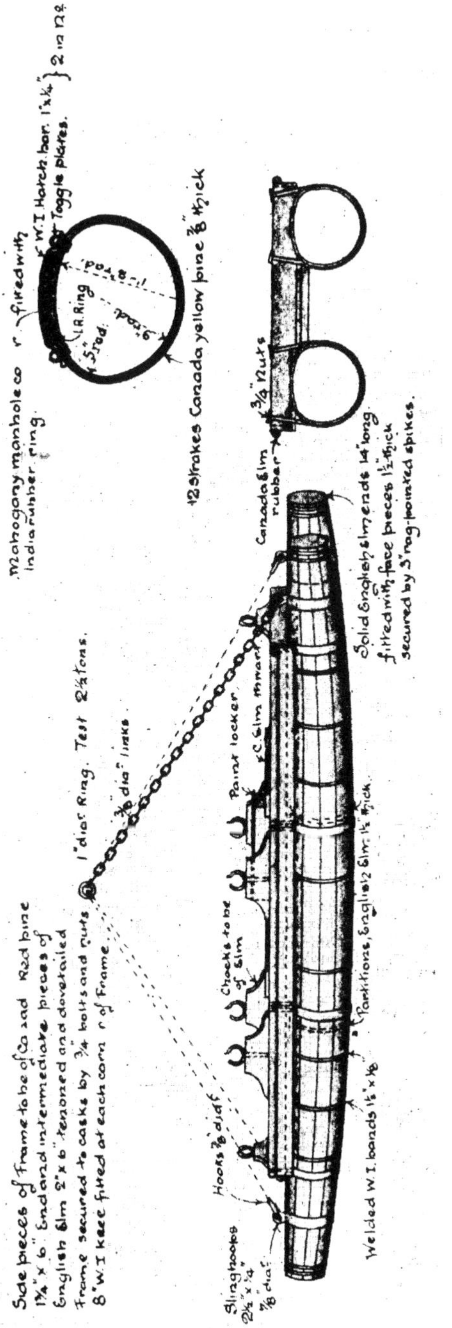
Side pieces of Frame to be of Canad Red pine
1¾" x 6". End and intermediate pieces of
English Elm 2" x 6" tenoned and dovetailed
Frame secured to casks by ¾ bolts and nuts
8" W.I knee fitted at each corner of Frame
1" dia. Ring. Test 2½ tons.
⅜" dia. links
Mahogany manhole cover fitted with
indiarubber ring
I.R. Ring
W.I. Hatch bar. 1" x ¼"
Toggle plates.
2 in No.
12 strakes Canada yellow pine ⅞" thick
Canada Elm rubber
¾" Nuts
Paint locker
C. Elm thwart
Chocks to be of Elm
Hooks ⅞ dia.
Sling hooks 2½" x ¾" ⅞ dia.
Welded W.I. bands 1½" x ⅛
Partitions, English Elm 1½" thick
Solid English Elm ends 14" long
fitted with face pieces 1½" thick
secured by 5" rag-pointed spikes.

Balsa Raft.

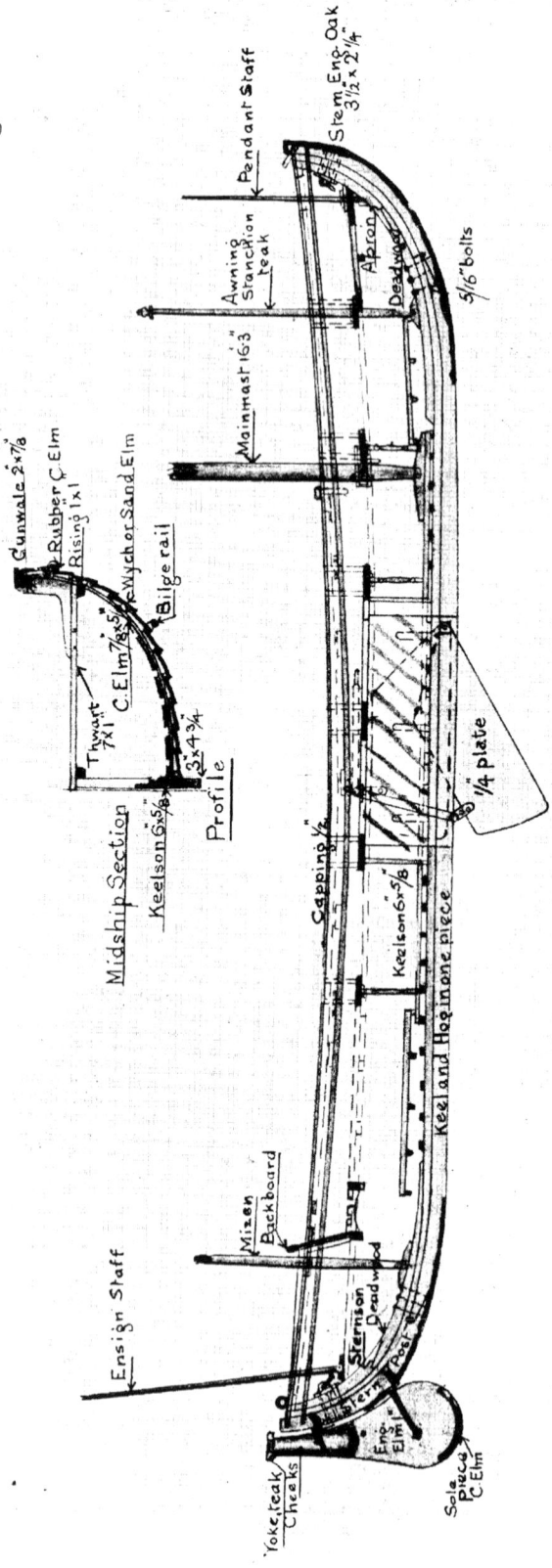

WHALER.

The diagram below enables the diameter at the collar of a boat's davit to be readily obtained when the working load and overhang are known, the stress assumed being 5 tons per square inch.

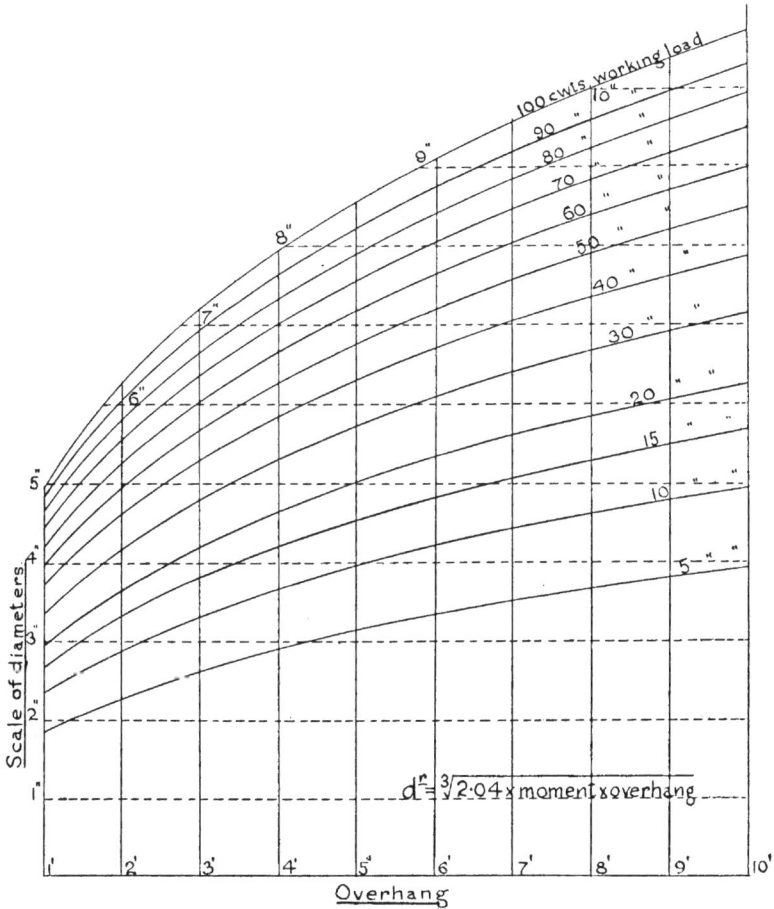

A similar diagram, on p. 316, gives the diameters for wood derricks, the factor of safety being taken as 10. For instance, if it were required to know the diameter of a derrick 20 feet long, and which has to carry a load of 4 tons, this would be obtained by measuring along the 20-feet ordinate to where this intersects the 4-ton curve of loads, thus giving a diameter of $8\frac{1}{2}$ inches.

In the formula A = sectional area, L = length, I = moment of inertia, f and c are constants, $f = 6000$ lbs., $c = 14666$.

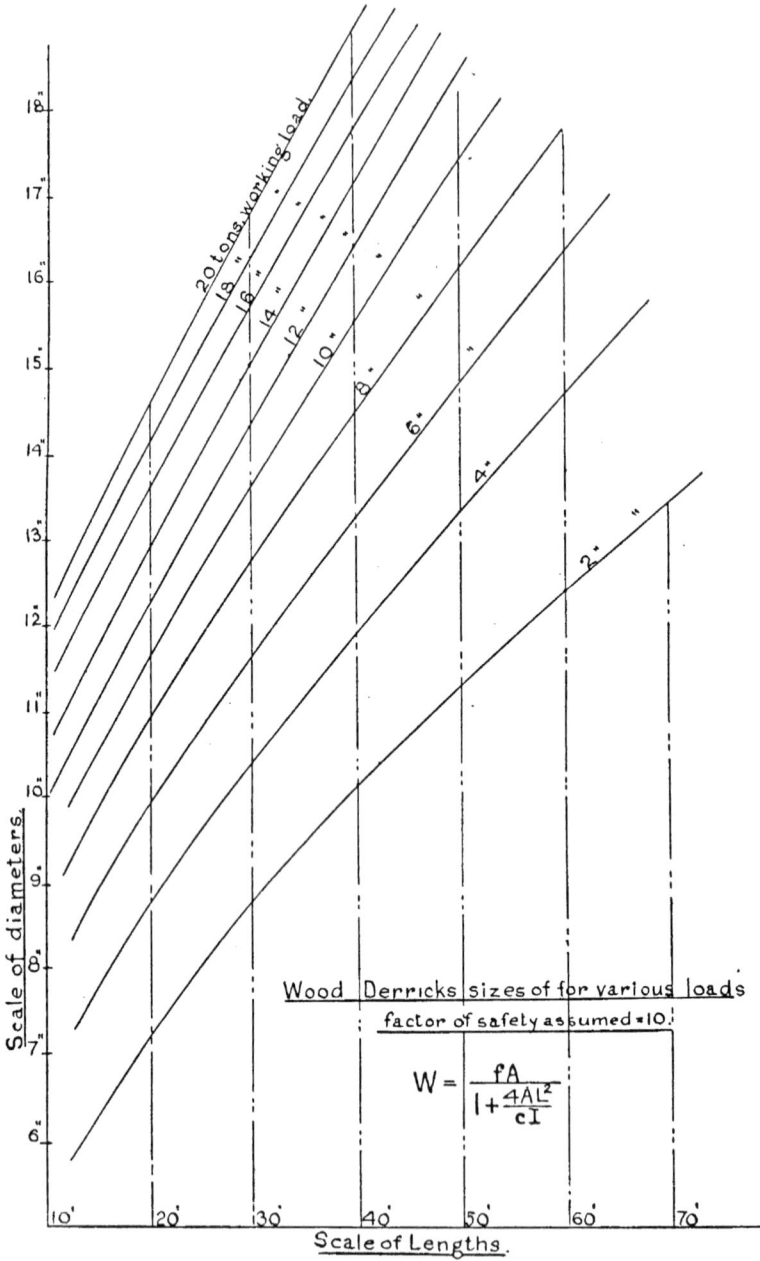

Wood Derricks sizes of for various loads
factor of safety assumed = 10.

$$W = \frac{fA}{1 + \frac{4AL^2}{cI}}$$

Scale of diameters,

Scale of Lengths.

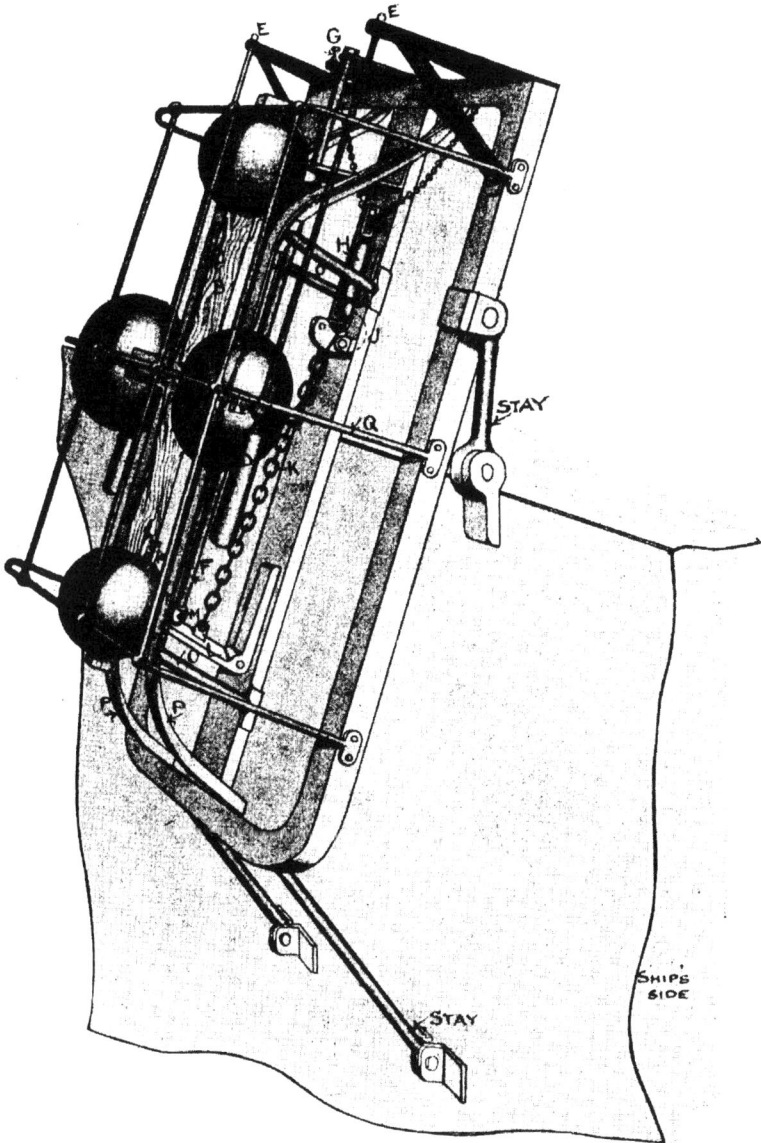

A. Copper globes.
B. Cross arms.
C. Man rope.
D. Tube for calcium light.
E. Extractor rod.
F. Guide rod.
G. Nob for releasing buoy.
H. Spring box.

I. Sheave.
J. Tumbler.
K. Holding chain.
L. Hinged bracket supporting buoy.
M. Lug and band attached to globe.
O. Stays.
P. Runners.
Q. Guard.

METHOD OF FITTING AWNINGS.

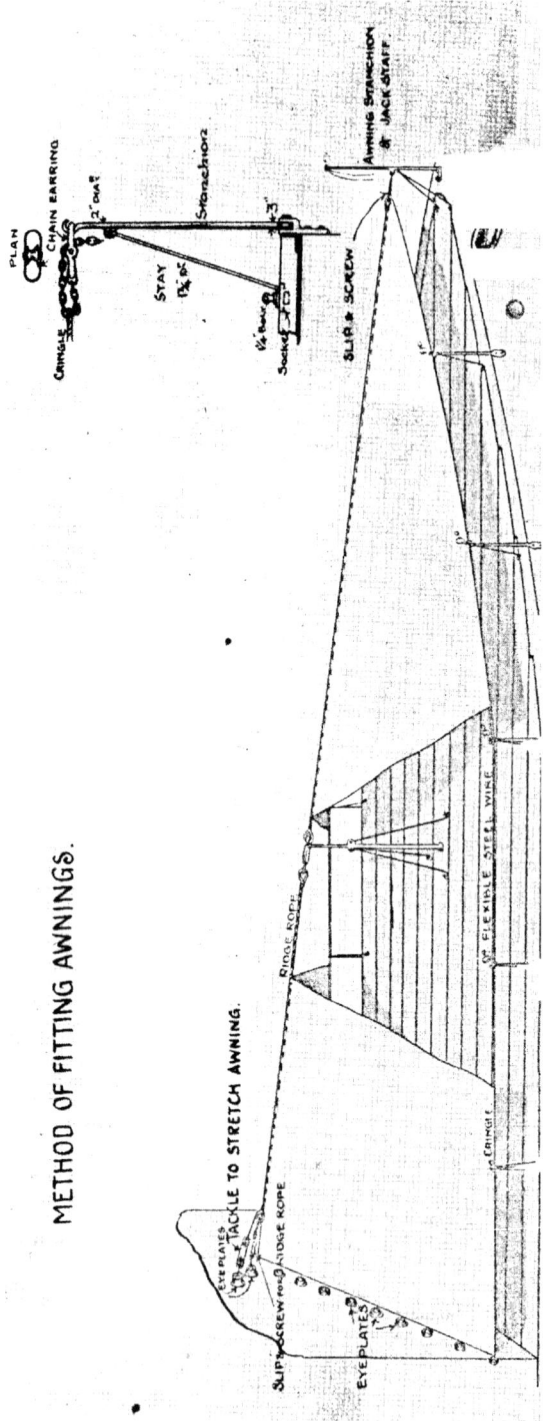

PLAN

CHAIN EARRING

CRINGLE

2" DIA

STAY
1¾ 6F

STANCHION

H-3

½ BAR

SOCKET 2F

SLI.P. SCREW

AWNING STANCHION
& JACK STAFF

TACKLE TO STRETCH AWNING.

EYEPLATES

RIDGE ROPE

SLIP.SCREW FOR RIDGE ROPE

EYEPLATES

RIDGE ROPE

CRINGLE

FLEXIBLE STEEL WIRE

CRINGLE

Description of boat.	Boat with permanent fittings.				Gear which can be removed.				Men.				Men.	Total weight.			
	tons.	cwts.	qrs.	lbs.	tons.	cwts.	qrs.	lbs.	tons.	cwts.	qrs.	lbs.	No.	tons.	cwts.	qrs.	lbs.
32-ft. steam barge	4	8	0	0	1	19	0	12	0	8	2	8		6	3	2	20
32-ft. „ cutter	4	9	0	7	1	19	0	2	0	8	2			6	8	0	8
23-ft. „ „	2	10	0		0	9			0	8	2			3	0	0	8
42-ft. sailing launch	6	16	0	1	1	11	0	1	1	8	2		2	9	1	1	15
36-ft. „ pinnace	3	7	0	1	1	1	0	1	1	5	2	2	1	6	0	0	22
32-ft. „ „	3	9	0	0	1	0	0	0	1	2	3		1	5	1	3	12
30-ft. „ „	3		0	6	0	18	0	1	1	0	0	0	1	5	2	1	16
34-ft. „ cutter	1	1	2	9	0	7			1	2	3	0	1	3	5	1	12
32-ft. „ „	1	1	0		0	6	0	2	1	2	3	1	1	3	0	3	18
30-ft. „ „	1	1	0	2	0	6	0		1	0	0	0	1	2	1	3	9
28-ft. „ „	1	0	0	21	0	5	0	0	0	17	0	½	1	2	10	3	16
26-ft. „ „	1	2	0	27	0	5	0	½	0	7	0	0	1	2	5	1	14
32-ft. galley or gig	0	13	0	14	0	3	0	0	0	1	1	0		1	8	2	9
30-ft. gig	0	4	0	4	0	3	0	0	0	1	1	20		1	9	3	2
28-ft. „	0	1	0	6	0	3	0	0	0	1	1	20		1	9	2	1
27-ft. whaler	0	2	0	0	0	3			0	0	0	0		1	0	3	
[1] 25-ft. „ L.S.	0	6	0	21	0	2	3	20	0	0	0	0		0	1	3	13
16-ft. skiff-dinghy	0	4	0	8	0	1	2	20	0	4	0	0		0	0	3	
13½-ft. dinghy	0	5	0	11	0	1	2	17	0	4	0	0		0	1	2	0
[1] 13½-ft. „	0	6	0	19	0	1	2	17	0	4	0	0		0	8	1	1
[2] 10-ft. „	0	0	0	0	0	1	1	4	0	4	0	0		0	1	0	0
13½-ft. balsa raft	0	0	0	4	0	0	0	24	0	2	0	1		0	9	1	0
10-ft. „	0	0	0	4	0	0	0	24	0	2	0	1		0	8	0	1
[1] 20-ft. Berthon	0	0	0	0	0	1	0	14	0	12	0	1		1	0	1	26

[1] Special boats for torpedo boat destroyers.
[2] Special boats for 1st class torpedo boats.

TABLE OF DIMENSIONS OF UNIVERSAL JOINTS.

Size of rod.	A	B	C	D	E	F	G	H	K	L	M	N	O	J
ins.														
1	$3\frac{1}{2}$	$1\frac{3}{8}$	$1\frac{5}{8}$	1	$1\frac{1}{2}$	$\frac{1}{2}$	$\frac{3}{4}$	$1\frac{5}{8}$	$\frac{3}{8}$	7	2	$1\frac{1}{2}$	$\frac{9}{16}$	$1\frac{1}{2}$
$1\frac{1}{8}$	$3\frac{5}{8}$	$1\frac{3}{8}$	$1\frac{7}{8}$	$1\frac{1}{8}$	$1\frac{1}{2}$	$\frac{9}{16}$	$\frac{3}{4}$	$1\frac{3}{4}$	$\frac{3}{8}$	$7\frac{1}{4}$	2	$1\frac{9}{16}$	$\frac{5}{8}$	$1\frac{1}{2}$
$1\frac{1}{4}$	4	$1\frac{1}{2}$	$2\frac{1}{16}$	$1\frac{1}{4}$	$1\frac{5}{8}$	$\frac{5}{8}$	$\frac{7}{8}$	2	$\frac{1}{2}$	8	$2\frac{7}{16}$	$1\frac{5}{8}$	$\frac{11}{16}$	$1\frac{5}{8}$
$1\frac{1}{2}$	$4\frac{1}{2}$	$1\frac{5}{8}$	$2\frac{3}{8}$	$1\frac{1}{2}$	$1\frac{3}{4}$	$\frac{5}{8}$	$\frac{15}{16}$	$2\frac{1}{8}$	$\frac{9}{16}$	9	$2\frac{9}{16}$	$2\frac{1}{8}$	$\frac{11}{16}$	$1\frac{3}{4}$

All dimensions are in inches.

Iron Cleats.—The sketch gives the dimensions for a cleat in terms of the circumference of the rope.

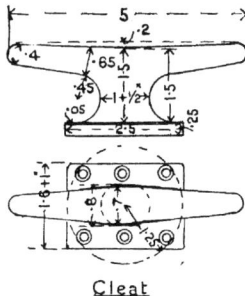

Cleat

TESTS OF MATERIALS.

Material.	U.T.S. in tons per square inch.	Elongation.	Form test piece.	Bending.	Remarks.
M.S. plates .	26 to 30	20 % in 8″	(1)	Double over radius = $1\frac{1}{2}$ × thickness plate.	Elastic limit ⊀ 20 tons.
H.T S. plates . .	34 to 38	—	—	—	
H.H.T.S. plates .	$\begin{cases} 3\frac{1}{2} \text{ lbs. and above—} \\ 37 \text{ to } 43 \\ \text{under } 3\frac{1}{2} \text{ lbs.—} \\ 35 \text{ to } 45 \end{cases}$	20 % in 8″	(1)	—	$\begin{cases} \text{Elastic limit} ⊀ \\ \text{one-half U.T.S.} \\ \text{with minimum} \\ 20 \text{ tons.} \end{cases}$
Nickel	36 to 40	18% in 8″	(1)	—	
Rivets (ordinary) .	26 to 30	25 %	(2)	$\begin{cases} \text{Cold} \\ \text{Hot} \end{cases}$	
,, (H.T.) . .	34 to 38	20 %	(2)	—	
,, (H.H.T.) .	37 to 43	18 %	(2)	—.	
,, (Nickel) .	36 to 40	20 %	(2)	—	
Steel forgings . .	⊀ 28	27 %	(3)	Double over radius ⊀ $\frac{3}{4}$″.	
,, casting (A) .	⊀ 26	$13\frac{1}{2}$ %	(3)	Bent through 45° over a radius 1″.	Drop tests for these in addition to other tests.
,, .. (B) .	⊀ 26	10 %	(3)	Do.	
,, ,, (C) .	—	—	—	—	Drop test only.
Pillars . . .	24 to 27	$\left\{ \begin{array}{c} \text{At least} \\ 33 \text{% in 2″} \end{array} \right\}$	—	Strips cut from tube bent double over radius $\frac{1}{2}$″.	End of tube expanded hot to increase $\frac{1}{20}$ diameter. Internal pressure, 100 lbs. per sq. in.
Wrought iron . .	⊀ 22 min.	25 % min.	(2)	Bent double cold over radius = $1\frac{1}{2}$ thickness test piece.	
Cable iron . .	$\begin{cases} ⊀ 23 \\ \text{for sizes below } 2\frac{1}{4}″ \\ ⊀ 22\frac{1}{2} \\ \text{for sizes } 2\frac{1}{4} \text{ to } 2\frac{8}{16}″ \\ ⊀ 22 \\ \text{for sizes above } 2\frac{9}{16}″ \end{cases}$	⊀ 22 %	(2)	—	
Cast iron . . .	⊀ 9	—	(3)	—	Pieces 1″ square section supported 12″ apart stand a load ⊀ 2000 lbs. in the middle.
Malleable cast iron	18	$4\frac{1}{2}$ % in 3″	—	—	Bent cold through 90° over radius 1″.
Naval brass— Bars and plates $\frac{3}{4}$″ and under .	⊀ 26	⊀ 30 %	(3)	(a) Hammered hot to a fine point.	Composition : Copper . 62 Zinc . . 37 Tin . . 1
Bars above $\frac{3}{4}$″, and square, flat, hexagonal . .	⊀ 22	—	—	(b) Bent cold over radius = thickness bar through 75°.	$\underline{100}$
Gun metal . .	⊀ 14	$7\frac{1}{2}$ %	(3)	—	Composition : Copper . 88 Tin . . 10 Zinc . . 2 $\underline{100}$

TESTS OF MATERIALS—*continued.*

Material.	U.T.S. in tons per square inch.	Elongation.	Form test piece.	Bending.	Remarks.
Naval brass—*cont.* Phosphor bronze	$\nless$ 17	15% in 6″	(3)	Bent double over 2″ bar.	Composition: Copper . 89·5; Tin . . 10·0; Phosphorus 0·5; 100·0
Copper bars . .	$\nless$ 13	30% in 4″ or 35% in 2″	—	Bent double over a radius = diameter of bar.	
Special steel for pins of sheaves	40 to 50	15% in 2″	—	Bent through 90° over radius 1″.	

2½D
D

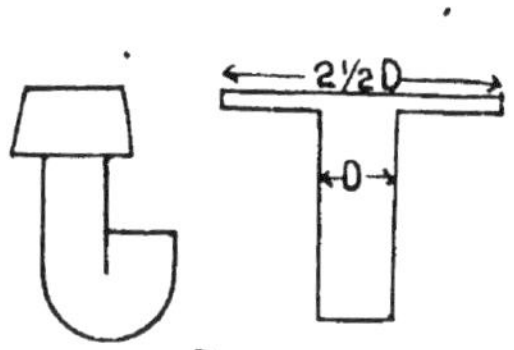

Rivets

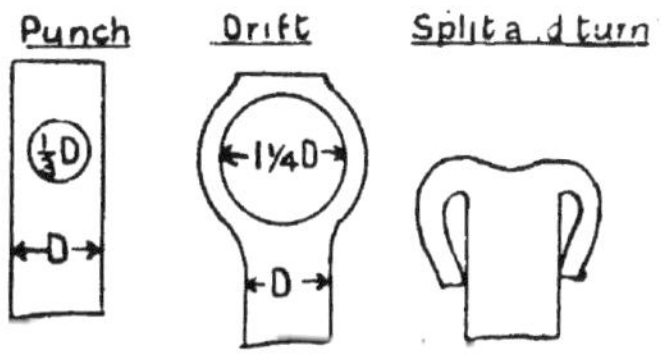

Punch
Drift
Split and turn
⅓D
D
1¼D
D

Cold tests

1" and above
Below 1"
radius of link

Cable iron

Width
over 35 lbs, 1½ (max^m)
15 to " 2" "
under 15, 2½ "
(1)
8"
Parallel portion, not less than 9"
ends enlarged
(2)
8 diameters
diameter
9 diameters (min^m) parallel portion
(3)
4 diam^rs
4½ diameters

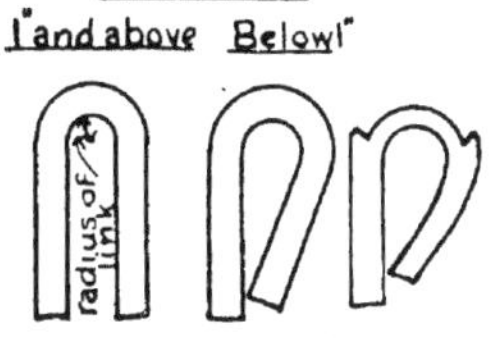

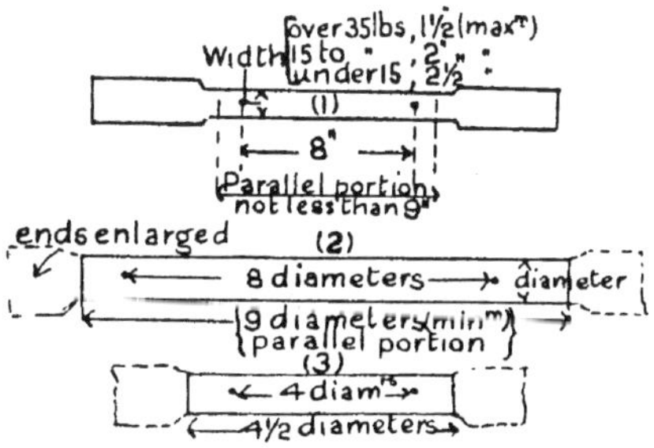

Forms of Test Pieces.

Fitting.	Proof strain.
Rings and ring bolts . . .	$4d^2$
Eyeplates	$12\frac{1}{2}d^2$
Shackle straight with forelock .	$9d^2$
,, ,, ,, screw .	$7d^2$
,, bow	$5d^2$
Chain cable, common link .	$12d^2$
,, ,, stud link	$18d^2$

d = diameter of iron in inches.

APPENDIX

ANOTHER method of arranging the fore part of the launch cradle is shown below, and differs from that described on pages 91, 92, in that the whole is free to come away from the ship when the latter is off the ways.

Sling plates, as shown, consisting of a $1\frac{1}{4}$-inch plate to fit under the keel, to which are bolted two $\frac{5}{8}$-inch plates, the upper ends of which are secured to stringer plates, S, of $\frac{5}{8}$-inch plate 4 feet wide. In the diagram seven of these plates are shown, the foremost ones being 5 feet wide and the remainder 4 feet.

The housing plates or shelf P for the heads of the poppets fit in between deep brackets connected to the sling plates by double angles $4'' \times 3\frac{1}{2}'' \times \frac{1}{2}''$, an angle bar being also worked on the underside of the shelf. Wood packing is fitted between the sling plates and the hull to prevent damage to the latter.

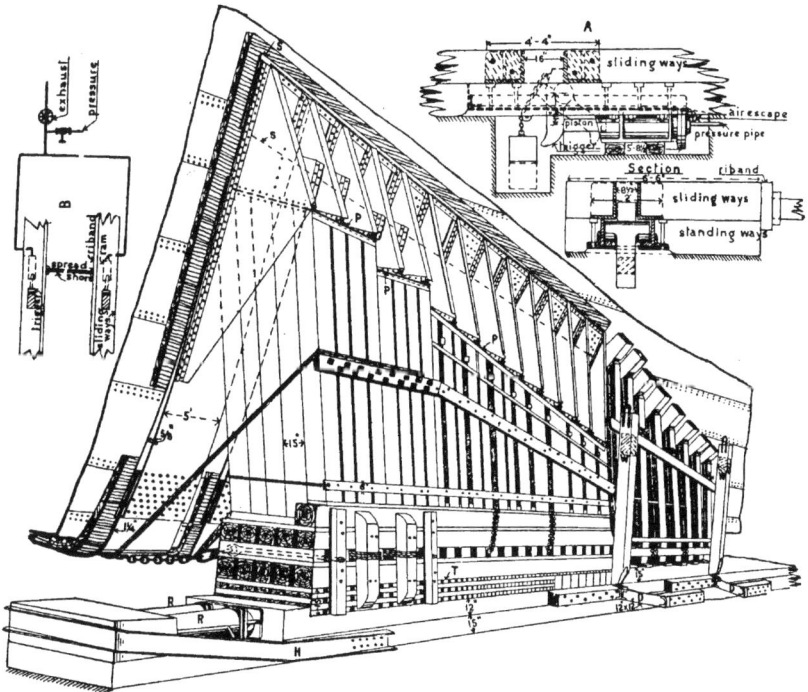

To keep the poppets as short as possible and also to permit of the fore poppet being carried as far forward as possible, they are inclined as shown, any tendency of the heels slipping outwards being guarded against by wire strops and chains carried across under the keel of the ship.

The wood marked T is soft wood plank spaced about 3 inches apart at the foremost portion gradually closing for the after portion, and is for the purpose of distributing the great pressure, which comes on the fore poppets when the stern lifts, over as large an area as possible through the crushing of the soft wood. R, R are hydraulic rams for starting the ship if necessary, the thrust on the standing ways being taken by the channel bars H.

The ribands are fitted on the inside edge of the sliding ways, this method doing away with the necessity for shoring the ribands from the building slip (see page 93).

The release gear consists of dogshores actuated by falling weights, W, about 6 cwt. each, and four hydraulic triggers, two on each side of the ship. The triggers are secured to the standing ways as in Fig. A, the upper part of the trigger pressing on the front of a steel shoe let into the sliding ways. Each trigger is kept in place by a ram working in a hydraulic cylinder which is attached to the underside of the standing ways. Up to the time of launching pressure is on the piston, and on opening the exhaust the trigger is released and the movement of the sliding ways turns it into the ticked position clear of the ways, this movement being aided by the counterbalance weight. Fig. B shows in plan the general arrangement of triggers, pressure and exhaust pipes, ways, etc.

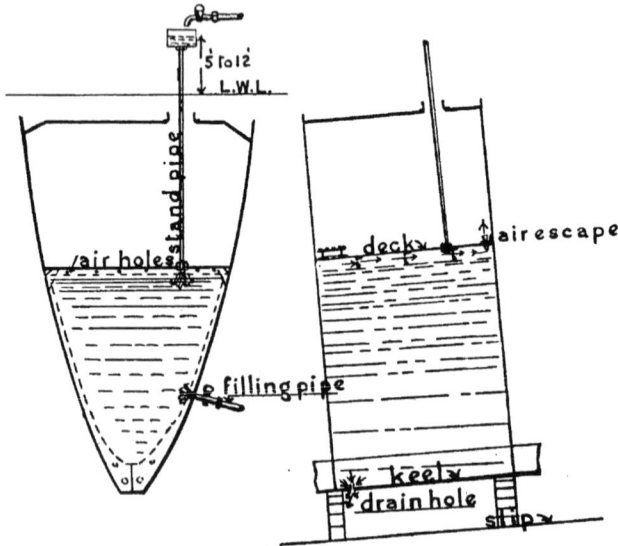

WATERTESTING OF COMPARTMENTS.

INDEX

THE END

PRINTED BY WILLIAM CLOWES AND SONS, LIMITED, LONDON AND BECCLES.

Printed in Great Britain
by Amazon